FAILURE ANALYSIS

WILEY SERIES IN QUALITY & RELIABILITY ENGINEERING

*and related titles**

Electronic Component Reliability:
Fundamentals, Modelling, Evaluation and Assurance
Finn Jensen

Measurement and Calibration Requirements
For Quality Assurance to ASO 9000
Alan S. Morris

Integrated Circuit Failure Analysis:
A Guide to Preparation Techniques
Friedrich Beck

Test Engineering
Patrick D. T. O'Connor

Six Sigma: Advanced Tools for Black Belts and Master Black Belts*
Loon Ching Tang, Thong Ngee Goh, Hong See Yam, Timothy Yoap

Secure Computer and Network Systems: Modeling, Analysis and Design*
Nong Ye

Failure Analysis:
A Practical Guide for Manufacturers of Electronic Components and Systems
Marius Bâzu and Titu Băjenescu

Reliability Technology:
Principles and Practice of Failure Prevention in Electronic Systems
Norman Pascoe

FAILURE ANALYSIS

A PRACTICAL GUIDE FOR MANUFACTURERS OF ELECTRONIC COMPONENTS AND SYSTEMS

Marius Bâzu

National Institute for Microtechnologies, IMT-Bucharest, Romania

Titu Băjenescu

University Professor, International Consultant, C. F. C., Switzerland

A John Wiley and Sons, Ltd., Publication

This edition first published 2011
© 2011 John Wiley & Sons, Ltd.

Registered office
John Wiley & Sons Ltd, The Atrium, Southern Gate, Chichester, West Sussex, PO19 8SQ, United Kingdom

For details of our global editorial offices, for customer services and for information about how to apply for permission to reuse the copyright material in this book please see our website at www.wiley.com.

The right of the author to be identified as the author of this work has been asserted in accordance with the Copyright, Designs and Patents Act 1988.

All rights reserved. No part of this publication may be reproduced, stored in a retrieval system, or transmitted, in any form or by any means, electronic, mechanical, photocopying, recording or otherwise, except as permitted by the UK Copyright, Designs and Patents Act 1988, without the prior permission of the publisher.

Wiley also publishes its books in a variety of electronic formats. Some content that appears in print may not be available in electronic books.

Designations used by companies to distinguish their products are often claimed as trademarks. All brand names and product names used in this book are trade names, service marks, trademarks or registered trademarks of their respective owners. The publisher is not associated with any product or vendor mentioned in this book. This publication is designed to provide accurate and authoritative information in regard to the subject matter covered. It is sold on the understanding that the publisher is not engaged in rendering professional services. If professional advice or other expert assistance is required, the services of a competent professional should be sought.

Library of Congress Cataloging-in-Publication Data
Bâzu, M. I. (Marius I.), 1948-
Failure Analysis : A Practical Guide for Manufacturers of Electronic Components and Systems / Marius Bâzu, Titu-Marius Băjenescu.
p. cm. – (Quality and Reliability Engineering Series ; 4)
Includes bibliographical references and index.
ISBN 978-0-470-74824-4 (hardback)
1. Electronic apparatus and appliances–Reliability. 2. Electronic systems–Testing. 3. System failures (Engineering)–Prevention. I. Băjenescu, Titu, 1938- II. Title.
TK7870.23.B395 2011
621.381–dc22

2010046383

A catalogue record for this book is available from the British Library.

Print ISBN: 978-0-470-74824-4 (HB)
E-PDF ISBN: 978-1-119-99010-9
O-book ISBN: 978-1-119-99009-3
E-Pub ISBN: 978-1-119-99000-0

Typeset in 9/11pt Times by Laserwords Private Limited, Chennai, India

With all my love, to my wife Cristina.

Marius I. Bâzu

To my charming dear wife Andrea – an unfailing source of inspiration – thankful and grateful for her love, patience, encouragement, and faithfulness to me, for all my projects, during our whole common life of a half-century.

To my descendants, with much love.

Titu-Marius I. Băjenescu

Contents

Series Editor's Foreword

During my eventful career in the aerospace and automotive industries (as well as in consulting and teaching) I have had plenty of opportunities to appreciate the contribution failure analysis makes to successful product development. While performing various engineering design functions associated with product reliability and quality, I have often found myself or a team member rushing to the failure analysis lab calling for help. And help we would always receive.

Reliability Science combines two interrelated disciplines: 1. Reliability Mathematics including probability, statistics and data analysis and 2. Physics of Failure. While the literature on the former is plentiful with a significant degree of depth, literature on the latter is somewhat scarce. The book you are about to read will successfully reduce that knowledge gap. It is not only a superb tutorial on the physics of failure, but also a comprehensive guide on failure mechanisms and the various ways to analyze them.

Engineering experience accumulated over the years clearly indicates that in reliability science Physics trumps Mathematics most of the time. Too often, jumping directly to statistical data analysis leads to flawed results and erroneous conclusions. Indeed, certain questions about the nature of failures need to be answered before life data analysis and probability plotting is carried out. Failure analysis is the key tool in providing answers to those questions. Physics of Failure is also vital to the emerging field of design for reliability (DfR). DfR utilizes past knowledge of product failure mechanisms to avoid those failures in future designs. No question that failure analysis requires sizable expenditures both in equipment and people, nevertheless done properly the return on this investment will be quick and substantial in terms of improved design and future cost reduction.

In this work, Marius Bâzu and Titu Băjenescu effectively demonstrate how critical it is to many engineering disciplines to understand failure mechanisms, and how important it is for the engineering community to continue to develop their knowledge of failure analysis. In addition, the authors clearly raise the bar on connecting theory and practice in engineering applications by putting together an exceptional collection of case studies at the end of the book.

Bâzu and Băjenescu successfully capture the essence and inner harmony of reliability analysis in this work. It undoubtedly will present a wealth of theoretical and practical knowledge to a variety of readers; from college students to seasoned engineering professionals in the fields of quality, reliability, electronics design, and component engineering.

Dr Andre Kleyner,
Global Reliability Engineering Leader at Delphi Corporation,
Adjunct Professor at Purdue University

Foreword by Dr Craig Hillman

Common sense.

In some respects, this is a simple, but powerful phrase. It implies grounding, a connection to the basics of knowledge and the physical world. But, it can also suggest a lack of insight or intelligence when thrown in an accusatory way. And, of course, common sense isn't always so common.

This phrase came to me as I was reading Marius' and Titu's manuscript, as it brought me back to my first days performing failure analysis on and predicting reliability of electronic components, boards, and systems. My background had been in Metallurgy and Material Science and I had experienced the revolution propagating through those disciplines as scientific approaches to processing and material behavior had resulted in dramatic improvements in cost and performance. These approaches had been germinating for decades, but demonstrated successes, economic forces, and the need for higher levels of quality control had forced their implementation throughout the materials supply chain.

So, imagine my surprise and amusement when I was informed that 'Physics of Failure' was the future of electronics reliability prediction and failure analysis. The future? Shouldn't this have been the past and present as well? How else could you ensure reliability? Statistics were useful in extrapolating laboratory results to larger volumes (either in quantity or size), but the fundamental understanding of materials reliability was always based on physics-based mechanisms. Isn't Physics of Failure, also known as Reliability Physics, common sense?

However, as Marius and Titu elaborate so well in their tome, the approaches that have evolved so well in other single-disciplinary fields did not always seamlessly integrate into the world of electronics. Even back in the 1940's and 1950's, electronics were complex (at least compared to bridges and boilers), involving multiple materials (copper, steel, gold, alumina, glass, etc.), multiple suppliers, multiple assembly processes, and multiple disciplines (electrical, magnetic, physics, material science, mechanics, etc.). The concept of assessing each mechanism and extrapolating their evolution over a range of conditions must have seemed mind-boggling at the time. As described succinctly, "*a physics-of-failure-like model developed by for small-scale CMOS was virtually unusable by system manufacturers, requiring input data (details about design layout, process variables, defect densities, etc.) that are known only by the component manufacturer*". Even users of the academic PoF tools out of the University of Maryland have admitted to the weeks or months necessary to gather the information required, which results in analyses that are more interesting exercises then a true part of the product design process.

Despite these frustrations, the path forward to physics-based failure analysis and reliability prediction is clear. Studies by Professor Joseph Bernstein of Bar-Ilan University and Intel have clearly demonstrated that wearout behavior will dominate the newer technologies being incorporated into electronics today. In addition, standard statistical approaches will fumble over trying to capture the increasing complexities in silicon-based devices (what is the FIT rate of a system-in-package consisting of a DRAM, SRAM, microcontroller, and RF device? The same as discrete devices? Is there a relevant lambda here?).

By laying out a methodical process of Why/When/How/What, Marius and Titu are not only laying out the menu for incorporating these best practices into the design and manufacturing process that are elemental to electronic components, products, and systems, but they are also laying out the arguments on why the electronics community should be indulging in this feast of knowledge and understanding.

With knowledge, comes insight. With insight, comes success. And isn't that just Common sense?

Dr Craig Hillman, PhD
CEO
DfR Solutions
5110 Roanoke Place
College Park, MD 20740
301-474-0607 x308 (w)
301-452-5989 (c)

Series Editor's Preface

The book you are about to read re-launches the Wiley Series in Quality and Reliability Engineering. The importance of quality and reliability to a system can hardly be disputed. Product failures in the field inevitably lead to losses in the form of repair cost, warranty claims, customer dissatisfaction, product recalls, loss of sale, and in extreme cases, loss of life.

As quality and reliability science evolves, it reflects the trends and transformations of the technologies it supports. For example, continuous development of semiconductor technologies such as system-on-chip devices brings about unique design, test and manufacturing challenges. Silicon-based sensors and micromachines require the development of new accelerated tests along with advanced techniques of failure analysis. A device utilizing a new technology, whether it be a solar power panel, a stealth aircraft or a state-of-the-art medical device, needs to function properly and without failure throughout its mission life. New technologies bring about: new failure mechanisms (chemical, electrical, physical, mechanical, structural, etc.); new failure sites; and new failure modes. Therefore, continuous advancement of the physics of failure combined with a multi-disciplinary approach is essential to our ability to address those challenges in the future.

The introduction and implementation of Restriction of Hazardous Substances Directive (RoHS) in Europe has seriously impacted the electronics industry as a whole. This directive restricts the use of several hazardous materials in electronic equipment; most notably, it forces manufacturers to remove lead from the soldering process. This transformation has seriously affected manufacturing processes, validation procedures, failure mechanisms and many engineering practices associated with lead-free electronics. As the transition continues, reliability is expected to remain a major concern in this process.

In addition to the transformations associated with changes in technology, the field of quality and reliability engineering has been going through its own evolution developing new techniques and methodologies aimed at process improvement and reduction of the number of design- and manufacturing-related failures.

The concepts of Design for Reliability (DfR) were introduced in the 1990's but their development is expected to continue for years to come. DfR methods shift the focus from reliability demonstration and 'Test-Analyze-Fix' philosophy to designing reliability into products and processes using the best available science-based methods. These concepts intertwine with probabilistic design and design for six sigma (DFSS) methods, focusing on reducing variability at the design and manufacturing level. As such, the industry is expected to increase the use of simulation techniques, enhance the applications of reliability modeling and integrate reliability engineering earlier and earlier in the design process.

Continuous globalization and outsourcing affect most industries and complicate the work of quality and reliability professionals. Having various engineering functions distributed around the globe adds a layer of complexity to design co-ordination and logistics. Also moving design and production into regions with little knowledge depth regarding design and manufacturing processes, with a less robust quality system in place, and where low cost is often the primary driver of product development affects a company's ability to produce reliable and defect-free parts.

The past decade has shown a significant increase of the role of warranty analysis in improving the quality and reliability of design. Aimed at preventing existing problems from recurring in new products, product development process is becoming more and more attuned to engineering analysis of returned parts. Effective warranty engineering and management can greatly improve design and reduce costs, positively affecting the bottom line and a company's reputation.

Several other emerging and continuing trends in quality and reliability engineering are also worth mentioning here. Six Sigma methods including Lean and DFSS are expected to continue their successful quest to improve engineering practices and facilitate innovation and product improvement. For an increasing number of applications, risk assessment will replace reliability analysis, addressing not only the probability of failure, but also the quantitative consequences of that failure. Life cycle engineering concepts are expected to find wider applications to reduce life cycle risks and minimize the combined cost of design, manufacturing, quality, warranty and service. Reliability Centered Maintenance will remain a core tool to address equipment failures and create the most cost-effective maintenance strategy. Advances in Prognostics and Health Management will bring about the development of new models and algorithms that can predict the future reliability of a product by assessing the extent of degradation from its expected operating conditions. Other advancing areas include human reliability analysis and software reliability.

This discussion of the challenges facing quality and reliability engineers is neither complete nor exhaustive; there are myriad methods and practices the professionals must consider every day to effectively perform their jobs. The key to meeting those challenges is continued development of state-of-the-art techniques and continuous education.

Despite its obvious importance, quality and reliability education is paradoxically lacking in today's engineering curriculum. Few engineering schools offer degree programs or even a sufficient variety of courses in quality or reliability methods. Therefore, a majority of the quality and reliability practitioners receive their professional training from colleagues, professional seminars, publications and technical books. The lack of formal education opportunities in this field greatly emphasizes the importance of technical publications for professional development.

The main objective of Wiley Series in Quality & Reliability Engineering is to provide a solid educational foundation for both practitioners and researchers in quality and reliability and to expand the readers' knowledge base to include the latest developments in this field. This series continues Wiley's tradition of excellence in technical publishing and provides a lasting and positive contribution to the teaching and practice of engineering.

Dr Andre Kleyner,
Editor of the Wiley Series in Quality & Reliability Engineering

Preface

The goal of a technical book is to provide well-structured information on a specific field of activity, which has to be clearly delimited by the title and then completely covered by the content. In accomplishing this *right book*, two conditions must be fulfilled simultaneously: the book must be written *at the right moment* and it must be written *by the right people*.

If the potential readers represent a significant segment of the technical experts, and if they have already been prepared by the development of the specific field, one may say that the book is written *at the right moment*. Is this the case with our book? Yes, we think so. In the last 40 years, failure analysis (FA) has received growing attention, becoming since 1990 the central point of any reliability analysis. Due to its important role in identifying the design and process failure risks and elaborating corrective actions, one may say that FA has been the motor of development for some industries with harsh reliability requirements, such as aeronautics, automotive, nuclear, military, and so on.

We have chosen to focus the book mainly on the field of electronic components, with electronic systems seen as direct users of the components. Today, in this domain, it is almost impossible to conceive a serious investigation into the reliability of a product or process without FA. The idea that failure acceleration by various stress factors (the key to accelerated testing) could be modelled only for the population affected by the same failure mechanism (FM) greatly promoted FA as the only way to segregate such a population damaged by specific FMs.

Moreover, the simple statistical approach in reliability, which was the dominant one for years, is no longer sufficient. The physics-of-failure approach is the only one accepted at world level, being the solution for continuously improving the reliability of the materials, devices and processes. Even for the modelling of FMs, the well-known models based on distributions like Weibull or Lognormal have today been replaced by analytical models that are elaborated based on an accurate description of the physical or chemical phenomena responsible for degradation or failure of electronic components and materials.

In FA, a large range of methods are now used, from (classical) visual inspection to expensive and modern methods such as transmission electron microscopy, secondary ion mass spectroscopy and so on. Nice photos and clever diagrams, true examples of scientific beauty, represent a sort of siren's song for the ears of the specialist in FA. It is easy to fill an FA report with the attractive results of sophisticated methods, without identifying in the end the root causes of the failure. The FA specialist has to be as strong as Ulysses, continuing to drive the 'boat' of FA guided only by the 'compass' of a logic and coherent analysis! On the other hand, the customers of FA, manufacturers of electronic components and systems, have to be aware that a good FA report is not necessarily a report with many nice pictures and 3D simulations, but a report with a logic demonstration of FA, based on results obtained with necessary FA techniques, and, very importantly, with solid conclusions about the root causes of the failure for the studied item. It is the most difficult route, but the only one with fruitful results.

Consequently, for all the above reasons, we think that a practical guide to FA of electronic components and systems is a necessary tool. Today, a book of this kind, capable of orienting reliability engineers in the complicated procedures of FA, is sadly lacking. It is our hope to fill this void.

The second question is: are we the *right people* to write this book? Yes, we think so. In answering this question, we have to put aside the natural modesty of any scientist and show here the facts that support this statement. We are two people with different but perfectly complementary backgrounds, as anyone can see from the attached biographies. Marius Bâzu has behind him almost 40 years of activity in the academic media of Romania, as leader of a reliability group from an applied research institute focused on microtechnologies, and author of books and papers on the reliability of components. Titu Băjenescu has had an outstanding career in industry, responsible for the reliability domain in many Swiss companies and authoring many technical papers and books – written in English, French, German and Romanian. He is also a university professor and visiting lecturer or speaker at various European universities and other venues. Both authors are covering the three key domains of industry, research and teaching. Moreover, we have already worked together, publishing two books on the reliability of electronic components.

We hope we have convinced you that this is indeed a book written at the right moment and by the right people.

While writing this book, we were constantly asking ourselves who might be its potential readers. As you will see, the book is aimed to be useful to many people.

If you are working as a component manufacturer, the largest part of the book is directly focused on your activity. Even if you are not a reliability engineer, you will find significant information about the possible failure risks in your current work, with suggestions about how to avoid them or how to correct wrong operations.

You will also be interested in this book if you are a component user. This includes reliability engineers and researchers, electrical and electronic engineers, those involved in all facets of electronics and telecommunications product design and manufacturing, and those responsible for implementing quality and process improvement programmes.

If you are from the academic media, including teachers and students of the electrical and electronic faculties, you will find in this book all the necessary information on reliability issues related to electronic components. The book does not contain complicated scientific developments and it is written at a level allowing an easy understanding, because one of its goals is to promote reliability and failure analysis as a fascinating subject of the technical environment. Moreover, the book's companion website contains Q&A sessions for each chapter.

If you are a manufacturer of electronic components, but not directly involved in the reliability field, you will be interested in this book. By gathering together the main issues related to this subject, a fruitful idea is promoted: all parts contributing to the final product (designers, test engineers, process engineers, marketing staff and so on) participate together in its quality and reliability (the so-called 'concurrent engineering' approach).

If you are the manager of a company manufacturing electronic components, you may use this book as a tool for convincing your staff to involve itself in all reliability issues.

A prognosis on the evolution of FA in the coming years is both easy and difficult to make. It is easy, because everyone working in this domain can see the current trend. FA is still in a 'romantic' period, with many new and powerful techniques being used to analyse more and more complex devices. Still, the need for new tools is pressing, especially for the smallest devices, at nano level, but it is difficult to predict the way these issues will be solved. On the other hand, we think that procedures for executing FA will be stabilised and standardised very soon, allowing any user of an electronic component to verify the reliability of the product. In fact, our book is intended to be a step on this road!

Marius I. Bâzu, Romania
Titu-Marius I. Băjenescu, Switzerland
September 2010

About the Authors

Marius Bâzu received his BE and PhD degrees from the Politehnica University of Bucharest, Romania. Since 1971 he has worked at the National Institute for Research and Development in Microtechnology, IMT-Bucharest, Romania. He is currently Head of the Reliability Laboratory and was Scientific Director (2000–2003) and Vice-President of the Scientific Council of IMT-Bucharest (2003–2008). His past work involved the design of semiconductor devices and in semiconductor physics. Recent research interests include design for reliability, concurrent engineering, methods of reliability prediction in microelectronics and microsystems, a synergetic approach to reliability assurance and the use of computational intelligence methods for reliability assessment. He developed accelerated reliability tests, building-in reliability and concurrent engineering approaches for semiconductor devices. He was a member of the Management Board and chair of the Reliability Cluster of the Network of Excellence Design for Micro and Nano Manufacture – PATENT-DfMM (2004–2008 project in the Sixth Framework Program of the European Union) and leader of a European project (Phare/TTQM) on building-in reliability technology (1997–1999). He currently sits on the Board of the European network EUMIREL (European Microsystem Reliability), established in 2007, which aims to deliver reliability services on microsystems.

Dr Bâzu has authored or co-authored more than 130 scientific papers and contributions to conferences, including two books (co-authored with Titu Băjenescu, published in 2010 by Artech House and in 1999 by Springer Verlag) and is the referent of the journals *Microelectronics Reliability*, *Sensors*, *IEEE Transactions on Reliability*, *IEEE Transactions on Components and Packaging* and *Electron Device Letters* and Associate Editor of the journal *Quality Assurance*. Recipient of the AGIR (General Association of Romanian Engineers) Award for the year 2000, he was chairman of and presented invited lectures to several international conferences: CIMCA 1999 and 2005 (Vienna, Austria), CAS 1991–2009 (Sinaia, Romania) and MIEL 2004 (Niš, Serbia and Montenegro), amongst others. He is currently Invited Professor for Post-Graduate Courses at the Politehnica University of Bucharest.

Contact: 49, Bld. Timisoara, Bl. CC6, ap.34, Sector 6, 061315 Bucharest (Romania);
Tel: +40(21)7466109; mariusbazu@yahoo.com, marius.bazu@imt.ro

Titu-Marius I. Băjenescu received his engineering training at Politehnica University of Bucharest, Romania. He designed and manufactured experimental equipment for the army research institute and for the air defence system. He specialises in QRA Management in Switzerland, the USA, the UK and West Germany and was a former Senior Member of the IEEE (USA). In 1969 he moved to Switzerland, joining first Asea Brown Boveri (as research and development engineer, involved in the design and manufacture of equipment for telecommunications), then – in 1974 – Ascom, as Reliability Manager (recruitment by competitive examination), where he set up QRA and R&M teams, developed policies, procedures and training and managed QRA and R&M programmes. He also acted as QRA manager, monitoring and reporting on production quality and in-service reliability. As Swiss Official, he contributed to the development of new ITU and IEC standards. In 1985, he joined Messtechnik und Optoelektronik (Neuchâtel, Switzerland and Haar, West Germany), a subsidiary of Messerschmitt-Bölkow-Blohm (MBB) Munich, as quality and reliability manager, where he was product assurance manager of 'intelligent cables' and managed applied research on reliability (electronic components, system analysis methods, test methods and so on). Since 1986 he has worked as an independent consultant and international expert on engineering management, telecommunications, reliability, quality and safety.

He has authored many technical books, on a large range of technical subjects, published in English, French, German and Romanian. He is a university professor and has written many papers and articles on modern telecommunications and on quality and reliability engineering and management; he lectures on these subjects as invited professor, visiting lecturer and speaker at various European universities and other venues. Since 1991, he has won many awards and distinctions, presented by the Romanian Academy, Romanian Society for Quality, Romanian Engineers Association and so on, for his contribution to the reliability of science and technology. Recently, he received the honorific title of *Doctor Honoris Causa* from the Romanian Military Technical Academy and from the Technical University of the Republic of Moldavia. He is Invited Professor at the Romanian Military Technical Academy and at the Technical University of the Republic of Moldavia.

His extensive list of publications includes: *Initiation à la fiabilité en électronique moderne*, Masson, Paris, 1978; *Elektronik und Zuverlässigkeit*, Hallwag-Verlag, Berne and Stuttgart, 1979; *Problèmes de la fiabilité des composants électroniques actifs actuels*, Masson, Paris, 1980; *Zuverlässigkeit elektronischer Komponenten*, VDE-Verlag, Berlin, 1985; *Reliability of Electronic Components. A Practical Guide to Electronic System Manufacturing* (with M. Bâzu), Springer, Berlin and New York, 1999; *Component Reliability for Electronic Systems* (with M. Bâzu), Artech House, Boston and London, 2010.

Contact: 13, Chemin de Riant-Coin, CH-1093 La Conversion (Switzerland);
Tel: ++41(0)217913837; Fax: ++41(0)217913837; tmbajenesco@bluewin.ch.

1

Introduction

1.1 The Three Goals of the Book

In our society, which is focused on *success* in any domain, *failure* is an extremely negative value. We all still remember the strong emotion produced worldwide by the crash of the Space Shuttle *Challenger*. On 28 January 1986, *Challenger* exploded after 73 seconds of flight, leading to the deaths of its seven crew members. The cause was identified after a careful failure analysis (FA): an O-ring seal in its solid rocket booster failed at lift-off, causing a breach in the joint it sealed and allowing pressurised hot gas from the solid rocket motor to reach the outside and impinge upon the attachment hardware. Eventually, this led to the structural failure of the external tank and to shuttle crash. This is a classical example of failure produced by the low quality of a part used in a system. Other examples of well-known events produced by failures of technical systems include the following:

- On 10 April 1912, RMS *Titanic*, at that time the largest and most luxurious ship ever built, set sail on its maiden voyage from Southampton to New York. On 14 April, at 23 : 40, the *Titanic* struck an iceberg about 400 miles off Newfoundland, Canada. Although the crew had been warned about icebergs several times that evening by other ships navigating through the region, the *Titanic* was travelling at close to its top speed of about 20.5 knots when the iceberg grazed its side. Less than three hours later, the ship plunged to the bottom of the sea, taking more than 1500 people with it. Only a fraction of the passengers were saved. This was a terrible failure of a complex technical system, made possible because the captain had ignored the necessary precautions; hence the failure was produced by a human fault. The high casualty rate was further explained by the insufficient number of life boats, which implies a design fault.
- On 24 April 1980, the US President Jimmy Carter authorised the military operation Eagle Claw (or Evening Light) to rescue 52 hostages from the US Embassy in Tehran, Iran. The hostages had been held since 4 November 1979 by a commando of the Iranian Revolutionary Guard. Eight RH-53D helicopters participated in the operation, which failed due to many technical problems. Two helicopters suffered avionics failures en route and a sand storm damaged the hydraulic systems of another two. Because the mission plan called for a minimum of six helicopters, the rest were not able to continue and the mission was aborted. This is considered a typical case of reliability failure of complex technical systems.

Obviously, we have to fight against failures. Consequently FA has been promoted and has quickly become a necessary tool. FA attempts to identify root causes and to propose corrective actions aimed at avoiding future failure.

Failure Analysis: A Practical Guide for Manufacturers of Electronic Components and Systems, First Edition.
Marius I. Bâzu and Titu-Marius I. Băjenescu.
© 2011 John Wiley & Sons, Ltd. Published 2011 by John Wiley & Sons, Ltd.

Given the large range of possible human actions, many specific procedures of FA are needed, starting with medical procedures for curing or preventing diseases (which are failures of the human body) and continuing with various procedures for avoiding the failure of technical systems. In this respect, the word 'reliability' has two meanings: first, it is the aim of diminishing or removing failures, and second, it is the property of any system (human or artefact) to function without failures, in some given conditions and for a given duration. In fact, FA is a component of reliability analysis. This idea will be detailed in the following pages.

From the above, one can see that the first goal of this book is to present the basics of FA, which is considered the key action for solving reliability issues. But there is a second purpose, equally important: to promote the idea of reliability, to show the importance of this discipline and the necessity of supporting its goals in achieving a given level of reliability as a key characteristic of any product.

Unfortunately, the first reliability issues were solved by statisticians, which led to a mathematical approach to reliability, predominant in the first 25–30 years of the modern history of the domain. Today other disciplines, such as physics and chemistry, are equally involved. All this issues are detailed in Section 1.2, where a short history of reliability as a discipline is presented.

The mathematical approach was restrictive and created the incorrect impression that the aim of reliability analysis was to impede the work of real specialists, forcing them to undertake redesigns due to cryptic results that nobody could understand. Today this misapprehension has generally been overcome, but its after-effects are still present in the mentality of some specialists. We want to persuade component manufacturers that reliability engineers are their best friends, simulating the behaviour of their product in real-life conditions and then recommending necessary improvements before the product can be sold. Manufacturers and reliability engineers must form a team, with information flowing in both directions.

Even more importantly, this book is aimed at showing to industry managers the reasons for taking reliability issues into account from the design phase onwards, through the whole cycle of development of a product. It has been proved that the only way to promote reliability requirements is top-down, starting with the manager and continuing down to every worker.

The third goal of the book starts from our subjective approach to reliability. We think reliability is a beautiful domain, offering immense satisfaction to any specialist, and involving a large range of knowledge, from physics, chemistry and mathematics to all engineering disciplines. That is why strong interdisciplinary teams are needed to solve reliability issues, which are difficult challenges for the human mind.

We want to show to young readers the beauty of reliability analysis, which can be compared to a simple mathematical demonstration. Another approach is to consider a reliability analysis to be similar to the activity of a detective: we have a 'dead component' and, based on the information gathered from those involved, we have to find out why this happened and 'who did it'. This is possible because failures follow the law of cause and effect [1].

Our focus on the above three goals has imposed the structure of this first chapter. As you can see, we want not only to deliver a high quantity of information, but to convince the specialists who manufacture electronic components and systems how important FA is, and, more generally, to attract the reader to the 'charming land of reliability'. This first chapter has a huge importance, as our main attractor. Consequently, we have tried to structure it as straightforwardly as possible. We have presumed the subject is a new one for the reader, so we thought it best to begin with a short history of reliability as a discipline, with a special emphasis on FA. A section on terminology will furnish definitions of the most important terms. Finally the state of the art in FA will be described, including a short description of the main challenges for the near future.

This first chapter will thus show the past, present and future of FA, together with the main terminology. With this knowledge acquired, we think the reader will be ready to learn the general plan of the book, which is given in the final part of this chapter.

1.2 Historical Perspective

There is a general consensus that reliability as a discipline was established during World War II (1939–1945), when the high number of failures noticed for military equipment became a concern, requiring an institutional approach. However, attempts to design a fair quality into an artefact or to monitor the way this quality is maintained during usage (i.e. reliability concerns) were first made a long time ago.

1.2.1 Reliability Prehistory

This story may begin thousands of years ago, during the Fifth Dynasty of the Ancient Egyptian Empire (2563–2423 BCE), when the pharaoh Ptah-hotep stated (in other words, of course, but this was the idea) that good rules are beneficial for those who follow them [2]. This is the first known remark about the quality of a product, specifically about the design quality. Obviously, during the following ages, many other milestones in quality and reliability history occurred:

- In Ancient Babylon, the Code of Hammurabi (1760 BCE) said: 'If the ship constructed for somebody is damaged during the first year, the manufacturer has to re-build it without any supplementary cost.' This could be considered the first specification about the reliability of a product!
- In China, during the Soong dynasty (960–1279 CE), there were six criteria for the quality and reliability of arches: to be light and elastic, to withstand bending and temperature cycles, and so on. Close enough to modern specifications!
- At the same time in Europe, the guilds (associations of artisans in a particular trade) elaborated principles of quality control, based on standards. Royal governments promoted the control of quality for purchased materials; for instance, King John of England (1199–1216) asked for reports on the construction of ships.
- In the 1880s mass production began and F.W. Taylor proposed the so-called 'Scientific Management': assembly lines, division of labour, introduction of work standards and wage incentives. Latter, he wrote two basic books on management: *Shop Management* (1905) and *The Principles of Scientific Management* (1911).
- On 16 May 1924, W.A. Shewhart, engineer at the Western Electric Company, prepared a little memorandum about a page in length, containing the basics of the control chart. He later became the 'father' of statistical quality control (SQC): methods based on continual on-line monitoring of process variation and the concepts of 'common cause' and 'assignable cause' variability.
- In 1930, H.F. Dodge and H.G. Romig, working at Bell Laboratories, introduced the so-called Dodge–Romig tables: acceptance sampling methods based on a probabilistic approach to predicting lot acceptability from sampling results, centred on defect detection and the concept of acceptable quality level (AQL).

All these contributions (and many others) have paved the way for current, modern approaches in quality and have prepared the development of reliability as a discipline (see Sections 1.2.2 and 1.2.3).

Following World War II, in parallel with the rise of reliability as a discipline, the quality field has continued to be developed, mainly in the USA. Two eminent Americans, W. Edwards Deming and Joseph Juran, alongside the Japanese professor Kaouru Ishikawa, were successful in promoting this field in Japan. Another name has to be mentioned, Philip B. Crosby, who initiated the quality-control programme named 'Zero Defects' at Martin Company, Orlando, Florida, in the late 1960s. In 1983, Don Reinertsen proposed the concept of concurrent engineering as an idea to quantify the value of development speed for new products.

Due to the efforts of the above, a new discipline, called quality assurance, was born, aimed at covering all activities from design, development and production to installation, servicing, documentation, verification and validation.

1.2.2 The Birth of Reliability as a Discipline

The following two events are considered the founding steps of reliability as a discipline:

1. During World War II, the team led by the German rocket engineer Wernher Magnus Maximilian Freiherr von Braun (later the father of the American space programme) developed the V-1 rocket (also known as the Buzz-Bomb) and then the V-2 rocket. The repeated failures of the rockets made a safe launch impossible. Von Braun and his team tried to obtain a better device, focusing on improving the weakest part, but the rockets continued to fail. Eric Pieruschka, a German mathematician, proposed a different approach: the reliability of the rocket would be equal to the product of the reliability of its components. That is, the reliability of all components is important to overall reliability. This could be considered the first modern predictive reliability model. Following this approach, the team was able to overcome the problem [3].
2. In 1947, Aeronautical Radio Inc. and Cornell University conducted a reliability study on more than 100 000 electronic tubes, trying to identify the typical causes of failures. This could be considered the first systematic FA. As a consequence of this study, on 7 December 1950, the US Department of Defense (DoD) established the *Ad Hoc Group on Reliability of Electronic Equipment* (AHGREE), which became in 1952 the *Advisory Group on the Reliability of Electronic Equipment* (AGREE) [4]. The objectives proposed by this group are still valid today: (i) more reliable components, (ii) reliability testing methods before production, (iii) quantitative reliability requirements and (iv) improved collection of reliability data from the field (including failure analyses). Later, in 1956, AGREE elaborated the first reliability handbook, titled 'Reliability Factors for Ground Electronic Equipment'. This report is considered the fundamental milestone in the birth of reliability engineering.

1.2.3 Historical Development of Reliability

The reliability discipline has evolved around two main subjects: reliability testing and reliability building. In this discussion, we think the history of *prediction methods* (which are based on FA) is the relevant element, being deeply involved in both subjects: as input data for reliability building and as output data for reliability testing.

Following the issue of the first reliability handbook, the TR-1100 'Reliability Stress Analysis for Electronic Equipment', released by RCA in November 1956, proposed the first models for computing failure rates of electronic components, based on the concept of activation energy and on the Arrhenius relationship.

On 30 October 1959, the Rome Air Development Center (RADC; later the Rome Laboratory, RL) issued a 'Reliability Notebook', followed by some other basic papers contributing to the development of knowledge in the field: 'Reliability Applications and Analysis Guide' by D.R. Earles (September 1960); 'Failure Rates' by D.R. Earles and M.F. Eddins (April 1962); and 'Failure Concepts in Reliability Theory' by Kirkman (December 1963).

From the early 1960s, efforts in the new reliability discipline focused on one of the RADC objectives: developing prediction methods for electronic components and systems. Two main approaches were followed:

1. '*The statistical approach*', using reliability data gathered in the field. The first reliability prediction handbook, MIL-HDBK-217A, was published in December 1965 by the US Navy. This was a huge success, being well received by all designers of electronic systems, due to its flexibility and ease of use. In spite of its wrong basic assumption, an exponential distribution of failures [5], the handbook became the almost unique prediction method for the reliability of electronic systems, and other sources of reliability data gradually disappeared [6].

2. *'The physics-of-failure (PoF) approach'*, based on the knowledge of failure mechanisms (FMs) (investigated by FA) by which the components and systems under study are failing. The first symposium devoted to this topic was the 'Physics of Failure in Electronics' symposium, sponsored by the RADC and the IIT Research Institute (IITRI), in 1962. This symposium later became the 'International Reliability Physics Symposium' (IRPS), the most influential scientific event in failure physics. On 1 May 1968, the MIL-HDBK-175 Microelectronic Device Data Handbook appeared (revised in 24 October 2000), with a section focused on FA ('Reliability and Physics of Failure').

The two approaches seemed to be diverging; system engineers were focused on the 'statistical approach' while component engineers working in FA were focused on the PoF approach. But soon both groups realised that the two approaches were complementary and attempts to unify the two methods have been made.

This has been facilitated by the fact that, in 1974, the RADC, which was the promoter of the PoF approach, became responsible for preparing the second version of MIL-HDBK-217, and of the subsequent successive versions (C...F), which tried to update the handbook by taking into account new advances in technology. However, instead of improved results, more and more sophisticated models were obtained, considered 'too complex, too costly and unrealistic' by the user community [6]. Another attempt, executed by RCA, under contract to the RADC, which tried to develop PoF-based models, was also unsuccessful. This was because the model users did not have access to information about the design and construction of components and systems.

In the 1980s, various manufacturers of electronic systems tried to develop specific prediction methods for reliability. Examples include the models proposed for automotive electronics by the Society of Automotive Engineers (SAE) Reliability Standards Committee and for the telecommunication industry (Bellcore reliability-prediction standards).

For the last version (F) of MIL-HDBK-217, issued on 10 July 1992, two teams (IIT/Honeywell and CALCE/Westinghouse (Centre for Advanced Life Cycle Engineering)) were commissioned by the (RL) RADC to provide guidelines. Both teams suggested the following conclusions:

- The constant-failure-rate model (based on an exponential distribution) is not valid in real life.
- Electromigration and time-dependent dielectric breakdown could be modelled with a lognormal distribution.
- Arrhenius-type formulation of the failure rate in terms of temperature should not be included in the package failure model.
- Temperature change and humidity must be considered as key acceleration factors.
- Temperature cycling is more detrimental for component reliability than the steady-state temperature at which the device is operating, so long as the temperature is below a critical value.

Similar conclusions were supported by the studies [7, 8]. In fact, these conclusions are preparing a unified approach on prediction method, which has not been issued yet.

During the first 30 years of the reliability discipline, military products acted as the main drivers of reliability developments. However, starting from the 1980s, commercial electronic components became more and more reliable. In June 1994, the so-called 'Acquisition Reform' took place: the US DoD abolished the use of military specifications and standards in favour of performance specifications and commercial standards in DoD acquisitions [9]. Consequently, in October 1996, MIL-Q-9858, Quality Program Requirements, and MIL-I-45208 A, Inspection System Requirement, were cancelled without replacement. Moreover, contractors were henceforth required to propose their own methods for quality assurance, when appropriate. The DoD policy allows the use of military handbooks only for guidance. Many professional organisations (e.g. IEEE Reliability Society) attempted to produce commercial reliability documents to replace the vanishing military standards [10]. A number of international standards were also produced, including IEC TC-56, some NATO documents, British documents and Canadian documents. In addition to the new standardisation activities, the RL is also undertaking a number of research programmes to help implement acquisition reform.

However, some voices, such as Demko [11], consider a logistic and reliability disaster to be possible, because commercial parts, standards and practice may not meet military requirements. For this purpose, in June 1997 the IITRI of Rome (USA) developed SELECT, a tool that allows users to quantify the reliability of commercial off-the-shelf (COTS) equipment in severe environments [12]. Also, beginning from April 1994, a new organisation, the Government and Industry Quality Liaison Panel (GIQLP), made up of government agencies, industry associations and professional societies, was intimately involved in the vast changes being made in the government acquisition process [13].

MIL-HDBK-217 was among the targets of the Acquisition Reform, but it was impossible to replace it, because no other candidates (prediction methods) were available. However, attempts at elaborating a new handbook for predicting the reliability of electronic systems were made. Supplementary to the existing handbook, this document, called the 'New System Reliability Assessment Method', has to manage system-level factors [6]. The system has to take into account previous information about the reliability of similar systems (built with similar technologies, for similar applications and performing similar functions), as well as test data about the new system (aimed at producing an initial estimate of the system's reliability).

On the other hand, a well-known example of commercial standards replacing the old military standards is given by the ISO 9000 family, first issued in 1987, with updates in 2000, 2004 and 2008. Basically, the IS0 9000 standards aimed to provide a framework for assessing the management system in which an organisation operates in relation to the quality of the furnished goods or services. The concept was developed from the US Military Standard for quality, MIL-Q-9858, which was introduced in the 1950s as a means of assuring the quality of products built for the US military services. But there is a fundamental new idea promoted by ISO 9000: the quality management systems of the suppliers are audited by independent organisations, which assess compliance with the standard and issue certificates of registration. The suppliers of defence equipment were assessed against the standards by their customers. In a well-documented and convincing paper about ISO 9000 standards, Patrick O'Connor elucidated the weak points of this approach [14]:

- The philosophy of 'total quality' demands close partnership between supplier and purchaser, which is destroyed if the 'third party' has to audit the quality management of the supplier.
- It is hard to believe that this 'third party' is able to have the appropriate specialist knowledge about the products to be delivered.
- In fact, the ISO 9000 standards aim only to verify that the personnel of the supplier are strictly observing the working procedures and not whether the procedures are able to ensure a specified quality and reliability level. Of course, some organisations have generated real improvements as a result of registration, but it is not obvious that this will happen in all cases. Moreover, the high costs required to implement ISO 9000 may have detrimental effect on a real improvement, by technical corrective actions, in the manufacturing process.
- Two explanations are proposed for wide adoption of these standards, in spite of the solid arguments of many leading teachers of quality managers: (i) the tendency to believe that people perform better when told what to do, rather than when they are given freedom and the necessary skills and motivation to determine the best ways to perform their work and (ii) working only with 'registered' suppliers is the easy way for many bureaucrats to select the most appropriate suppliers for their products. The main responsibility is transferred to the 'third party'.

As one can see, the ISO 9000 approach seeks to 'standardise' methods that directly contradict the essential lessons of the modern quality and productivity revolution (e.g. 'total quality'), as well as those of the new management.

Some other important contributors to the domain of reliability include the following:

- Genichi Taguchi, who proposed robust design, using fractional factorial design and orthogonal arrays in order to minimise loss by obtaining products with minimal variations in their functional characteristics.
- Dorian Shainin, reliability consultant for various companies, including NASA (the Apollo Lunar Module), who supported the idea of discovering and solving problems early in the design phase, before costly manufacturing steps are taken and before customers experience failures in the field.
- Wayne B. Nelson, who developed a series of methodologies on accelerated testing, based on FA.
- Larry H. Crow, independent consultant as well as an instructor and consultant for ReliaSoft Corporation, who made contributions in the areas of reliability growth and repairable system data analysis.
- Gregg K. Hobbs, the inventor of the highly accelerated life test (HALT), a stress-testing methodology, aimed at obtaining information about the reliability of a product.
- Michael Pecht, the founder of CALCE and the Electronic Products and Systems Consortium at the University of Maryland, which made essential contributions to the study of FMs of electronic components.
- Patrick O'Connor, who made contributions to our understanding of the role of failure physics in estimating component reliability; he also proposed convincing arguments against ISO 9000.

1.2.4 Tools for Failure Analysis

Initiated in 1947 for electronic tubes, FA was developed mainly for other microelectronic devices (transistors, integrated circuits (ICs), optoelectronic devices, microsystems and so on), but also for electronic systems. Since 1965, the number of transistors per chip has been doubling every 24 months, as predicted by Moore. Today, ICs are made up of hundreds of millions of transistors grown on a single chip. Key FA tools have been developed continuously, driving the growth of the semiconductor industry by solving difficult test problems.

The search for physical-failure root causes (especially with physical inspection and electrical localisation) is aimed at breaking through any technology barrier in each stage of chip development, package development, manufacturing process and field application, offering the real key to eradicate the error. Testing provides us with information on the electrical performance; FA can discover the detractors for the poor performance [15]. In today's electronic industry FA is squeezed between the need for very rapid analysis to support manufacturing and the exploding complexity of the devices. This requires knowledge of subjects like design, testing, technology, processing, materials science, physics, chemistry and even mathematics [16]!

As can be seen in Figure 1.1, a number of key FA tools and advanced techniques have been developed. These play essential roles in the development of semiconductor technology.

1.3 Terminology

The lack of precision in terminology is a well-known disease of our times. Very often, specialists with the same opinion about a phenomenon are in disagreement because they are implicitly using different definitions for the same term, or because various terms having the same meaning. Of course, standards for the main terms of any technical field have been elaborated, but it is difficult to read a book with an eye to one or many standards. This is why we felt that this book needs a glossary. The glossary is located within the back matter and contains only the basic terms referring to the subject of the book, **failure analysis for electronic components and systems**, divided into two main sections:

- Terms related to electronic components and systems.
- Terms related to FA.

Other more specific terms will be explained within the main body of the text, when necessary.

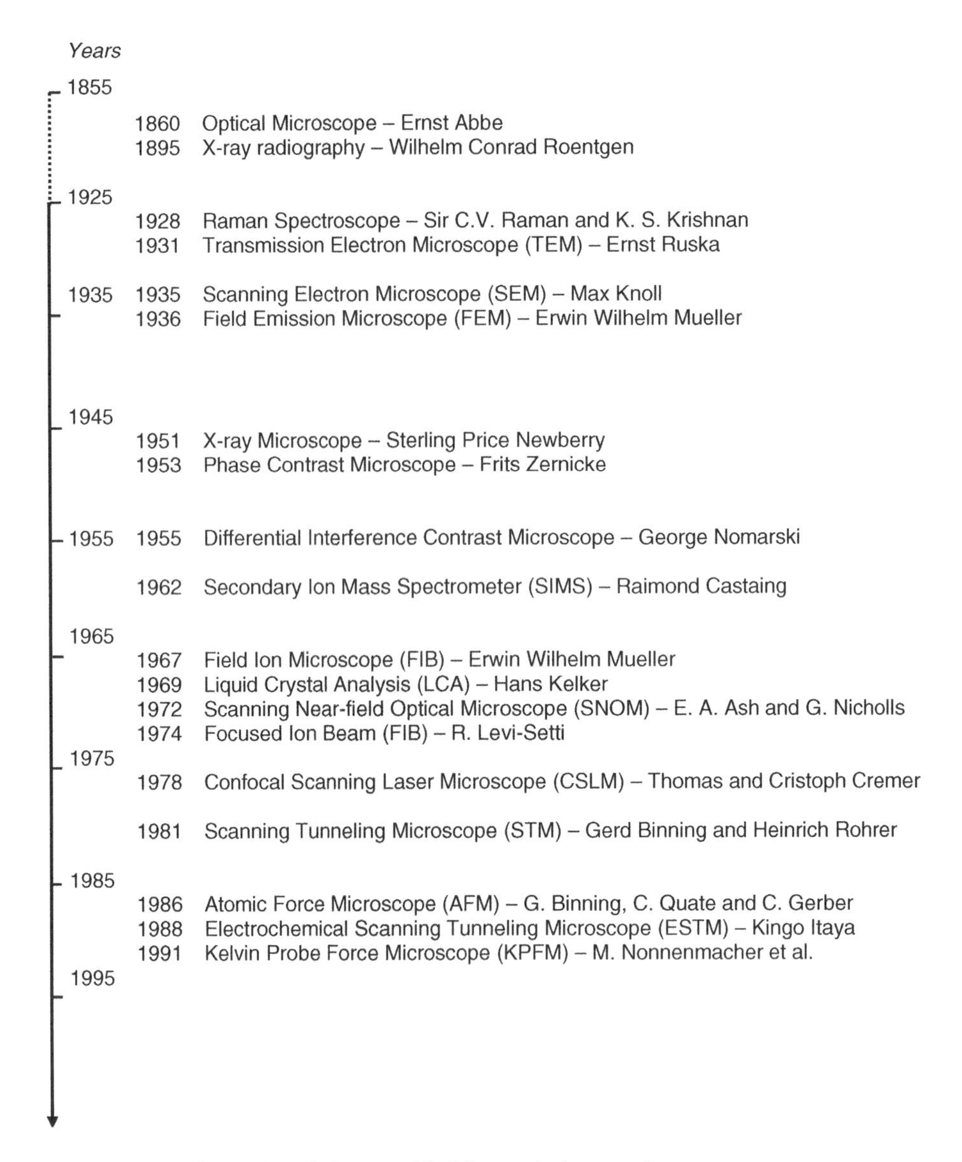

Figure 1.1 History of the main techniques used in failure analysis: year of appearance, name, acronym, name of inventor (After [14])

1.4 State of the Art and Future Trends

The current issues related to FA of electronic components and systems can be structured around three main areas:

- techniques of FA;
- FMs;
- models for the PoF.

In this section, the most important subjects of each area will be discussed.

1.4.1 Techniques of Failure Analysis

Today, when FA is performed at *system level*, we have to analyse not only discrete (active and passive) components but a large range of ultra-high-density ICs, with a high design complexity that exceeds 300 million gates, manufactured by a huge variety of technologies (bipolar silicon, CMOS, BiCMOS, GaN, GaAs, InP, GaN, SiC, complex heterojunction structures and microelectromechanical systems, MEMSs), which will be detailed in Chapter 5. This is why the today's analyst faces complex equipment sets (curve tracers, optical microscopes, decapsulation tools, X-ray and acoustic microscopies, electron and/or optical and/or focused ion beam (FIB) tools, thermal detection techniques, the scanning probe atomic force microscope, surface science tools, a great variety of electrical testing hardware and so on) that are necessary to realise a spatial and complex FA. FA is a highly technical activity with increasingly complex, sophisticated and costly specialised equipments. It is very difficult to achieve a balance between customer satisfaction, cost-effectiveness and future challenges. Very often the analyst must use a limited set of tools, as the cost of all the required tools exceeds the budget of current operations. FA techniques are used to confirm and localise physical defects. The final objective is to find the root cause.

Failure modes and effects analysis (FMEA) is the systematic method of studying failure, formally introduced in the late 1940s for military usage by the US armed forces [3]. Later, FMEA was used for aerospace/rocket development to avoid errors in small sample sizes of costly rocket technology. Now FMEA methodology is extensively used in a variety of industries, including semiconductor processing, food service, plastics, software and health care. It is integrated into advanced product quality planning (APQP) to provide primary risk-mitigation tools and timing in the preventing strategy, in both design and process formats. FMEA is also useful at component level, especially for complex components (ICs, MEMS, etc.). In FMEA, failures are prioritised according to how serious their consequences are, how frequently they occur and how easily they can be detected. Current knowledge about risks of failures and preventive actions for use in continuous improvement are also documented. FMEA is used during the design stage in order to avoid future failures and later for process control, before and during ongoing operation of the process. Ideally, FMEA begins during the earliest conceptual stages of design and continues throughout the life of the product or service. The purpose of FMEA is to take action to eliminate or reduce failures, starting with the highest-priority ones.

At *component level*, a broad definition of FA includes: collection of background data, visual examination, chemical analysis, mechanical properties, macroscopic examination, metallographic examination, micro-hardness, scanning electron microscopy (SEM) analysis, microprobe, residual stresses and phases, simulation/tests, summary of findings, preservation of evidence, formulation of one or more hypotheses, development of test methodologies, implementation of tests/collection of data, review of results and revision of hypotheses. Each time, the customer will be notified.

First, the causes of a failure can be classified according to the phase of a product's life cycle in which they arise – design, materials processing, component manufacturing or service environment and operating conditions. Then two main areas of FA enable fast chip-level circuit isolation, circuit editing for quick diagnostic and problem-solving, helping bring forward semiconductor development:

- **Physical inspection**, represented by three important tools: SEM, emission microscopy and transmission electron microscopy (TEM).
- **Electrical localisation**, executed mainly with liquid crystal analysis (LCA), photo electron microscopy (PEM) and FIB.

The package global localisation tool infra-red lock-in thermography (IR-LIT) became widely available in 2005 and is the most popular tool for global localisation for complex packages, such as system-in-package (SiP) and system-on-chip (SoC). Today the tool support for SoC development is X-ray CT, due to a significant resolution and speed improvement. FA has given and gives a continuous contribution to technological innovation in the whole history of semiconductor development.

When ICs are analysed, a number of tools and techniques are used due to device-specific issues: additional interconnection levels, power distribution planes and flip-chip packaging completely eliminate the possibility of employing standard optical or voltage-contrast FA techniques without destructive intervention. The defect localisation utilises techniques based on advanced imaging, and on the interaction of various probes with the electrical behaviour of devices and defects.

The thermal interaction between actively operated electronic components and applied characterisation tools is one of the most important interactions within FA and reliability investigations. It allows different kinds of thermal interaction mechanism to be utilised, which would normally have to be separated–for instance into classes with respect to thermal excitation and/or detection, spatial limitations and underlying physical principle. Although they all have in common the ability to link the thermo-electric device characteristic to a representing output signal, they have to be interpreted in completely different ways. Recently, the complementarity of the methods for localisation and characterisation as well as the according industrial demands and related limitations have been shown [17]; techniques such as IR-LIT and thermal induced voltage alteration (TIVA), case studies and the capability of non-established techniques like scanning thermal microscopy (SThM), thermal reflectance microscopy (TRM) and time domain thermal reflectance (TDTR) are also presented and their impact on reliability investigations is discussed.

Over the last few years, the increased complexity of devices has scaled the difficulty in performing FA. Higher integration has led to smaller geometry and better wire-to-cell ratios, thus increasing the complexity of the design. These changes have reduced the effectiveness of most of the current FA techniques; over the past few years, a variety of techniques and tools, such as electron-beam (E-beam) probers, FIB, enhanced imaging SEM and field emission SEM (FESEM), have been developed to determine the defects at wafer level. All these tools improve FA capabilities, but at substantial cost, running into hundreds of thousands of dollars. Some other examples of new techniques are given below:

- A strategy was derived for FA in random logic devices (such as microprocessors and other VLSI chips) where the electrical scheme is not known. This strategy is based on the use of a test tool composed of an SEM allied to a voltage contrast, an exerciser, an image processing system and a control and data processing system [18].
- Three new FA techniques for ICs have been developed recently using localised photon probing with a scanning optical microscope (SOM) [19]. The first two are light-induced voltage alteration (LIVA) imaging techniques that (i) localise open-circuited and damaged junctions and (ii) image transistor logic states. The third technique uses the SOM to control logic states optically from the IC backside. LIVA images are produced by monitoring the voltage fluctuations of a constant current power supply as a laser beam is scanned over the IC. High selectivity for localising defects has been demonstrated using the LIVA approach. Application of the two LIVA-based techniques to backside FA has been demonstrated using an infrared laser source.
- It is critical to develop improved analysis techniques that are easier to use, less damaging, more sensitive and provide better spatial resolution. One example is 'passive' techniques, which are non-invasive, in the sense that the normal operation of the IC provides the information or energy being measured. Recently, dynamic photoelectric laser-stimulation techniques were applied to mixed-mode ICs, where the major difficulty is their considerable intrinsic sensitivity [20]. Indeed, the analogue circuitry is more sensitive than the digital circuitry since a slight change in an electrical parameter can trigger a functionality failure. This property limits the defect localisation because of the complex interpretation of the results: the laser-stimulation mapping. In this case, dynamic laser-stimulation mapping is coupled with photoelectric impact simulations run on a previously analysed structure. The goal is to predict and interpret the laser-sensitivity mapping and to isolate the defective areas in the analogue devices.
- A technique used for decapsulating the device for FA is the ultra-short-pulse laser-ablation-based backside sample-preparation method [21]. This technique is contactless, nonthermal, precise, repetitive and adapted to each type of material present in IC packages. However, it can create thermal

stresses to the device. In order to minimise these stresses, a new method was proposed for controlling the thermal effect of the laser on the component [22].

- Various methods of preparation to repack dice with a nondestructive access to the backside are shown in [23]. With new processes, and where backside milling is not possible, this work is mandatory for any fault localisation technique. Various mountings – to be chosen depending on the original package (mainly BGA) and on the requirements of the emission techniques – and how these techniques can be used for a new product family (SiP, with multiple dice inside a package) are described. In this specific case the challenge is to extract the failed die without destroying the module.
- As new technologies in the electronic environment develop from 2D IC to 3D complex packages, it becomes necessary to find new techniques to detect and localise the different kinds of failure. A solution to localise defects for SiP devices is to measure the magnetic field that is generated by the current flowing through the device with a magnetic microscope (Magma C20) and compare it with several simulated faults in order to choose the most probable one [24].

1.4.2 Failure Mechanisms

From a technical perspective, failure can be defined as the cessation of function or usefulness. FA is the process of investigating such a failure. Basically, FA analysis the failure modes (FMos) with the aim of identifying the FMs, by using optical, electrical, physical and chemical analysis techniques.

Reliability is built into the device at the design and manufacturing process stages. In most practical cases, the final damage rarely reveals a direct physical FM; often the original cause (or complete scenario of failure) is hidden by secondary post-damage processes. On the other hand, it is impossible to eradicate failures during the manufacturing process and upon field use. Therefore, FA must be performed to provide timely information and prevent the recurrence of similar failures.

The wafer fabrication and assembly process involves numerous steps using various types of material. This, combined with the fact that devices are used in a variety of environments, requires a wide range of knowledge about the design and manufacturing processes. This explains why FA of semiconductor device is becoming increasingly difficult as VLSI technology evolves towards smaller features and semiconductor device structures become more complex. Since it is usually not possible to repair faulty component devices in a VLSI, each device in a chip can become a single point of failure unless some redundancy is introduced. Therefore, VLSIs have to be designed based on the characteristics of the worst devices rather than on those of average devices. Even if a chip is equipped with some redundant device, today's scale of integration is so high that the yield requirement will still be very severe. The final chip yield is governed by the device yield.

A recent report [25] has demonstrated that, for any item, once the major cause of failure is somehow identified or assumed, the Monte Carlo method may be used to study yield problems. This method was applied to the analysis of leakage current distribution for double-gate MOSFETs; the microscopic FM that limits the final yield was identified. This explains experimental data very well. The insight into the FM gives clear guidelines for yield enhancement and facilitates device design alongside the quantitative yield prediction. It is useful for yield prediction and device design. Transistors should be designed such that I_t (the maximum current generated by a single trap) is very much lower than the tolerable leakage current at the specified cumulative probability. The method does not have any convergence problems, unlike the conventional Monte Carlo approach.

At system level, a general study [6] performed by the Reliability Analysis Centre (RAC) into the predominant causes of failure in electronic systems led to the results shown in Table 1.1. As one can see, only 22% of the failures are directly linked to the reliability of the components. An important number of failures have noncomponent causes, such as defects in design and manufacturing of the system. This could be considered to go against the general belief that the reliability of the components is decisive for the reliability of the system. However, the results from Table 1.1 have to be understood as follows: when appropriately selected components are used, they are responsible

Table 1.1 Distribution of failure causes for electronic systems

Failure cause	Details	Percentage (%)
Parts	Failure of components (e.g. transistors, diodes, ICs, resistors, etc.)	22
Mysterious cause	System failures not reproduced upon further testing; uncertain that there are actual failures	20
Manufacture	Anomalies of the manufacturing process (e.g. wrong manipulation, faulty solder joints, etc.)	15
Induced	Produced by applied stress (e.g. electrical overstress, maintenance-induced failures)	12
Design	Inadequate design (e.g. nonrobust design for environmental stress)	9
Wearout	Wearout-related failure mechanisms of parts and interfaces	9
Software	System failures produced by software fault	9
Management	Wrong management (e.g. wrong interpretation of system requirements, failure to provide required resources)	4

After [6].

for less than a quarter of system failures. If inappropriate components are used, they can determine failures beyond the 22% already mentioned (see Table 1.1): for example, failures from induced causes, design, wear-out and so on. Actually, the total percentage of possible failures linked to the parts is higher that 50%.

1.4.3 Models for the Physics-of-Failure

For electronic components, models describing the action of FM versus time and specific stresses arose early in reliability history (as mentioned at Section 1.2.3). This PoF approach has to be used because the statistical processing of reliability data is significant only for a population affected by a single FM; otherwise the extrapolation of data beyond the duration of the reliability tests is no longer valid. Of course, for such models, the reliability engineer has to perform extensive FA in order to identify accurately the populations affected by each FM. In the last few years, these have been called 'empirical' models, because the new tendency at component level is to develop 'physical' models; that is, models based on the physical or chemical phenomena responsible for degradation or failure of electronic components and materials (details are provided in Chapter 2, Section 2.3 and in Chapter 5).

However, the 'statistical' approach is still used, but mainly for electronic systems, offering the only possible solution for assessing the reliability of such a system. This is so because the 'physical' approach is difficult, if not impossible, to use at system level. For example, a PoF-like model developed by for small-scale CMOS was virtually unusable by system manufacturers, requiring input data (details about design layout, process variables, defect densities and so on) that are known only by the component manufacturer [6]. So, in spite of their lower accuracy, 'statistical' models (e.g. MIL-HDBK-217) are used to evaluate the reliability of electronic systems.

Obviously, the solution is a unified approach, trying to get the best features of each model. It is easy to say this, but more difficult to accomplish. So, such a unified approach is still a dream. The new tool that has to be created from this new approach must offer the possibility of an interactive design (unlike MIL-HDBK-217 models), allowing a trade-off between the physical performances of the device and the reliability implications thereof [6].

Meanwhile, many models were developed for the PoF of FM in semiconductor technology. No details will be furnished in this section, as Chapter 5 is dedicated to the presentation of the typical FMs for various categories of electronic components. As an example of such models, we note here the important contribution of the CALCE Electronic Packaging Research Center, a research team led by

Michael Pecht of Maryland University (USA); a series of tutorial papers on FM linked to packaging issues were published since 1991 by IEEE Transactions on Reliability.

1.4.4 Future Trends

Today, FA is the key method in reliability analysis. It is impossible to conceive of a serious investigation into the reliability of a product or process without FA. The idea that failure acceleration by various stress factors (which is the key to accelerated testing) could be modelled only for the population affected by a single FM greatly promoted FA as the only way to separate populations damaged by specific FMs.

A large range of methods are now used, from (classical) visual inspection to such expensive and sophisticated methods as atomic force microscopy and scanning near-field optical microscopy. Many others are still waiting to be created.

Recently, through device shrinking and complexity growing, it has become more and more difficult to carry out FA of semiconductor devices. A recent prediction [26] of the Semiconductor Industry Association (SIA), made in the International Technology Roadmap for Semiconductor (ITRS), says that silicon technology will continue its historical rate of advancement predicted by the Moore's law. So the challenges for FA in the next few years will be extended to broad new aspects (design for analysis or design for test; physical limit – tools for chip, tools for package; chip–package co-design; and organisational issues like FA cost and FA cycle time). Above all, it is clear that failure diagnosis and failure position analysis have to increase in accuracy. Hitachi High Technologies, Ltd. has developed an extremely fine SEM-type mechanical proving system [27], for example; a high-precision probe and stage mechanism corresponding to the fine devices, a six-probe mechanism expandable to applications including inverter testing and a high-precision unit transistor testing were all investigated, together with an in-vacuum probing, sample-exchanging mechanism and a computer aided design (CAD) navigation system. As this system was applicable to 65 nm devices, it seems likely it will be possible to apply it to any device of such a scale in the future.

The microsystems are relatively new devices, containing, on a single chip, a mixture of components–a sensor, an actuator (a mechanical component) and the electronics–which creates new challenges for their reliability [28]. The package should protect the chip from an often harsh and demanding environment (as for the 'classical' microelectronic devices: transistors, ICs, etc.), but is also an interface between the sensor and that environment. The small dimensions of the mechanical elements of the actuator produce new FMs and the interactions between mechanical, electrical and material reliability must be taken into account. Moreover, the third dimension (the depth) of the structure cannot be ignored, as occurred for microelectronic devices, where all the simulations are basically two-dimensional. This collection of reliability risks has to be taken into account when FA is performed for any type of microsystem. It should be noted that the subject is common in papers today, but we still have limited knowledge about how microsystems fail. The main explanation is a lack of specific tools for studying microsystems. Fabricating multiple devices on the same chip will have to deal with more FMos. Complex interactions of cross-domain signals, interference and substances induce new FMos and FMs.

Another challenge for FA comes from a new domain, called nanotechnology. Here everything is new and the FMs for nanomaterials, which are different from those for the same materials at micro level, have to be studied. Supplementary issues are induced by organic materials, which is a new trend in this field. Also, at nano level, new techniques for FA have to be created [29]. As one can see, nano-reliability (study of the reliability of nano-devices) offers a huge range of subjects for FA. The near future will see an important step forward in this field.

In conclusion, it is both easy and difficult to predict the future evolution of FA. Easy because everyone working in this domain can see the current trend. FA is still in a 'romantic' period, with fabulous pictures and smart figures smashing customers, convinced by such a 'scientific' approach.

Seldom do these users of electronic components understand the essence of the FA procedure, because the logic is frequently missing. But this situation is only a temporary one. Very soon, the procedures for executing FA will be stabilised and standardised, allowing any user of an electronic component to verify the reliability of the purchased product.

But it is also difficult to predict the evolution of FA, because the continuous progress in microelectronics and microtechnology makes it almost impossible to foresee with good accuracy the types of electronic component that will be most successful on the future market. And FA must serve this development, being one step ahead and furnishing manufacturers with the necessary tools for their researches. However, with sufficiently high probability, one may say that nano-devices (or even nanosystems) will become a reality in the next five years, so we must be prepared to delve deeper into the matter, with more and more expensive investigation tools.

1.5 General Plan of the Book

At this point, having delivered a lot of information about the past, the present and the future of FA, we shall explain to the reader the plan of the rest of the book.

FA is a vast subject, being used for almost all manufactured product. Thus procedures for performing FA have been developed in all technical domains. Obviously, in order to be useful, a book about FA must focus on a limited range of subjects; otherwise it will be impossible for the reader to understand many points. In the present volume, we are firmly focused on *electronic components*, which are, in fact, the domain in which FA was initiated. Of course, *electronic systems* will be discussed too, because their reliability is highly dependent on the reliability of electronic components.

So we have defined the subject, which is still a broad one, but will be comprehensible in a single volume, we hope!

The book is divided into four main parts, providing possible answers to four questions:

- **Why** is it so important to use FA (Chapter 2)? Eight possible reasons are presented: (i) forensic investigation, (ii) reliability modelling, (iii) reverse engineering, (iv) controlling critical input variables, (v) design for reliability, (vi) process improvement, (vii) saving money by early control and (viii) a synergetic approach.
- **When** is it appropriate to use FA (Chapter 3)? It is recommended that FA be used during the whole life cycle of any electronic components, starting with design (based on previous knowledge about FMs of similar or quasi-similar components, as part of the concurrent engineering approach; that is, participation from the first phase until the last one of all persons responsible for product achievement), continuing with prototyping, fabrication and (most importantly) post-fabrication, and on into reliability testing and operational life.
- **How** should FA be used (Chapter 4)? This guide contains a large variety of methods of FA (electrical methods, thermal methods, optical methods, electron microscopy, mechanical methods, X-ray methods, spectroscopic methods, acoustic methods, laser methods and so on).
- **What** is the result of using FA (Chapter 5)? Typical FMs for the most important technologies used for manufacturing electronic components are detailed (silicon bipolar and MOS technologies, optoelectronic and photonic technologies, nonsilicon technologies, hybrid technologies and microsystem technologies). The main technologies are discussed, not the main products, because the FMs are more or less dependent on technologies–and, of course, on the main applications of the products!

Chapter 6 provides 12 complex case studies, covering the main technologies and using various methods of FA. This is a synthesis of the previous chapters, containing some practical lessons about using FA.

Finally, Chapter 7 provides some conclusions, which will be of value to manufacturers and users of electronic components.

The website of the book offers Q&A sessions about each chapter, as an effective learning tool.

The book as a whole aims to serve as a reference work for those involved in the design, fabrication and testing of electronic components, but also for those who are using these components in complex systems and want to discover the roots of the reliability flaws for their products.

References

1. Kirkman, R.A. (1963) Failure concepts in reliability theory. *IEEE Trans. Reliab.*, **51**, 1–10.
2. Motoiu, R. (1994) *Quality Engineering*, Chiminform Data, Bucharest.
3. DeVale, J. (1998) Traditional Reliability, Carnegie Mellon University Internal Reports, Spring 1998, http://www.ece.cmu.edu/~koopman/des_s99/traditional_reliability/index.html. (Accessed 2010).
4. Băjenescu, T.I. and Bâzu, M. (1999) *Reliability of Electronic Components. A Practical Guide to Electronic Systems Manufacturing*, Springer, Berlin and New York.
5. Nash, F.R. (1993) *Estimating Device Reliability: Assessment of Credibility*, Chapter 6, Kluwer, Boston, MA.
6. Denson, W. (1998) The history of reliability prediction. *IEEE Trans. Reliab.*, **47** (3), 321–328.
7. Kopanski, J., Blackbum, D.L., Harman, G.G. and Berning, D.W. (1991) Assessment of reliability concerns for wide temperature operation of semiconductor device circuits. Transactions of the 1st International High Temperature Electronics Conference, Albuquerque, 1991, pp. 137–142.
8. Pecht, M., Lall, P. and Hakim, E. (1992) Temperature dependence on integrated circuit failure mechanisms, in *Advances in Thermal Modeling III*, Chapter 2 (eds A. Bar-Cohen and A.D. Kraus), IEEE and ASME Press, New York, pp. 61–152.
9. Yates W. and Johnson, R. (1997) Total quality management in U. S. DoD electronics acquisition. Proceedings of the Annual Reliability and Maintainability Symposium, Philadelphia, January 13–16, 1997, pp. 571–577.
10. Ermer, D. (1996) Proposed new DoD standards for product acceptance. Proceedings of the Annual Reliability and Maintainability Symposium, Las Vegas, January 22–25, 1996, pp. 24–29.
11. Demko, E. (1996) Commercial-off the shelf (COTS): challenge to military equipment reliability. Proceedings of the Annual Reliability and Maintainability Symposium, Las Vegas, Nevada, January 22–25, pp. 7–12.
12. Nicholls, D. (1996) Selection of equipment to leverage commercial technology (SELECT). Proceedings of the Annual Reliability and Maintainability Symposium, Las Vegas, Nevada, January 22–25, 1996, pp. 84–90.
13. Schneider, C. The GIQLP-product integrity's link to acquisition reform. Proceedings of the Annual Reliability and Maintainability Symposium, January 13–16, 1997, pp. 26–28.
14. O'Connor, P.D.T. (2000) Reliability past, present and future. *IEEE Trans. Reliab.*, **49** (4), 335–341.
15. Xue, M., Görlich, S. and Quek, J. Semiconductor Failure Analysis for 60 Years and Beyond http://www.semiconsingapore.org/ProgrammesandEvents/cms/groups/public/documents/web_content/ctr_022432.pdf. (Accessed 2010).
16. Henderson, C.L. Advanced to Failure and Yield Analysis. Overview of a two days course organized by Semitracks Inc., www.semitracks.com/courses/fa-course.htm. (Accessed 2010).
17. Altes, A. *et al.* (2008) Advanced thermal failure analysis and reliability investigations–industrial demands and related limitations. *Microelectron. Reliab.*, **48** (8–9), 1273–1278.
18. Bergher, L. *et al.* (1986) Towards automatic failure analysis of complex ICs through E-Beam testing. Proceedings of International Test Conference, Testing's Impact on Design and Technology, Cat. No. 86CH2339-0, Washington, DC, 1986.
19. Cole, E.I. Jr. *et al.* (1994) Novel failure analysis techniques using photon probing with a scanning optical microscope. Proceedings of Annual IEEE Reliability Physics Symposium, pp. 388–398.
20. Sienkiewicz, M. *et al.* (2008) Failure Analysis Enhancement by evaluating the Photoelectric Laser Stimulation impact on mixed-mode ICs. *Microelectron. Reliab.*, **48** (8–9), 1529–1532.
21. Beaudoin, F. *et al.* (2000) New non-destructive laser ablation based backside sample preparation method. *Microelectron. Reliab.*, **40** (8–10), 1425–1429.
22. Aubert, A., Dantas de Morais, L. and Rebrassé, J-P. (2008) Laser decapsulation of plastic packages for failure analysis: process control and artefact investigations. 19th European Symposium on Reliability of Electron Devices, Failure Physics and Analysis (ESREF 2008); Microelectronics Reliability, Vol. 48, Issues 8-9, August-September 2008, pp. 1144–1148.
23. Barberan, S. and Auvray, E. (2005) Die repackaging for failure analysis. *Microelectron. Reliab.*, **45** (9–11), 1576–1580.

24. Infante, F. (2007) Failure analysis for system in package devices using magnetic microscopy. PhD thesis, CNES. http://www.tesionline.com/intl/pdfpublicview.jsp?url=../__PDF/22186/22186p.pdf. (Accessed 2010).
25. Amakawa, S., Nakazato, K. and Mizuta, H. (2002) A new approach to failure analysis and yield enhancement of very large integrated systems. Proceedings of 32th European Solid-State Device Research Conference, Firenze, Italy, September 2002, pp. 147–150.
26. International Technology Roadmap for Semiconductor (ITRS), http://www.itrs.net/reports.html. (Accessed 2010).
27. Munetoshi, F. (2006) Invisible failure analysis system by nano probing system. *Hitachi Hioron*, **88** (3), 287–290.
28. Bâzu, M. (2007) Quantitative accelerated life testing of MEMS accelerometers. *Sensors*, **7**, 2846–2859.
29. Jeng, S.-L., Lu, J.-C. and Wang, K. (2007) A review of reliability research on nanotechnology. *IEEE Trans. Reliab.*, **56** (3), 401–410.

2

Failure Analysis – Why?

2.1 Eight Possible Applications

Failure analysis (FA) is the key method in any reliability analysis, because fighting against failure is the reason reliability exists as a discipline [1]. In short, FA is a scientific method which aims to discover the root cause for the incorrect behaviour of a product; that is, the failure of a device or a component to meet user expectations. For modern products (e.g. semiconductor devices), the causes and mechanisms of failures are complex and involve various factors: design issues, process environment and process parameters during product manufacture, but also environmental and exploitation stresses corresponding to a specific application of the product. FA is the process of determining the cause of failure, collecting and analysing data and developing conclusions to eliminate the failure mechanism (FM) causing the specific device or system failures. The reason it is so important to use FA – that is, to discover the cause of product failure – is what we intend to discuss in this chapter.

Reliability analysis is by no means the only 'customer' of FA. Other fields, such as business management and military strategy, also use this term [2]. In order to offer the reader a more complete picture, we have identified eight possible applications of FA in various fields (industry, research, etc.), to be detailed in this chapter:

- forensic engineering;
- reliability modelling;
- reverse engineering;
- controlling critical input variables;
- design for reliability (DfR);
- process improvement;
- saving money through early control;
- a synergetic approach.

If the reader is able to identify other reasons for using FA, we would be grateful to receive suggestions for future editions.

Failure Analysis: A Practical Guide for Manufacturers of Electronic Components and Systems, First Edition.
Marius I. Bâzu and Titu-Marius I. Băjenescu.
© 2011 John Wiley & Sons, Ltd. Published 2011 by John Wiley & Sons, Ltd.

2.2 Forensic Engineering

Historically, the first goal of FA was to identify the failure of a product, in order to take the necessary corrective actions to avoid it. This is called forensic investigation[1] or failure diagnosis.

In this book, we are focused on electronic systems, where failures can have dramatic consequences, including the *loss of human life*. There are many possible examples, including devices used for health care (pacemakers, various sensors, etc.) or in transportation (redundancy systems in aeroplanes, car security circuits such as airbags, etc.). On a less serious level, failures can lead to *important economic damage*, such as loss of satellites (with costs in the millions of dollars) or degradation of industrial equipment. Even *failures in domestic equipment* could produce significant damage: cost of product return by the client, cost of repairing the defective object, decrease of supplier credibility and so on. Consequently, the best policy is to avoid failures. In doing this, one has to be aware of the possible FMs affecting a product and initiate corrective actions to diminish or even eliminate the failures. The basis for all corrective actions is the knowledge generated by FA.

Forensic engineering is the investigation of materials, products, structures or components that fail or do not operate/function as intended, causing personal injury or damage to property [3]. As in any detective investigation, this investigation starts by analysing information about the 'dead' or 'wounded' product, by looking at reject reports or examples of previous failures of the same type of product. To avoid progress of failure, the sample is handled and stored with great care so that environmental (temperature and humidity), electrical and mechanical damage to the device is prevented. The failed components are evaluated to verify that they have indeed failed before being stored in a failed parts archive for possible future analysis. Then the 'body' is investigated (the failed product itself), using specific methods such as electrical and mechanical characterisation. In most practical cases the final damage doesn't reveal a direct physical FM and the original cause or complete scenario of failure is hidden by secondary post-damage processes. Thus it is extremely difficult to restore the failure scenario and particular physical mechanism in order to understand step by step the events resulting in catastrophe. In many cases this problem is excessive even for the experienced researcher [4]. Usually, a typical FA photograph presents somewhat useful information related to the location of the structural damage, which identifies the failed device, especially in the case of integrated circuits. Additional information can be derived about the damage location inside the device itself that might reveal or provide an idea of some topology defects of the device. Methods based on more sophisticated tools (such as optical microscopes, electron microscopes, spectroscopes, etc.) have to be used. All these methods are FA ones. Witness statements are also analysed, in order to discover the possible involvement of human factors in failure. Of course, if the failed product has led (or may lead in the future) to loss of human life or to important economic damage (e.g. aircraft accidents), all products of the same type are stopped from operating until the investigation is completed. The discipline that deals with the consequences of failures is *product liability*, which is involved in any proceedings concerning accidents in the operation of vehicles or machinery. Another discipline that may be involved in forensic investigation is *intellectual property*, which refers principally to patents.

The final goal of an FA in a forensic-engineering investigation is to locate the 'root cause' of failure. Then a solution to the immediate problem is given, followed by some valuable guidelines (corrective actions) aimed at preventing similar future failures. The root cause can be defined as 'events and conditions existing immediately before the failure which directly resulted in its occurrence and which, if eliminated or modified, would have prevented that occurrence' [5]. This is the correct procedure for any FA. An example of successful corrective action based on FA is given by the Apollo Programme. On 27 January 1967, the whole crew of Apollo 1 was lost in a fire inside the command module during a simulation run aboard an unfueled Saturn 1-B launch vehicle [6]. The forensic investigation

[1] 'Forensic', as an adjective, means pertaining to, connected with, or used in courts of law or public discussion and debate (from Latin: belonging to the forum, public). In modern language, the meaning has been enlarged towards relating to the use of science or technology in the investigation and establishment of facts or evidence.

proved to be efficient, and just two years, five months and twenty-three days later, on 20 July 1969, the Lunar Module Eagle of Apollo 11 landed on the moon. Today this is considered a classical case of well-implemented FA. A significant detail is that during this FA, a second capsule, the next in line, was sacrificed in order to gain a better understanding the failure and determine its root cause [6].

Very often, an FA is stopped after the primary cause (or the simple physical cause) of the failure is established, instead of being continued until the root cause is discovered [7]. Because the root cause is not identified, the conclusions of the FA do not provide long-term solutions for the problems. Bhaumik has found the root cause for limiting an FA before the root cause of a failure is evidenced: human error. Bhaumik agrees with Zamanzadeh [8], based on examining 1100 cases of service failures, that all failures are eventually caused by human errors, which may be classified into three categories [7]:

- errors of knowledge;
- errors of performance (e.g. negligence);
- errors of intent (e.g. acts of sabotage).

Moreover, in most cases, multiple causes contribute to a failure: bad design, poor-quality materials, negligence during the operation of a process, poor use and so on. It is not easy to discern amongst all these factors, so FA is stopped before the root cause is found, because the human investigators are not able to comprehend the complicated set of human interactions within a company, manufacturer or user. This often results in the truth being left undiscovered. Corporate cultural changes are needed, which are hard to implement in the short term. But the conclusion is simple: FA must be pushed as close as possible to its final goal: identifying the root cause of failure.

Because the basic methods used for FA are different for systems and for components, some details of both will be given below. The modelling and simulation activities linked to FA will be detailed too.

2.2.1 FA at System Level

For electronic systems, the common method for assessing and improving reliability, based on FA, is called failure modes and effects analysis (FMEA). Initially developed by the military (US Armed Forces, in the late 1940s), FMEA was first used in aerospace development (e.g. the Apollo space programme) and then in the automotive industry (Ford Motor Company, late 1970s) [9]. Today, FMEA is supported by the American Society of Quality, which provides detailed guides on the use of FMEA [10], mainly in industries with special reliability requirements. FMEA complies with Automotive Industry Action Group (AIAG), QS-9000, SAE J 1739, IEC 60812, JEP131 and other standards, and is sometimes called 'automotive' or 'AIAG' FMEA. It is required by many more standards, such as ISO 14971 (medical devices risk management) and others [11].

The main goal of FMEA is to identify the potential failure modes (FMos) of the components contained in the system under analysis and their possible effects on system performance [12]. It also aims to prioritise the failures according to how serious their consequences are, how frequently they occur and how easily they can be detected [44]. Basically, any FMEA contains three main phases:

- **Severity** – all the possible FMos are analysed and a *severity number* (S) is assigned to each, from 1 (no danger) to 10 (critical).
- **Occurrence** – the causes and frequencies of all possible failures are identified and for each FMo an *occurrence ranking (O)* is assigned, from 1 (low) to 10 (extremely high).
- **Detection** – all the existing techniques that could be used to 'fight' against a failure are identified and the ability of planned tests and inspections to remove defects or detect FMos in time is evaluated. Each failure receives a *detection number (D)*, ranked from 1 (low) to 10 (extremely high), which measures the risk that it will *escape detection*.

- After accomplishing these three basic steps, *risk priority numbers* (RPNs) are calculated, by multiplying the three rankings (S, O and D). The highest RPN is $10 \times 10 \times 10 = 1000$, which indicates that the failure is very severe, that occurrence is almost certain and that it is not detectable by inspection. The RPN is used to decide the order in which FMos must be diminished (or eliminated) by corrective actions.
- **Criticality** is a fourth phase that is added in FMECA (failure modes, effects and criticality analysis), an extension of FMEA. Criticality highlights FMos with a relatively high probability and severity of consequences, allowing remedial effort to be directed where it will produce the greatest value [13]. Many standards and regulations for aerospace, defence, telecommunications, electronic and other industries require that FMECA analysis be performed for all designed/manufactured/acquired systems, especially if they are mission- or safety-critical. FMECA includes FA, criticality analysis and testability analysis. It analyses different FMos and their effects on the system, and classifies and prioritises the end-effect level of importance based on failure rate and the severity of the effect of failure [11].

FMEA is a bottom-up approach, the basic rule being to treat each FMo as an independent event. However, very often this rule is not valid, which leads to the use of other reliability tools, such as fault-tree analysis (FTA) and reliability block diagram (RBD).

FMEA can be used at various stages of product development: during the design phase (in order to avoid future failures), during manufacture (as process control) and during operation, furnishing real-life data to the manufacturer (via the service network).

There are various degrees of detail for executing FMEA, from a simple analysis of system functions only, to the most complex FMEA, which uses physics of failure (PoF) to push the research towards identifying and evaluating, for each system component, the probability of occurrence of all FMs that produced an FMo [12]. Such analysis has to be executed if reliability requirements are mandatory. In this case, additional information is needed, such as models of failure-rate dependence on stress level and data from previously executed FTAs. This is a hard task, but with fruitful results: eventually, remedial actions can be taken to avoid or diminish certain FMs.

FMEA can also be used to plan reliability-centred maintenance operations. For example, a virtual maintenance system based on computer-aided FMEA has been reported [14]. This system uses an extended product model, where possible machine-failure information is added to describe machine status. By applying generic behaviour simulation to extended product models, it is possible to detect abnormal or mal-behaviour of machines under use conditions. Based on FMEA results, the life-cycle operations of machines can be simulated and reliability during operation can be predicted, and a maintenance programme can be planned accordingly. The proposed computer-aided FMEA method has been validated by several experiments.

There are several drawbacks to the traditional FMEA approach, as described in the literature, associated with the fact that it is only a *qualitative* reliability prediction method: low accuracy, possible mistakes in identifying FMos (important FMos can be neglected), a subjective approach to risk ranking, the impossibility of distinguishing between FMos with equal RPNs and so on. Examples are given in [15, 16].

Recently, a new procedure has been proposed to avoid one of these issues: omission of important FMos [17]. A checklist with 10 types of FMo, to be used by the team executing the FMEA, has been drawn up:

(1) The intended function is not performed.

The intended function is performed, but: (2) there is a safety problem or a problem in meeting a regulation associated with the intended function performance; (3) at a wrong time (availability problems); (4) at a wrong place; (5) in a wrong way (efficiency problems); (6) the performance level is lower than planned; (7) its cost is higher than planned (maintenance problems).

(8) An unintended (unplanned) and (or) undesirable function is performed.

(9) Period of intended function performance (life time) is lower than planned (reliability problems).
(10) Support for intended function performance is impossible or problematic (maintenance, reparability, serviceability problems).

The risk of using FMEA as the only method for reliability prediction is illustrated in [18], where a complete reliability-prediction approach called ROMDA is proposed, which adopts FMEA to predict product reliability but uses other methods too.

The ROMDA concept attempts to establish a relationship between product reliability and product design, and investigates the failure of products with respect to a dominant performance characteristic, where 'performance characteristic' is defined as a measure expressing how well a product fulfils its function. The concept expresses the characteristics of product reliability as a function of the dominant design parameters, which are degrading over time. The degradation profiles are superimposed on the design parameters in the functional or mechanistic model, by which the degradation of the performance characteristic as a function of time and the design parameter values can be explained. The degradation of the performance characteristic is then used to derive reliability characteristics, such as mean time to failure (MTTF). The most important result is that FMEA does not identify the dominant FM. Hence, the application of FMEA may cause problems in real-life business processes. In the new approach, the FMEA results are compared with error field data or product service data from the same or a comparable product. Two paths that are performed in parallel: on the one hand, the time-dependent FMEA process is executed as described above; on the other, field errors of the same or a comparable product are presented in a Pareto diagram in order to indicate the main causes of failure. For the dominant FM, an FTA is performed, which gives the main time-dependent causes of product failure. The results of both paths are compared and combined to provide the root cause of failure.

FTA is a graphical representation of events in a hierarchical, tree-like structure. It is used to determine various combinations of hardware, software and human-error failures that could result in a specified risk or system failure [45]. Fault trees are used to investigate the consequences of multiple simultaneous failures or events. This is the main advantage of FTA over FMEA/FMECA, which investigate single-point failures. A variant of FTA, called *event-tree analysis* (*ETA*), is an inductive FA performed to determine the consequences of a single failure for the overall system risk or reliability. ETA uses similar logic and mathematics to FTA, but the approach is different – FTA employs a deductive approach (from a system failure to its causes) and ETA an inductive approach (from a basic failure to its consequences).

The main differences between FMEA and FTA are outlined in Table 2.1. Each method has its limitations: FMEA does not take into account combined failures, while FTA is very dependent on personal knowledge and can miss some FMos. Unfortunately, FA at system level is currently based on either FMEA or FTA. When, rarely, both FMEA and FTA are performed, they are separated, with activities executed one after another. Recently, a methodology for connecting both methods was proposed [19]. Called bouncing failure analysis (BFA), this method 'bounces' between both approaches, from top-down to bottom-up, and from FTA diagram to FMEA table and back, changing the presentation and the direction of the analysis for convenience's sake at any point in the process. FMEA is extended by taking into account combinations of FMos (as in FTA, instead of looking at just one FMo at a time as in traditional FMEA). BFA replaces the traditional top-down FTA process with

Table 2.1 Main differences between FMEA and FTA

Method	Boundaries of analysis	Direction of analysis	Result presentation
FMEA	Single-point failures	Bottom-up	Tables
FTA	Combination of failures	Top-down	Logic diagram

a bottom-up approach, which is far more intuitive and easy to use for most engineers, but bounces to top-down and back from time to time to ensure analysis verification and subsequent update. BFA results in a highly efficient, systematic study of FMos and a dramatic decrease of time to acceptable analysis (TTAA); that is, it decreases the period from the beginning of analysis to a satisfactory report. The result is a complete coverage of all FMos accompanied by testability and detectability analyses.

2.2.2 FA at Component Level

At system level, failures can have two causes: incorrect function of a component and a fault in the system design and/or manufacture. If a component is 'guilty', the problem is transferred to the component's manufacturer, who has to solve it: that is, find out why the component failed (identify the FM) and take the necessary measures to prevent it failing again (prevent the FMo). At component level, the FMo is the status of the failed component and the FM is the physico-chemical process leading to failure. It is important to make this distinction clear because many papers claiming to study FMs actually refer to FMos and vice versa.

But both 'FM' and 'FMo' are also used at system level. In this case, the FMo is the incorrect function of one circuit of the system, while the FM might be the failure of a component of this circuit (e.g. a transistor) or a design fault in the circuit. In the former case, the FM at system level becomes an FMo at component level.

It is obvious that FA requires complex expertise in various fields. This is true for electronic systems, where electronics, mathematics and physics are involved, but especially for electronic components, where chemistry, material science and characterisation techniques have to be added to the above. Because the problem of identifying and avoiding the action of FMs at component level is a complex one, a large part of this book is dedicated to this subject (Chapter 5). Therefore we will not offer further details on the subject in this section.

Finally, it is necessary to establish a link between the reliability of components and of systems. This might be done by analysing the reliability of circuits based on the reliability of their components. One approach is presented in [20], which extrapolates circuit/product reliability based on transistor-lifetime distributions and transistor-bias distributions; that is, the number of transistors in each bias condition in the circuits. The impact of the statistical nature of transistor lifetime on circuit reliability is integrated to give a more realistic circuit-reliability projection. Although this approach is demonstrated for low-temperature applications, focusing on a hot-carrier ageing FM, it can easily be applied and extended to other FMs for any varying temperature operating conditions.

2.3 Reliability Modelling

Modelling the time and stress dependence of the failure rate (for components) or MTBF (for systems) is one of the main goals of reliability analysis, for three reasons:

- The existence of such models makes it possible to extrapolate the results obtained by reliability testing beyond the test duration (which is the basic element of reliability assessment through laboratory tests, as shown in Section 3.4).
- One may estimate (even at the design phase) the potential reliability of a product in various applications (for various stress factors and stress levels), which is necessary for modern manufacturing actions, such as concurrent engineering (CE), virtual prototyping and so on. More details are given in Sections 3.1–3.3.
- Information contained in reliability models can be used to design efficient accelerated reliability tests (which aim to simulate real function, but at higher stress levels, and to accelerate the action of various FMs). This subject will be detailed in Section 3.4.1.

For electronic systems, the first attempt to model reliability can be traced to November 1956, with the publication of the RCA release TR-1100, 'Reliability Stress Analysis for Electronic Equipment', which presented models for computing rates of component failure. In October 1959, the 'RADC Reliability Notebook' was published, followed by a military reliability-prediction handbook format known as MIL-HDBK-217, which was based on field data. For years, this so-called 'statistical approach' (because there was a statistical processing of field data) was largely used for any reliability prediction concerning electronic systems. Many editions of this standard were issued, from A to F (the last one being published in February 1995), each trying to update the previous issue with new technologies and new field results. However, many studies now indicate that this approach is obsolete.

The alternative approach, named 'physics-of-failure' (PoF), develops models based on root-cause analysis of FMs, FMos and failures causing stresses [21], which are more effective at preventing, detecting and removing the failures. Historically, the PoF approach was first used in the design of mechanical, civil and aerospace structures, as a mandatory approach to buildings and bridges. This new approach has already replaced MIL-HDBK-217 and is used by many US commercial electronics companies, by the UK Ministry of Defence and in East and South-East Asia. It is based on the fact that FMs are governed by fundamental mechanical, electrical, thermal and chemical processes. By understanding the possible FMs, potential problems in new and existing technologies can be identified and solved, sometimes even before they occur. In most PoF approaches, the basic tool is FA of the failed components, which is the only way to identify and explain scientifically the action of FMs. The final result is a set of corrective actions aimed to diminish and, eventually, avoid the action of a given FM. The failure rate for a batch of products is obtained by summing the contribution of each FM.

In 2005 a new PoF approach to predicting the reliability of systems was proposed, named FIDES. This was elaborated by a French consortium formed by big companies such as Airbus France, Eurocopter, GIAT Industries and Thales Group. FIDES proposed a standard for reliability prediction, called UTE C 80 811,[2] available in both French and English. The FIDES methodology is based on PoF (but does not require FA of the failed components) and supported by the analysis of test data, field returns and existing modelling. A large range of factors are taken into account: the manufacturing technology, the development cycle, the application type, the operational conditions and the over-stresses.

At component level, the PoF approach has some specific features. Of course, FA is indispensable before modelling, because models are valid only for a population affected by a single FM. A model for a population mixing two or more FMs has no relevance. So the correct procedure is to analyse all the failed products, to separate the populations affected by each FM and then to process the data for elaborating models. Two main models can be used to describe the actions of FMs:

- 'Empirical' models, which use statistical distributions (exponential, lognormal, Weibull, etc.) to describe the time and stress behaviour of FMs. In this case, FA's only role is to separate the populations affected by each FM. For electronic systems, these models are widely used, as the only possible method for modelling the reliability of electronic components.
- 'Physical' models begin with an accurate description of the physical or chemical phenomena responsible for the degradation or failure of electronic components and materials. In other words, these models use the microstructural information acquired by FA. In recent years, many such models have been elaborated. Details are given in Chapter 4 of the typical FMs encountered in the functioning of electronic components (and materials). Unfortunately, such models for electronic components are virtually unusable by system manufacturers, as they require input data that is not accessible to component users.

The role of FA in reliability modelling may not be easily understood from the above. We have therefore synthesised the relevant information in Figure 2.1.

[2] UTE = Union Technique de l'Electricité.

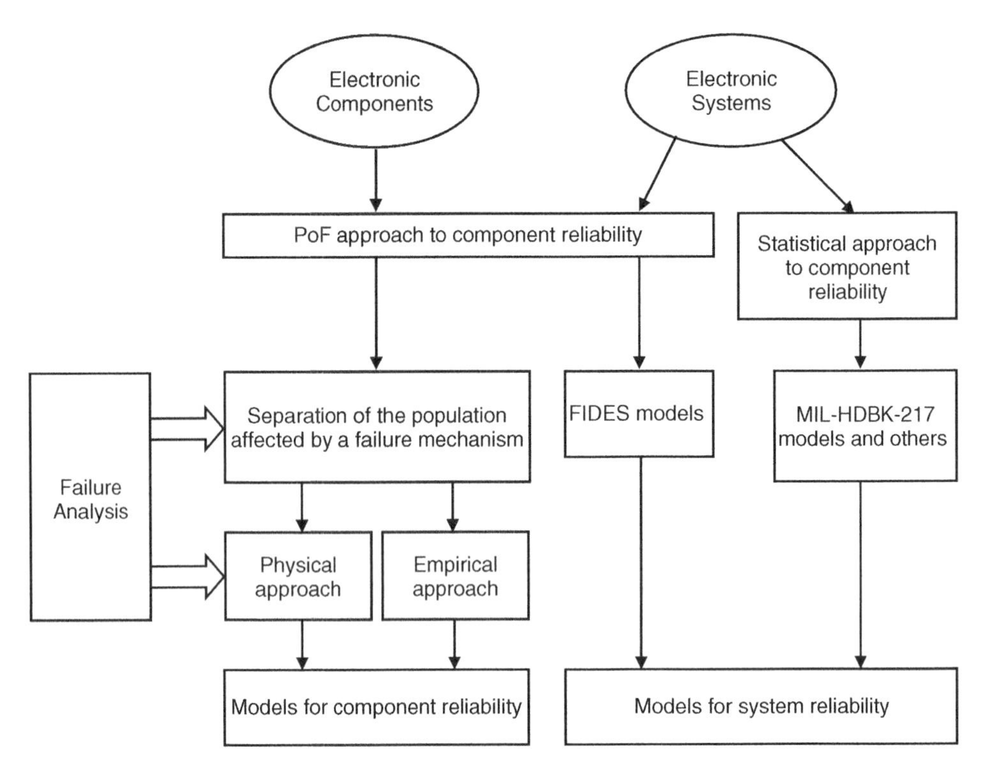

Figure 2.1 Possible approaches to modelling the reliability of electronic components and systems. The role of FA is emphasised

In conclusion, one may say that FA is now integrated in the concept of PoF, which is well established as the central tool for building reliability models, covering the following [1, 21]:

- Identifying the FMs of a batch of devices undergoing reliability testing, by using various methods of FA; statistical processing of data is valid only for a population affected by a single FM.
- Finding the dominant root cause of failure in each case and elaborating corrective actions for diminishing the action of the specific FM.
- Modelling the FMs versus time and stress (thermal, mechanical, etc.) in order to create simulation tools for the actions of FMs (e.g. computer-aided design of microelectronic packages, CADMP-2, and computer-aided life-cycle engineering, CALCE [22]) and to design procedures for accelerated testing of the devices.
- Helping suppliers measure how well they are fulfilling customer demands and what kinds of reliability assurance (RA) they can give to the customer.
- Helping customers determine that the suppliers know what they are doing and that they are likely to deliver what is desired.

Many of the leading international commercial electronics companies have now abandoned the traditional (statistical) methods of reliability prediction. Instead, they use reliability-assessment techniques, which are based on the root-cause analysis of FMs, FMos and failures causing stresses. The US Army has discovered that the problems with traditional reliability-prediction techniques are enormous and have cancelled the use of MIL-HDBK-217 in Army specifications. Instead they have developed Military Acquisition Handbook-179A, which recommends the best commercial practices, including PoF [21].

PoF has proven to be effective in the prevention, detection and correction of failures associated with the design, manufacture and operation of a product. By understanding the possible FMs, potential problems in new and existing technologies can be identified and solved before they occur. Using such science-based reliability techniques early in the design process can yield great cost savings in the manufacture, testing, fielding and sustainment of a new system. A PoF approach is also needed to assess the reliability cost impact of utilising commercial and new technologies in system design.

2.3.1 Economic Benefits of Using Reliability Models

The use of models to predict the reliability of a component or system has important financial implications. A recent paper [23] shows the economic benefits that can be obtained by the use of models. Even if exponential failure distribution is used (the simplest model, but the worst choice, deviating widely from the real world), important savings in human and material resources are achieved. The explanation is simple: if models are used, maintenance can be planned, and carried out at a time and place convenient to the owner or operator. In this paper, a quantitative assessment of the benefits of prognostics is made, by subtracting the maintenance savings (achieved during the whole operational period) from the expenses of purchasing and installing the prognostic instrumentation (paid once only, at the commencement of manufacture). The main issue to be solved in achieving high economic benefits is an accurate identification of the possible FMs. So even from an economic approach, FA is essential to developing reliability models. For electronic components, the savings from using models for reliability prognostic and maintenance scheduling are particularly important in aircraft applications. Of course, electronic equipment that is essential to safety of flight or mission completion always eventually becomes redundant, and this is not expected to change even if prognostics are shown to be effective for some FMs. Thus the benefits claimed for prognostics should not include avoidance of loss of aircraft, but they might include improved aircraft availability. Similar considerations apply to electronics installed in other vehicles (trains, buses, boats, etc.) as well as fixed electronic equipment in inaccessible locations, such as remote pumping stations and nuclear power plants.

2.3.2 Reliability of Humans

We feel it necessary to underline that reliability models can also be used to describe the ageing of humans. In October 2004, two biologists from the University of Chicago, Leonid Gavrilov and Natalia Gavrilova, tried to use reliability engineering to describe living organisms [24]. They noted that there are similarities between a human (described as a machine composed of redundant components) and a technical system. In both cases, the failure rate (death rate for humans) follows a 'bathtub' curve (see Figure 2.2). An initial period of 'infant mortality' (10–15 years for humans), where the failure rate is initially high and then quickly decreases, is followed by a relatively long 'useful life' period and, eventually, by a final period in which the failure rate rapidly increases again ('wear-out', or 'ageing' for humans).

According to Gavrilov and Gavrilova, if the failure (death) risk of humans could be kept at the level attained after 'infant mortality', people could expect to live about 5000 years on average. Unfortunately, this is not possible, because the failure risk increases exponentially with age. But a strange thing was noticed: as humans approach the age of 100, the risk of death stops increasing exponentially and instead begins to plateau. 'If you live to be 110, your chances of seeing your next birthday are not very good, but, paradoxically, they are not much worse than they were when you were 102', the authors write. 'There have been a number of attempts to explain the biology behind this in terms of reproduction and evolution, but since the same phenomenon is found not only in humans, but also in

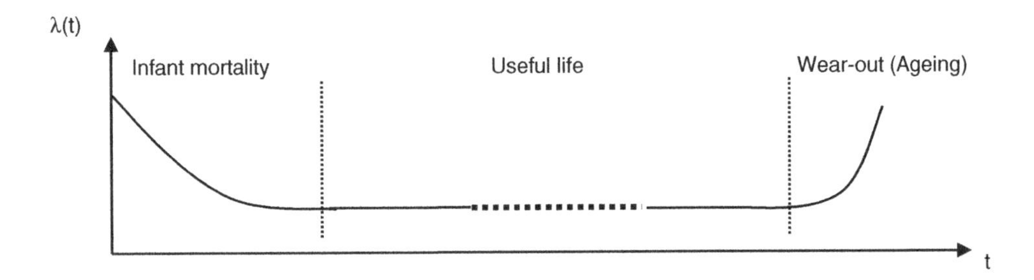

Figure 2.2 Model of the time variation of failure rate (bathtub shape)

such man-made stuff as steel, industrial relays and the thermal insulation of motors, reliability theory may offer a better way' [24].

The most interesting conclusion of this research is the following: the use of reliability theory for biological ageing of humans as a collection of redundant parts that do not age indicates that humans begin life with many defective parts. It follows that even small improvements in the processes of early human development, able to increase the numbers of initially functional elements, could result in a remarkable fall in mortality and a significant extension of human life.

2.4 Reverse Engineering

Initially, reverse engineering was a tool for forensic investigation; that is, for activity aimed at discovering the causes of product failure. The concept was developed for electronic components, but it could be used for electronic systems too. The failed product is carefully analysed and the failure causes are identified. This is a *reverse engineering* because it starts with the failure of the finished product and proceeds upstream, under the fabrication flaw, to identify the causes of failure (e.g. design faults, process faults or application faults). Obviously, the main tool in this activity is FA.

The success of reverse-engineering activity in reliability analysis has made it attractive in other fields, such as military and commercial espionage. It became clear that any hardware or software system could be reverse-engineered in order to discover the technological principles (design, processes, controls, etc.) behind it. Later, other possible goals of reverse engineering were discovered, and today many small companies are focused on reverse engineering, especially of electronic components. This activity covers a broad spectrum of applications, such as [25] recovering lost documentation for the manufacture of a product (e.g. integrated circuits designed on obsolete systems have to be reverse-engineered and re-designed on existing systems), analysing possible patent infringement, competitive technical intelligence (understanding what your competitor is actually doing rather than what they say they are doing, creating unlicensed/unapproved duplicates, avoiding the same mistakes that others have already made and subsequently corrected), academic/learning purposes and so on.

2.5 Controlling Critical Input Variables

The inputs to a manufacturing process are different at system and at component level.

For electronic systems, the main input variables are the electrical parameters of the components used. First, the manufacturer of an electronic system has to choose appropriate types of component. In this process, knowledge about the possible FMs of each component (acquired using FA) is essential, because any decision to use a low-reliability component in manufacturing a high-reliability system will be expensively paid later, as shown in Section 2.8, where the economic issues are discussed.

The main input variables to electronic components are the parameters of the materials used. The input control of the materials is carried out by verifying the parameters of the specifications, but a deeper understanding of the possible FMs induced by a material (necessary when high-reliability products are to be manufactured) requires the use of a special procedure: reliability tests are performed on materials with high reliability risks, the failed samples are analysed and the possible FMs are identified. This procedure is able to give the component manufacturer the necessary degree of confidence in a given material. Details about the procedures used for FA of materials are given in Section 3.3.

2.6 Design for Reliability

DfR is a modern concept in product design, which means taking reliability issues into account at the design stage, or 'pro-acting rather than reacting' [1], based on previous knowledge of the failure risks of similar (or close enough) products. The design team[3] must be aware of the typical FMs that could arise in a given application targeted by the product being designed. Obviously, FA is the main tool for acquiring the necessary knowledge. If no previous results are available, reliability tests, followed by careful FAs, have to be executed on the materials used in product manufacture and on test structures. The idea is to elaborate the design based on a collection of design rules intended to ensure maximum product reliability [1]. Moreover, predictive methods, able to assess the reliability of the future device, based on design data and on models describing the time and stress behaviour of similar products, are also needed (these models were discussed in Section 2.3).

The DfR approach starts by capturing the customer voice, translated in an engineering function [26]. Then a design immune to the action of perturbing factors (so-called 'robust design') is created. Kuo and Oh [27] describe a procedure for this: (i) develop a metric able to capture the function while anticipating possible deviations downstream and (ii) design a product that ensures the stability of the metric in the presence of deviation. Finally, the design team must use reliable prediction methods. In principle, DfR means passing from evaluation and repair to anticipation and design.

The design of any product has three phases: conceptual design, preliminary design and detail design [28]. Conceptual design does not focus on reliability issues, just on combining technologies in order to determine the feasibility of the product. At system level, the preliminary design and conceptual design are largely based on FA, each phase using one FA method (together with other methods): FTA and FMEA, respectively (see Section 2.2). At component level, FA is also deeply involved in design, the term used being 'failure physics-based design', which is integrated in the larger concept of PoF, a science-based approach to reliability that uses modelling and simulation to design-in reliability [29]. A methodology for PoF was elaborated in the 1970s, but its use in semiconductor devices was not possible until the early 1990s [28]. Today PoF is well established as the central element of DfR. The role of the PoF approach was underlined by Pecht [21].

2.7 Process Improvement

Once designed (as above), a process is used to manufacture a product. The PoF approach qualifies design and manufacturing processes to ensure that the nominal design and manufacturing specifications meet or exceed reliability targets. During production-volume manufacturing, a system of *reliability monitors* promotes a continuous reliability improvement in order to achieve RA. A reliability monitor consists of one or more samples withdrawn from a given node of the fabrication process (or test structures processed in the same conditions as the products) and subjected to previously designed reliability tests. The failed items are analysed by specific FA methods (electrical characterisation, as

[3] We speak of a 'design team' and not a 'designer', because today a team is necessary in designing a product. The team contains not only designers, but test engineers, process engineers, reliability engineers and even marketing specialists (according to the concurrent engineering concept).

well as other, more sophisticated methods) and, if necessary, corrective actions are elaborated. More details about reliability monitors are given in Section 3.3.

The basic concept of reliability monitoring is *reliability assurance*, which is detailed below.

2.7.1 Reliability Assurance

The first steps towards RA were made in early 1990s, when the approach used in reliability was changed from final inspection to prevention and elimination of root causes by corrective action. The key to this new RA paradigm was the feedback process, containing corrective and preventive actions. Corrective actions aim to eliminate reliability problems found during the production of an item by modifying the design, process, control and testing instructions and parameters, or even the marketing programme. Preventive actions deal with the response given by the manufacturer to the corrective action and aim to eliminate generic causes of product unreliability [1]. All these actions are the 'bricks' of RA, included in a quality and reliability assurance (Q&RA) programme. Besides the manufacture of electronic components, another industry that leads the way in Q&RA is the nuclear one. In 1996, the IAEA initiated a task to develop an RA guidebook to support its implementation within advanced reactor programmes and to facilitate the next generation of commercial nuclear reactor to achieve a high level of safety, reliability and economy [30]. In the early 2000s, the ISO 9000 standards (descending from MIL-Q-9858) were launched, which describe the Q&RA system for any given product: a set of well-established procedures for every facet of product design and manufacture. These standards were widely accepted as the best tool for Q&RA. Customers required manufacturers to have implemented systems for Q&RA, as a necessary condition to being accepted in the market.

However, many eminent engineers and scientists (e.g. Michael Pecht [29] and Patrick O'Connor [31, 32], well-known 'fighters' against the procedures imposed by the ISO 9000 series) have shown that implementing the ISO 9000 standards leads to results that are far from those foreseen. In fact, the procedures described by the ISO 9000 series do not ensure a reliability level, but impose a fixed working mode established by the manufacturers themselves. The goal of the standards is to monitor whether the requisite procedures are observed, not whether they are effective. Moreover, this fixed system inhibits continuous improvement, the only real tool for Q&RA [31].

Manufacturers need to know how things fail, as well as how things work [33], by combining the PoF approach with the use of best practices. This alternative methodology was proposed in the early 1950s by Deming (explaining basic principles in Japan) and is a 'people-oriented philosophy'. The key concept is total quality management (TQM), which is the integration of all activities that influence the achievement of ever-higher quality [31]. In this context, 'quality' is all-embracing, covering customer perceptions, reliability, value and so on. In the engineering companies that use this approach we do not see any demarcation between 'quality' and 'reliability', in organisations or in responsibilities. Every person in the business becomes committed to a never-ending drive to improve quality and productivity. The drive must be led by top management, and it must be vigorously supported by intensive training, appropriate application of statistical methods and motivation for all to contribute. The total-quality concept links quality to productivity. It has been the prime mover of the Japanese industrial revolution, and it is fundamental to the survival of any modern manufacturing business competing in world markets. TQM has generated enormous gains in the reliability of complex products such as cars, machines and electronic devices and systems [32]. When production quality is well managed, complexity is not the enemy of reliability, as it used to be perceived.

The main difference between the ISO 9000 standards and Deming's approach is the principle of 'scientific management', which teaches that people work most effectively when given specific instructions [31]. Managers elaborate these work instructions and ensure that they are followed. In effect, workers are treated like machines. Scientific management led to work study, demarcation of work boundaries, the mass production line and the separation of labour and management. In the 1950s, Peter Drucker underlined the errors of this system [34]. Unlike machines, people want to make contributions to how their work is planned and performed, and to do their jobs well. In principle, there is no limit to the quality of work that people can produce, both as individuals and in teams, so continuous improvement is a realistic objective.

In our opinion, the use of FA represents a powerful tool for the continuous improvement of process Q&R, hence we support Deming's TQM approach. By using FA, people can visualise the results of their work and be stimulated to improve day by day the Q&R of the products obtained.

2.8 Saving Money through Early Control

The controls performed during manufacture aim to monitor Q&R as early in the process as possible. The reason for this is economical: the cost saved by identifying and eliminating a defective component is highest at input control and diminishes the later it occurs [35]. To put it another way, the later the control is executed, the higher the cost that must be paid for an item identified as being defective. A comparison between the costs of four selection levels and three categories of electronic systems is given in Table 2.2. The conclusion is that it is more economical to identify and eliminate a defective component through the input controls than the equipped PCB controls.

An empirical rule says that these costs grow by an order of magnitude at each successive control level. This is the so-called 'rule of ten': if the cost of a defective die identified at chip level is C, then when the fault is found at board level the cost is 10 C, and at system level the cost becomes 100 C.

This means the input control must be the most severe one, aiming to discover and remove all the possible risks related to process inputs. In [36], it is recommended that electronic systems utilise 100% input controls, this being justified by a detailed economical analysis.

Of course, beyond the economic issues, the consequences of a product failure can be very important. In some cases failure can result in the loss of human life, as for example in the case of onboard computers, motors, aeroplane redundancy systems, car security circuits (airbags, brake assistance) and pacemakers. In other, less serious cases, the economic damage might be considerable. With satellite applications, which can cost in the order of $200 million, a system breakdown might lead to the loss of the satellite. In other cases, a design fault, or a badly-manufactured electronic component, can cause failures and the consequent return of the product by the client, with a total cost that can be extremely high, including repair and modification costs. The damage produced by the loss of credibility of the supplier must also be taken into account.

The importance of FA in obtaining significant cost savings in a specific case (electronic equipment development for the US Army) is emphasised in [37]. First, the large range of software tools for PoF used by US Army Material System Analysis Activity is shown. There are such tools for: solid modelling, dynamic simulation, finite-element modelling, fatigue analysis and thermal fluid analysis, together with electronic circuit card and IC toolkits. An example of a complex software tool is UD CalcePWA, which contains modules for: architecture and environment modelling, thermal analysis, vibration analysis, failure assessment and sensitivity analysis, as well as reports and documentation. The results obtained by US Army Material System Analysis Activity in cost savings are quite impressive. Several success stories concerning the use of FA in electronics are synthesised in Table 2.3.

Table 2.2 Comparison between control costs (expressed in $) of a defective component

Products	Input control to components	Control of PCBs	System control in the factory	At the user, during the warranty period
General use	6	15	15	150
Industrial	12	75	135	645
Military	21	150	360	3000

After [35].

Table 2.3 Cost savings obtained by US Army Material System Analysis Activity using FA for electronics

System	Problems to be solved	Solutions	Results	Cost saving
Radar ground station	Determining whether a commercial processor circuit card could be used instead of a ruggedised circuit card in the ground station	Vibrational, thermal and solder-joint fatigue analyses using UD CalcePWA software performed	Commercial circuit card fatigue life is 11 years versus 23 years for ruggedised, which was acceptable	\$12 000/card; For 100 cards = \$1 200 000
Tri-service radio	Validating UD CalcePWA software through accelerated life testing; assessing reliability of the module in a military environment; improving reliability of the module	20-pin leadless chip carrier was weak in design; estimated time-to-failure during accelerated life test cycle; estimated life under operating conditions = 6.5 years	Operating and support cost avoidance	\$27 000 000
Army helicopter and aircraft	Aircraft and helicopter had common circuit cards. UD CalcePWA software used to determine if commercial integrated circuits could be used	Analysis shows commercial integrated circuits can be used without degrading reliability	Savings: circuit card #1: \$18 501/card Circuit Card #2: \$20 228/card Also, a 15% weight reduction per card	For a buy of 1292 vehicles, total savings = \$50 M

After [37].

2.9 A Synergetic Approach

FA allows a synergetic approach by taking into account not only the failure risks induced by each technological step, but also the synergies between subsequent steps. FA is also useful for analysing the synergy of environmental factors, which must be considered when the effect of the environment on product operation are to be studied. The necessity of FA in a synergetic approach to reliability is demonstrated below.

2.9.1 *Synergies of Technological Factors*

There is a large variety of technological factors involved in product reliability. The best example might be given for the manufacture of electronic components, because this technology is mature enough to be well controlled. In this case, which will be detailed in this section, the main issues are related to [1]: quality of materials, contamination, quality of chemicals and of the packaging elements, and so forth. It may be noted that these factors are interdependent and, consequently, the failure risks may be induced by each technological step or by the *synergy* of these steps [38]. In all cases, FA is used to identify these failure risks. An example showing the synergetic effect between some process steps is given in Table 2.4. The technological flow of electronic components is formed by a front-end (executed on wafers, also called wafer-level technology) and a back-end (mainly containing all packaging operations). During all technological flow, the failure risks are identified at each technological step by means of FA methods, as detailed in Section 3.3. Note that because some FA methods are destructive, at wafer level, *test structures* are used to analyse the processes without damaging the whole wafer.

Table 2.4 Synergetic effects between various process steps of the technology for the manufacture of electronic components with semiconductors

Process step (phenomenon)	Failure risks	Synergy with other process steps	Possible corrective actions
Photolithography (contamination with particles)	Dust particles reaching the transparent area of the masks could transfer their images on the wafer	Metallisation – parasitic bridges between metallic areas leading to short or open circuits	Contamination prevention, by identifying and avoiding the contamination sources Wafer cleaning to remove the particles reaching the wafer
Oxidation/diffusion	Contaminants containing alkali ions become active at the thermal processes (oxidation, diffusion)	Oxidation/diffusion – localised regions with ionic contamination with ions that might migrate to the active areas of the device, producing increased leakage currents	–

2.9.2 *Test Structures*

Test structures were initiated in the early 1970s, as the necessary tool for monitoring the technological process without interfering too much with the technological flow. They are built on the same wafer as the devices to be obtained; hence they are processed in exactly the same technological environment and using the same parameters.

First, the test structures are characterised on-line, during processing. Then, chips with test structures are packaged in the same conditions as the devices and undergo reliability tests to identify specific FMs. Note that many test structures are built to study specific FMs. For example, the test structures focused on electromigration contain various metal structures (with different lengths and widths), which can be used to study the dependence of the electromigration process on the geometry of the device. In most cases, accelerated tests (simulating the operational stresses, but at higher stress levels) are used, in order to minimise the duration of reliability tests. From the acceleration induced by the stress used (known from previous studies), the results are extrapolated for operational stress levels. However, one must be cautious when extrapolating such data, because the test structures are not exactly like the real devices.

For small and medium companies, which generally do not have enough money to develop their own group to design and manufacture test structures, specialised companies can offer such services. An example of extended services in the field is Coventor, which launched *Coventor Catalyst*, a design for manufacturability (DfM) methodology for developing micro-electro-mechanical systems (MEMSs) to ensure optimal manufacturing yield [39]. This methodology includes a software environment for designing MEMS plus a kit needed to qualify and optimise a MEMS manufacturing process. The kit includes test structures for characterising the centres and corners of MEMS manufacturing processes and for extracting material properties. Using the test structures, the fabrication and assembly variations resulting from misalignment or machine failures are evaluated and the most suitable materials and minimal steps for successful fabrication are identified. Coventor extracts relevant process variables to centre the design for the process and predict future performance, and then builds material properties and process data-sets to simulate the design within the boundaries of the target process. Coventor process engineers are also available to define process requirements, measure performance and recommend a standard process that can be reused successfully and economically for an entire product family.

2.9.3 Packaging Reliability

The back-end operations (including packaging) are extremely important to the reliability of the electronic component. Even if automated equipment reduces the handling of and subsequent damage to die and wafers, the failure risk is still high. The role of FA is increasing, because new trends in packaging represent important challenges to reliability.

- *IC packaging technologies* are evolving towards either packaging at wafer level, including wafer-level burn-in and test, or system-in-package (SiP) or three-dimensional packaging of ICs.
- *System-on-package* (SoP) was developed from traditional board packaging with discrete components assembled by SMT processes. The SoP concept proposes to improve the system integration by embedding passive and active components and including such components as filters, switches, antennas and embedded optoelectronics. The first product introduction of this concept was the embedding of RF components by many companies. One of the SoP technologies is embedding of optical components in the board by means of waveguides, gratings and couplers.
- Other predictable future trends are towards *nano-packaging for nano-systems*. Attempts to develop nano-structured connections in the short term to 20–100 μm pitch and nano-grain or nano-fibre connections in the long term to 1 μm pitch were made by a consortium formed by the Georgia Institute of Technology NUS and the Institute of Microelectronics (IME), Singapore.
- In the longer term, the transition is towards *nano-chips*, with features smaller than 100 nm. Some of these chips will have several hundred million transistors, which require I/Os in excess of 10 000 and power in excess of 200 W, providing computing speed in terabits per second and supporting 20–50 GHz digital/RF applications. These nano-chips require nano-packaging at the wafer and system-board levels. Wafer-level packaging provides unprecedented advances in electrical, mechanical and thermal properties as well as the smallest form factor and lowest cost.

In all these developments, the main goal is the FA study of material characterisation at nano-scale and the modelling of various FMs (e.g. fatigue).

2.9.4 Synergies of Operational Stress Factors

During operation, any electronic component has to undergo operational stress factors, which can be divided into two main categories: (i) electrical bias and (ii) environmental stresses (which are in fact a combination of many factors: temperature, air pressure, vibration, thermal cycling and mechanical acceleration). Experimentally, there is evidence that the combined effect of these factors is higher than the sum of the individual effects, because a synergy of the stress factors occurs [40]. Hence, for an accurate reliability analysis both the individual effects and the synergy of these factors must be studied. Certain stress factors are strongly interdependent, such as solar radiation and temperature, or functioning and temperature. Independent factors may also be outlined, such as mechanical acceleration and humidity.

In order to analyse the influence of the operational stress factors, FA must be used, because a correct determination of the dependence of reliability on the stress factors can only be obtained at the level of FMs (or degradation mechanisms, if the failure has not yet arisen).

For electronic components, there are many examples of such synergies arising from combinations of operational stress factors during specific applications of electronic systems, such as:

- **Temperature and humidity** – during storage of a large variety of electronic systems.
- **Temperature and vibration** – during operation of automotive equipment or aeroplanes.
- **Temperature and air pressure** – during operation of aeroplanes at high altitude.
- **Electrical bias and temperature** – for the majority of electronic systems during operation; in this case, the electrical bias involves not only the thermal effect produced by the current flow, but also an electrical effect (depending on the electrical current or either the electrical potential).

2.9.5 Synergetic Team

So far, we have demonstrated that a synergetic approach must cover the study of both technological (Section 2.9.1) and environmental (Section 2.9.2) factors. But this synergetic approach means more: even the team that is studying the reliability has to be synergetic. The approach involved is CE, and FA plays an important role in applying it.

CE means an integration of all specialists (in design, technology, testing, reliability, marketing, etc.) with possible involvement in the team that is bringing about the product, from the design stage and through all subsequent stages. The multidisciplinary nature of FA is obvious for any product. For example, if mechanical failures are being studied, engineering disciplines such as materials, mechanics, thermal- and fluid-mechanics engineering must be involved, but biology, human factors and statistics may also be included in some cases [41]. The most important role is played by the team leader, who has to apply the right expertise and the correct specialists to solve the task: identifying the root cause(s).

CE is also called 'parallel engineering', as opposed to 'sequential engineering' (based on the Taylor assembly-line principle), which is still used by most industrial companies. In sequential engineering, product development is viewed as a series of isolated 'do what you want' technical projects [42], which can lead to a mismatch between design and manufacturing, requiring a product to be redesigned after final testing.

If product development is managed in a concurrent manner rather than sequentially, the new product can be a good one 'right from the first shot', with a higher chance of quickly gaining a place in the market. A synergy of the whole team must be realised: the final result must overreach the sum of the individual possibilities. Moreover, an important change in mentality must take place: from 'toss it over the wall' (from one specialist to another) to everyone working together. Again, this is a task

Table 2.5 Involvement of FA in CE actions

CE action	Basic tools	Involvement of FA
Involving all specialists required for product development from the design phase	*Design for reliability* (DfR) – taking into account the reliability issues from the design phase and through the whole manufacturing process.	The reliability engineer participating in design-team activity has to know all the possible FMs (obtained by previous FA performed on similar products) and the design rules for avoiding them.
Moving metrics from back to front	*Building-in reliability* (*BiR*) – aims to define and monitor the reliability risk as early as possible in manufacturing flow [42]. Experimental results have proved that BiR is an effective approach, significantly increasing product reliability [43].	In defining and quantifying the reliability risks, the possible FMs must be studied by FA on similar products. At well-established milestones in the manufacturing flow, reliability monitors must be designed, offering the necessary information and allowing the best decision to be taken.
Taking into account the 'customer voice'	*Quality function deployment* (*QFD*) – provides the means for translating the voice of the user into the appropriate technical requirements for each stage of product development.	The user's operational conditions have to be translated into technical requirements concerning the operational stresses. Reliability tests, based on FA, have to verify that these requirements can be fulfilled by the product.

for the team leader, who has to bring every specialist together and create a smooth communication environment, with no unnecessary energy loss in internal debates with no significant results. If this goal is achieved, the benefits of this approach are important. With CE, the number of iterations needed to achieve a project is diminished and the time required to obtain a new product is significantly shortened [31]. A strong supporter of CE is the US Department of Defense (DoD), which encourages its contractors to lead the way in this field.

CE promotes three main actions, which are more or less linked with reliability issues and based on FA, as described in Table 2.5.

References

1. Băjenescu, T. and Bâzu, M. (2010) *Component Reliability for Electronic Systems*, Boston and London, Artech House.
2. http://en.wikipedia.org/wiki/Failure_analysis. (Accessed 2010).
3. http://en.wikipedia.org/wiki/Forensic_engineering. (Accessed 2010).
4. Venkataraman, S. (2004) Diagnosis meets physical failure analysis: what is needed to succeed? Proceedings of IEEE ITC International Test Conference 2004, pp. 1442–1445.
5. Office of NASA Safety & Mission Assurance, Chief Engineers Office. Root Cause Analysis: an Overview, July 2003, www.au.af.mil/au/awc/awcgate/nasa/rootcauseppt.pdf. (Accessed 2010).
6. Dennies, D.P. (2002) The organization of a failure investigation. *J. Fail. Anal. Prev.*, **2** (3), 11–41.
7. Bhaumik, S.K. (2009) A view on the general practice in engineering failure analysis. *J. Fail. Anal. Prev.*, **9**, 185–192.
8. Zamanzadeh, M. *et al.* A Re-examination of Failure Analysis and Root Cause Determination, www .matcoinc.com.(Accessed 2010).
9. http://en.wikipedia.org/wiki/Failure_mode_and_effects_analysis.(Accessed 2010).
10. http://www.asq.org/learn-about-quality/process-analysis-tools/overview/fmea.html. (Accessed 2010).
11. http://www.aldservice.com/en/fracas/failure-analysis-methods-and-tools.html. (Accessed 2010).
12. Krasich, M. Can failure modes and effects analysis assure a reliable product? Proceedings of Annual Reliability and Maintainability Symposium (RAMS), Orlando, January 22–25, pp. 277–281.
13. http://en.wikipedia.org/wiki/Failure_Mode,_Effects,_and_Criticality_Analysis. (Accessed 2010).
14. Kimura, F., Hata, T. and Kobayashi, N. (2002) Reliability-Centered maintenance planning based on computer-aided FMEA. The 35th CIRP-International Seminar on Manufacturing Systems, Seoul, May 12–15, 2002, pp. 1–6.
15. Montgomery, T.A., Pugh, D.R., Leedham S.T. and S.R. Twitchett (1996) FMEA automation for the complete design process. Proceedings of Annual Reliability and Maintainability Symposium (RAMS), Las Vegas, January 22–25, 1996, pp. 30–36.
16. Pillay, A. and Wang, J. (2002) Modified failure mode and effects analysis using approximate reasoning. *Reliab. Eng. Syst. Saf.*, **79** (1), 69–85.
17. Bluvband, Z. and P. Grabov (2009) Failure analysis of FMEA. Proceedings of Annual Reliability and Maintainability Symposium (RAMS), Fort Worth, January 26–29, 2009, pp. 344–347.
18. de Visser, I.M. and van den Bogaard, J.A. (2006) The risks of applying qualitative reliability prediction methods: a case study. Proceedings of Annual Reliability and Maintainability Symposium (RAMS), Newport Beach, January 23–26, 2006, pp. 532–538.
19. Bluvband, Z., Polak, R. and Grabov, P. (2005) Bouncing failure analysis (BFA): the unified FTA-FMEA methodology. Proceedings of Annul Reliabillity and Maintainability Symposium (RAMS), Alexandria, January 24–27, 2005, pp. 463–467.
20. Chen, Y. *et al.* (2006) Approach to extrapolating reliability of circuits operating in a varying and low temperature range. Proceedings of IEEE 44th Annual International Reliability Physics Symposium (IRPS), San Jose, March 26–30, 2006, pp. 306–312.
21. Pecht, M. (1996) Why the traditional reliability prediction models do not work – is there an alternative? *Electron. Cooling*, **2** (1), 10–12.
22. Anon. (1996) MEMS design for Manufacturability Coventor Catalyst, Coventor, July 2002.
23. Hecht, H. (2006) Prognostics for electronic equipment: an economic perspective. Proceedings of Annual Reliability and Maintainability Symposium (RAMS), Newport Beach, January 23–26, 2006, pp. 165–168.

24. Gavrilov, L. and Gavrilova, N. (2004) Why we fall apart engineering's reliability theory explains human ageing. *IEEE Spectr.*, **41** (9), 31–35, www.coventor.com/pdfs/sensorsexpo_2002.pdf.
25. http://en.wikipedia.org/wiki/Reverse_engineering. (Accessed 2010).
26. Batson, R. and Elam, M. Robust Design: An Experiment-based Approach to Design for Reliability, http://ie.eng.ua.edu/research/MRC/Elam-robustdesign.pdf. (Accessed 2010).
27. Kuo, W. and Oh, H. (1995) Design for reliability. *IEEE Trans. Reliab.*, **44** (2), 170–171.
28. Taguchi, G. (1995) Quality engineering (Taguchi methods) for the development of electronic-circuit technology. *IEEE Trans. Reliab.*, **44** (2), 225–229.
29. White, M. and Bernstein, J. (2008) *Microelectronics Reliability: Physics-of-Failure Based Modeling and Lifetime Evaluation*, JPL Publication 08-05, Jet Propulsion Laboratory, California Institute of Technology, Pasadena, CA, 2/08.
30. Anon. (2001) *Reliability Assurance Program Guidebook for Advanced Light Water Reactors*, International Atomic Energy Agency (IAEA), December 2001.
31. O'Connor, P.D.T. (1995) Achieving world class quality and reliability: science or art? http://www.pat-oconnor.co.uk/qrscorart.htm
32. O'Connor, P.D.T. (2000) Reliability past, present and future. *IEEE Trans. Reliab.*, **49** (4), 335–341, www-pub.iaea.org/mtcd/publications/pdf/te_1264_prn.pdf.
33. Pecht, M., Nash, F. and Lory, J. (1995) Understanding and solving the real reliability assurance problems. Proceedings of Annual Reliability and Maintainability Symposium (RAMS), Washington, DC, January 16–19, pp. 159–161.
34. Drucker, P.F. (1955) *The Practice of Management*, HarperCollins.
35. Băjenescu, T. and Bâzu, M. (1999) *Reliability of Electronic Components*, Springer, Berlin and New York.
36. Băjenesco, T.I. (1975) Quelques aspects de la fiabilité des microcircuits avec enrobage plastique. *Bull. ASE/UCS (Switzerland)*, **66** (16), 880–884.
37. Stadterman, T. (2002) Cost savings from applying physics of failure analysis during system development. National Defense Industrial Association 5th Annual Systems Engineering Conference, Tampa, October 22–24, 2002.
38. Bâzu, M. (1994) A synergetic approach on the reliability assurance for semiconductor components, PhD thesis, Politehnica University of Bucharest, Romania, March 1994.
39. Schröpfer, G. *et al.* (2004) Designing manufacturable MEMS in CMOS compatible processes–methodology and case studies. SPIE's Photonics Europe, conference 5455–MEMS, MOEMS, and Micromachining, Strasbourg, April 26–30, 2004.
40. Bâzu, M. (1995) A combined fuzzy logic and physics-of-failure approach to reliability prediction. *IEEE Trans. Reliab.*, **44** (2), 237–242.
41. Huet, R. (2002) The interdisciplinary nature of failure analysis. *J. Fail. Anal. Prev.*, **2** (3), 17–19 and 42–44.
42. Peck, D.S. and Trapp, O.D. (1987) *Accelerated Testing Handbook*, Technology Associates, Portola Valley, CA.
43. Flynn, A. and Millar, S. (2005) Investigation into the correlation of wafer sort and reliability yield using electrical stress testing. Proceedings of 43rd Annual IEEE Reliability Physics Symposium, San Jose, April 17–21, 2005, pp. 674–675.
44. http://en.wikipedia.org/wiki/Failure_Mode,_Effects,_and_Criticality_Analysis. (Accessed 2010).
45. http://www.aldservice.com/en/fracas/failure-analysis-methods-and-tools.html. (Accessed 2010).

3

Failure Analysis – When?

When should failure analysis (FA) be used? The answer is simple: in all phases of the product development cycle and during product manufacture and exploitation. Why? Because FA aims to identify the causes of failures, and failures may arise at any stage of development, fabrication and operation [1]. As an example, the potential failures at various stages of the life cycle of an electronic system are shown in Table 3.1.

For electronic components, the situation is the same: FA should be used at every phase of the development cycle. However, in this case there are some peculiarities, as shown in Table 3.2.

Of course, the information contained in the above two tables is purely illustrative, and will not fit every case. This chapter provides examples of how FA is used at various stages of the life cycle of both electronic components and systems. By combining the knowledge developed for various products, by various companies, we hope to convince you that, in the future, product manufacturing and exploitation will not be possible without FA.

3.1 Failure Analysis during the Development Cycle

The development cycle has to follow a standard process, which basically contains the following four stages: Design, Demonstrator, Experimental Model and Prototype.

In a charming paper[1] emphasising the lessons design specialists can learn from the truly sage advice given to Alice by the Cheshire Cat in Lewis Carroll's *Alice in Wonderland* [2], George Hazelrigg also says that *'concisely stated, the fundamental preference driving the design is: the company wants to make money, and more is better'*. And in making money, the main goal of any manufacturer of a new product is to find a market niche. This is always a tough battle, in which the time factor is one of the most important: the time-to-market (i.e. the time from the initial idea to the finished and ready-to-be-sold product) must be as short as possible. Companies which succeed in launching new products faster than their competitors can obtain many advantages, including the gain of a large market share and higher profit through premium prices. But in order to do this today, the product development processes must bring good products to the market much faster than ever before [3]. The product must also fulfil its intended functions, not only when initially purchased by the customer, but after even quite a long period (at least to the end of the warranty period). In other words, the reliability level

[1] The paper demonstrates that in *Alice's Adventures in Wonderland*, the Cheshire Cat provides to Alice the underlying tenets of modern normative decision theory. The principles of engineering design are discussed, the conclusion being that decision making is a very personal thing.

Failure Analysis: A Practical Guide for Manufacturers of Electronic Components and Systems, First Edition.
Marius I. Bâzu and Titu-Marius I. Băjenescu.
© 2011 John Wiley & Sons, Ltd. Published 2011 by John Wiley & Sons, Ltd.

Table 3.1 Possible failures during the life cycle of an electronic system

Stage	Possible failures	Role of failure analysis
Design	Wrong processes or wrong process parameters	Provides knowledge of possible failure risks
Prototype	Wiring and assembly; component failure	Provides knowledge of typical failure mechanisms
Fabrication	Wiring and assembly; component failure	Provides knowledge of typical failure mechanisms
Field operation	Component failure; operator errors; environmental factors	When performed on each failed product during field operation, can delineate the three possible failures

After [1].

Table 3.2 The role of FA in the life cycle of an electronic component

Stage	Actions	Role of failure analysis
Development cycle (design, demonstrator, experimental model, prototype)	The components undergo high stresses (circuit and mounting changes, inversion of the bias, too small supply voltages)	Identifies the weak points of the components and elaborates the necessary measures for their elimination
Before fabrication (input control)	The quality of the materials to be used for component fabrication is analysed, by performing reliability tests	Any deviation from normal behaviour may lead to the decision to return the whole batch to the supplier
During fabrication	At the various stages of the manufacturing flow, reliability monitors are established, where accelerated reliability tests are performed	Any deviation from normal behaviour may lead to the decision to stop the whole batch or to change the destination of the final products of the fabrication
After fabrication (in testing laboratory)	The reliability tests that are listed in the specifications are executed. If the number of failures is higher than the specified threshold, an alarm signal is pulled	The FA laboratory must give this stage priority in analysis, because a failure can cause an interruption in manufacture. Possible failure causes must be determined quickly, in order to restart the fabrication
During operation or storage	Based on data furnished by the customer, the failure rate is assessed continuously by the manufacturer of the component. In most cases, as a consequence of FA and with the aid of the manufacturer, it is possible to find a convenient solution for both sides	In the case of an obvious deviation, a careful analysis of the failures must be made. But an analysis of field failures is difficult to make because sometimes the relevant information is lost

required by the customer has to be obtained (which includes the target quality level), preferably right from the start. This means the 'long and winding road' from idea to finished product (from design and fabrication, through reliability testing, re-design and so on) has to be shortened drastically. If the desired reliability level is not obtained after a single development cycle, it is important to be as close to it as possible, so that it can be attained with only minor adjustments.

Hence, there are two basic requirements for the development of a new product:

- The time-to-market has to be minimised.
- The necessary quality and reliability has to be obtained (or close to) after a single development cycle.

3.1.1 Concurrent Engineering

It is difficult to fulfil both the above requirements: it may be relatively easy to minimise the development time, but maximising the reliability level normally requires time-consuming tests, increasing development time again. How to solve this?

Moreover, there is a circumstantial factor that makes this task even harder: the complexity of the technical content of products is increasing, but so is the diversity and the variety of these products. Companies have to deal with this increasing product complexity in their product creation processes by delivering products that satisfy customer requirements [4].

The conclusion: in order to achieve these two goals, a powerful tool is required. This seems to be the approach called 'concurrent engineering', where every specialist working on a product is involved at every stage of development, from the design phase through to commercialisation, significantly shortening time-to-market and producing highly reliable products.

Concurrent engineering involves not just reliability, but also manufacturability, testability and so on. But our main focus in this book will be on reliability issues; that is, on the discipline called design for reliability (DfR), which is aimed at developing the necessary procedures and methods to take reliability issues into account at the design phase [5]. DfR has to understand, predict, optimise and design up front for an optimal reliability, by taking into account materials, processes, interconnections, packaging and the application environment [6]. The necessary conditions for implementing a DfR approach are: (i) a broad range of qualitative and quantitative methods for predicting the reliability of the future product, even at design phase and (ii) some design guidelines for avoiding possible reliability risks. FA plays an important role in developing these two types of tool, as shown below.

3.1.2 Failure Analysis during the Design Stage

Some years ago, the use of FA during the design stage was considered rather peculiar, because FA was traditionally believed to be linked to reliability tests, which were performed only after a product was manufactured. Today this view is obsolete; by implementing the modern ideas of concurrent engineering, the reliability engineer is involved right from the design stage. The modern reliability engineer uses the function-failure design method (FFDM) [7], a methodology for performing FA in conceptual design, by using a function-based concept-generator approach [8] to streamline the design process. FFDM is a start-to-finish design method that utilises knowledge bases which link product function to likely failure modes and product function to possible concepts [9]. FFDM requires a knowledge base of historical failure occurrences linked to product function in order to generate the likely failure modes for new designs. Once a knowledge base is generated, it can be used by many different designers in a wide range of design cases.

A modelling tool set that can be used to analyse variability at any product level (from component to system) was proposed by VEXTEC [10]. At component level, the physics of failure (PoF) describes the process for fatigue crack initiation and development within the material microstructure. PoF models are used to predict failure rates based on measured statistical variations. In complex electronic systems, a constant failure rate is not assumed, as it is in most other systems. In fact, no assumptions are made about failure rate, but a virtual representation of the electronics product is used to simulate the product's real-world performance. Thus, reliability can be estimated while the product is still at the design phase.

Alam provides a nice tutorial on how to take process variations into account during the design phase of integrated circuits (ICs) [11]. He gathered the literature on reliability- and process-variation-aware VLSI design, together with solutions and solution strategies. Case studies on logic circuits, memories (e.g. SRAM) and thin-film transistors were analysed, starting with the physics of process variation and parameter degradation, in which FA plays an important role.

Because in real cases the PoF approach should focus on each failure mode separately, it is better to reduce the number of potential failure modes. For instance, failure modes with insufficient margins (meaning a high likelihood of occurrence in the time of interest) should be extensively analysed and tested to ensure that they do not contribute to an unacceptable number of failures in the field. The use of this method requires that the equipment manufacturer actively collects and lists failure modes for previous products, as well as for components, materials and processes [12].

In the case of electronic components, knowledge obtained by FA is used to evaluate the reliability of a product during actual operation. A failed component can provide important information to enhance the reliability of a device or product. Depending on the type of component failure, failure modes (FMos), failure mechanisms (FMs) and factors such as stresses can be identified, inducing the failure and initiating appropriate corrective measures. FA provides feedback to the product designers in order to improve design and correct minor faults that may have been overlooked in the initial design [13]. The main types of failure that can arise at the design stage are produced by [6]: mask design faults, incorrect dimensions of some device elements (too long, too short), incorrect assumptions about the models used, material properties not taken into account (e.g. large stress gradient) and so on.

An introduction to the methods used to represent and model expert judgment for reliability prediction in the early stages of the product development process is given in [14]. It provides a survey of the required inputs and resulting outputs of the single approaches. The problem of handling uncertainties in early design stages is exemplified, evaluating the reliability for the software portion in mechatronic systems. The influence of uncertainty is demonstrated and an approach which enables the comparison of several concepts in early development stages based on expert judgment is introduced.

For electronic systems, there are some well-known reliability prediction methods that can be used in DfR, such as MIL-HDBK-217, PRISM and Telcordia (Bellcore) SR-332. In June 2005, PRISM (which was produced by the Reliability Analysis Centre, RAC) was replaced by 217PlusTM (elaborated by the Reliability Information Analysis Centre, RIAC, the successor to the RAC) [15]. Without getting into details, none of these methods really use FA, because they treat failures globally, rather than analysing each population affected by an FM.

In 2004, a new reliability prediction method, *FIDES Guide 2004: Reliability Methodology for Electronic Systems*, was developed by a consortium of French industrialists from the aeronautics and defence fields (composed of Airbus, Eurocopter, GIAT ELECSYS, MBDA Missile Systems, Thales Airborne Systems, Thales Avionics, Thales Services and Thales Underwater Systems). It is based on the PoF method and supported by the analysis of test data, field returns and existing modelling. Most importantly, FA is involved more directly than in previous methods (MIL-HDBK-217, etc.), because technological and physical factors are considered, including mission profile, mechanical and thermal over-stresses and the possibility of distinguishing the failure rate of a specific supplier of a component. It also takes into account failures linked to development, production, field operation and maintenance processes [16]. Results have shown that the FIDES prediction methodology produces acceptable results.

One may say that FIDES is a step towards a paradigm shift in the approach to product quality: the integration of design-centric analysis technologies is required to facilitate virtual qualification using the PoF-based 'design–fix–build–test' methodology for reliability prediction. In this methodology, the thermal design plays an important role [17]. In the PoF approach, user-defined load and environmental conditions are used in combination with layout and other input data and physics-based FMs, resulting in a ranking of most probable failures. Also, and most significantly, the PoF approach requires that the root cause of the failure is identified. Hence FA becomes increasingly important. One of the leaders in this field is the CALCE Electronic Products and Systems Center at the University of Maryland, which promoted systematic research into the FMs of electronic components. The results were reported in a tutorial series, published by IEEE Transactions on Reliability in the period 1992–1994. In these papers, for each FM, design guidelines were proposed in order to avoid the possible reliability risks. Conceptual models of failure were proposed in the paper initiating this tutorial series [18]. The basic idea is that the real failures of a system are due to a complex set of interactions between the its materials and elements and the stresses acting on and within it. Hence, system reliability must be analysed by starting with the response of material/elements to stresses and the variability of each parameter. Four simple conceptual models for failure are defined, corresponding to various real-life situations:

- **Stress–strength:** The item fails if and only if the stress exceeds the strength and there is no damage to the item properties of non-failed items. One example is transistor behaviour when a voltage is applied between collector and emitter.
- **Damage–endurance:** Damage is accumulated irreversibly under stress, but the cumulative damage does not degrade performance. The item fails when and only when the damage exceeds the endurance. Accumulated damage does not disappear when the stresses are removed, although sometimes annealing is possible. Examples include corrosion, wear, fatigue and dielectric breakdown.
- **Challenge–response:** The system fails because one element is bad, but only when that element is solicited is the system failure produced. The most common example is computer program (software) failures, but telephone switching systems also resemble this failure model.
- **Tolerance–requirement:** System performance is satisfactory if and only if its tolerance remains within its requirement. In other words, failure occurs when something is nominally working, but not well enough. An example is a measuring instrument that shows an incorrect value for the measured parameter.

Recently, new ideas about the best way to manage reliability were proposed by the most dynamic domain of the technique: the space exploration programme. The United States National Aeronautics and Space Administration (NASA) is running a space exploration programme, called Constellation, intended to send crew and cargo to the international space station (ISS), to the moon, and beyond. In the frame of this programme, NASA proposed a new 'design for reliability and safety' approach, applied to the new launch vehicles ARES I and ARES V [19]. An integrated probabilistic functional analysis will be used to support the design analysis cycle (DAC) and a probabilistic risk assessment (PRA) to support the preliminary design (PD) and beyond. This approach addresses up-front some key areas in the support of ARES I, including DAC and PD. Named probabilistic functional failure analysis (pFFA), this is a probabilistic physics-based approach which combines failure probabilities with system dynamics and engineering failure impact models to identify key system risk drivers and potential system design requirements. It is a dynamic top-down scenario-based approach intended to identify, model and understand high system risk drivers in order to influence both system design and system requirements. Failures not initiated by the launch vehicle, other than those induced by the natural environment, and launch-vehicle software failures are not currently being considered. As mentioned above, in this approach FA plays a central role. The main managing tool of pFFA is PRA, which models what can go wrong with a system, predicts how often it might go wrong (the probability that specific undesired events will occur), identifies the consequences if something does go wrong, and engages the design and development community to the fullest extent. Within NASA the PRA uses as input, among other things, safety, reliability and quality models and analyses, which include hazard analyses, fault-tree analyses, failure modes and effects analyses, reliability predictions, and process characterisation and control analyses. PRA provides information on the uncertainty of the predictions and identifies which failures and therefore which systems, subsystems and components pose the most significant risk to the overall system.

3.1.3 Virtual Prototyping

As shown in Table 3.2, the development cycle of a product begins with the design stage and finishes with a prototype. In a concurrent engineering approach, the most important characteristics of the prototype must be taken into account even from the design stage, a concept called virtual prototyping [20]. Obviously, this concept requires extensive knowledge of the FMs involved and their dependence on technological factors and the functioning environment.

A detailed example of the use of virtual prototyping is presented in [21], for a tuneable Fabry–Perot cavity for optical applications, which is a resonant micro-cavity composed of two flat, parallel, high-quality mirrors separated by a variable gap. The movable membrane (mirror) is electrostatically actuated by applying a voltage between a fixed electrode and the movable mirror (deformable membrane). Elaboration of the reliability analysis of the virtual prototype is based on previous knowledge of typical FMs, as shown in Table 3.3.

The reliability analysis contained the following steps:

1. Simulating the mirror displacement versus Young's modulus and calculating the Young's modulus value corresponding to the failure criterion. The failure criterion is $\lambda/8$, which corresponds to a value of the displacement of 0.32 μm. For this displacement, the Young's modulus value E_d is 130 GPa.
2. Simulating the oscillation frequency versus Young's modulus and calculating the Young's modulus values for the oscillation frequency values chosen as failure criteria. The oscillation frequency values corresponding to the failure criteria (±15%) are 25 and 28 kHz, respectively. The Young's modulus values for these two values of oscillation frequency are $E_{f1} = 128$ GPa and $E_{f2} = 160$ GPa.
3. Comparing the values obtained at steps A and B in order to obtain the limiting Young's modulus values for the life of the device. For the resonant mode, the limiting value of the Young's modulus must be chosen as the higher value between E_d and E_{f1}, which is 130 GPa. For the switch mode, the failure criterion linked to the oscillation frequency could not be used, because the limiting value, 160 GPa, is already the initial value of the Young's modulus and this value decreases with cycling. Hence the limiting value is again $E_d = 130$ GPa. As one can see, in both modes (the resonant one and the switch one), the limiting value is 130 GPa.
4. Using the curve's Young's modulus versus the number of cycles to calculate the number of cycles corresponding to the limiting value of Young's modulus. If the initial Young's modulus value is 160 GPa and the limiting value is 130 GPa, the corresponding number of cycles to failure is 3×10^4.

Reliability tests were performed on prototypes of a microelectromechanical system (MEMS)-tuneable Fabry–Perot optical cavity, the results were statistically processed and the number of cycles to failure was obtained: a value close to the 3×10^4 obtained by the virtual prototype. Hence, virtual prototyping, based on FA, seems to be a useful tool for concurrent engineering, furnishing *at the design phase* accurate predictions about the reliability of the future device.

3.1.4 Reliability Testing during the Development Cycle

During the development cycle, any component or system has to be tested to ensure that its design is correct in relation to performance, reliability, safety and other requirements. In spite of its importance for product development, testing is still treated as a marginal subject in engineering degree courses. Consequently, as O'Connor said, 'much testing in industry, particularly in relation to reliability, is based upon inappropriate thinking and blind adherence to standards and traditions. Products are tested to "measure" reliability, when the proper objective of development testing should be to find opportunities for improving reliability, by forcing failures using accelerated stresses' [24].

Table 3.3 Reliability analysis of the typical failure mechanisms of the tuneable Fabry–Perot micro-cavity

Typical failure mechanisms	Main characteristics	Chance of occurrence	Detection method	Failure criterion
Fatigue damage	The stiffness of the membrane is influenced by the number of cycles [22]; a direct relationship links the stiffness degradation rate and the number of fatigue cycles	High – this is the most probable FM	The displacement of the mirror at a given voltage becomes slightly different when the elasticity is damaged. This phenomenon is detrimental to normal functioning when the displacement is higher than $\lambda/8$,[a], which is the failure criterion The time response of the device (when fixed voltage is applied) is modified as the movable structure's stiffness varies. The geometry of the structure and the elastic properties of the material define its stiffness. Also, the fundamental natural frequency is coupled with the stiffness of the movable structure, and is influenced by the value of the Young's modulus (E)	For the resonant mode: a decrease of 15% of the natural frequency. For the switch mode: an increase of 15% of the natural frequency
Formation of an oxide layer	An oxide layer is formed on the polysilicon and the cracks initiated in this oxide layer are propagated into the polysilicon layer [23]; this process is increased in a humid environment and is called reaction-layer fatigue	Low – this FM was not studied	–	–

[a] λ is the wavelength of the radiation filtered or modulated, depending of the usage of the device – as accordable filter or modulator, respectively.

As mentioned above, the development cycle generally contains four stages. The role of FA during the first stage (design) has already been described in Sections 3.1.2 and 3.1.3. It is summarised in Table 3.4, together with the characteristics of the other three stages.

Some comments must be made on the involvement of FA in Table 3.4. The first reliability tests are performed at the experimental model stage. These are aimed at checking the limits of the functioning domain for the component. Hence, a large range of stress types and stress values are used, not only by simulating the typical operational environment, but by combining stresses that might be encountered in operation. These are called stress tests (STs). FA must be performed carefully, for each failed component, because the resulting information is extremely valuable in identifying the possible failure risks and elaborating adequate corrective actions. In fact, this is an iterative process for improving the quality and reliability of the product, and continues at the prototype stage.

At the prototype stage, together with STs, all the tests from the general and particular specifications are performed, as well as the tests necessary for obtaining the reliability level of the batch of components (so-called quantitative life tests, QLTs). Frequently, these are accelerated tests (see Section 3.1.4.2), because the tests at normal stress level do not produce enough failures for the statistical processing of data. Again, FA has a central role, being aimed at identifying all the typical FMs, including the influence of environmental and working factors.

From the above, it becomes obvious that there are two main categories of reliability test:

- *Qualitative life tests* (or STs), as mentioned above at the experimental model stage, used to establish the operational limits of the components; that is, the behaviour at stress limits, for example failures or drift of significant electrical parameters. Examples of qualitative tests are structural and fatigue testing and environmental stress screening (ESS) of electronic assemblies; other highly accelerated life testing (HALT) and highly accelerated stress screening (HASS), developed by Hobbs [25], belong in the same category.
- *Quantitative life tests* (QLTs), as mentioned above at the prototype stage, used to assess the reliability indicators (or parameters) of the batch of components. The reliability indicator for a component is the failure rate and for a system is the mean time between failures (MTBF).

Table 3.4 The role of FA in the main stages of the development cycle of a product

Main stages of the development cycle	Characteristics and role of failure analysis
Design	Starting from an idea and the required functions, the processes required to enable product fabrication are imagined. The virtual prototype approach is used to estimate the potential reliability of the future product
Demonstrator	A limited number of items are fabricated and characterised, in order to demonstrate that the imagined product could be manufactured
Experimental model	A first batch of product, with known and monitored characteristics, is fabricated; reliability tests are performed to check the product at the limits of the functioning domain; FA is used to identify the possible reliability risks; corrective actions are proposed, leading to an iterative process for improving the quality and reliability of the product
Prototype	Subsequent batches of the product are fabricated, with the aim of monitoring the quality and reliability of the product; details about the manufacturing process are elaborated (procedures, monitors, etc.); all tests from the general and particular specifications are performed; (accelerated) tests to obtain the reliability level of the batch of components are also performed; FA aims to identify all the typical FMs, including the influence of environmental and working factors

For both methods, *corrective actions* are established after FA of the failed items and analysis of the test data; these actions refer to changes in manufacturing and control technology, such as:

- New/modified working procedures for the processes used on the manufacturing flow.
- New parameters to be measured at various control stages (or modifications of the measuring conditions/limits for the existing parameters).
- New control and/or monitoring points (or modifications of the requirements for the existing ones) of the manufacturing flow.

QLTs are more difficult to design and use than STs, because the real operation must be simulated in order to test the components in the same conditions as in real life. Consequently, QLTs have to solve many issues [5]:

- Choosing the dimension of the sample(s) used, given the desire to have statistically significant results, even for the most costly/difficult-to-obtain types of component.
- Measuring the most significant electrical parameters for component reliability.
- If a continuous inspection of the parameters cannot be performed, executing the measurements at some given times (so-called 'noncontinuous inspection'), which must be chosen appropriately.
- Establishing the right failure criteria (the limits for declaring a component as failed) corresponding to nonfunctionalities in operation (this depends on the application, which make the choice more difficult).
- Choosing the stress types characteristic for the application, because QLT is an application-driven test; in fact, this is the most difficult issue, because it is not a simple task to simulate the operational conditions.
- Choosing the appropriate criteria for test-ending (a test duration or a percentage of failed components), in order to obtain the most fruitful results.
- If no (or a small number of) failures are obtained after the whole test, establishing appropriate means for withdrawing significant results from the data, such as statistically processing the drift of the electrical parameters.

The first three issues are also valid for qualitative life tests.

3.1.4.1 Environmental Reliability Testing

The components have to function in an environment made up of a combination of factors, such as temperature, air pressure, vibration, thermal cycling and mechanical acceleration. Experimentally, it was demonstrated that the combined effect of these factors is higher than the sum of the individual effects, because a synergy of the stress factors occurs [26]. Hence, the individual effects, but also the synergy of these factors, must be studied. Some stress factors are strongly interdependent, such as solar radiation and temperature, or functioning and temperature. Independent factors are also found (mechanical acceleration, humidity). An analysis of the possible synergies must cover the main phases of the component's life: storage, transport and operation.

Storage and Transport

Normally, the storage and transport period, if correctly done, does not influence the component's reliability. However, for components manufactured for special military and industrial purposes (weapons, nuclear plants, etc.), the storage period becomes important to reliability, mainly because for such components no failure can be allowed when they are first used. The stresses involved in the transport and storage period are therefore carefully analysed and the possible FMs must be avoided. Hence, FA plays an important role.

As an example, weapons kept in storage by the US Army at Anniston experience a daily variation in temperature of up to 2 °C, and a humidity of 70% [27]. There are also storage areas in tropical and arctic zones. For systems exposed to solar radiation, temperatures higher than 75 °C were measured, with temperature variations exceeding 50 °C. To check the component reliability in these situations, studies of the behaviour at temperature cycling were performed (see Section 3.2.2). Other stress factors, such as rain, fog, snow and fungus or bacteria, may act and must be investigated. At transport, the same (temperature cycling, humidity) or other (vibration, mechanical shocks, etc.) specific stress factors may arise. For all these factors, studies about the involved stress synergies and FMs were performed. An example is given in [28], where the behaviour of temperature and vibrations of electronic equipment protecting an airplane against sol–air missiles is investigated. Operational data was collected, using a complex system (elaborated by the specialists from Westinghouse) containing 64 temperature sensors (AD 590 MF, from Analogue Devices) and 24 vibration sensors (PCB Piezotronics 303 A02 Quartz Accelerometers), mounted on two ALQ-131 systems used for the F15 fight plane. The tests were performed between December 1989 and August 1990. The data were processed and laboratory tests were built–based on the obtained information – for the components with abnormal behaviour. Eventually, corrective actions were made to improve component reliability. The result was that during the Gulf War (January 1991), the ALQ-131 equipment had a higher reliability than before.

Operation

The essential difference between the storage and transport environment and operation is the presence of electrical bias. At first sight, it seems that the only effect of the electrical factor is an increase in chip temperature, following the relation [5]:

$$T_j = T_a + r_{th_{j-a}} P_d \qquad (3.1)$$

where T_j is the junction temperature, T_a the ambient temperature, $r_{th_{j-a}}$ the thermal resistance junction – ambient and P_d the dissipated power. If the effect of the electrical factor is only a temperature increase, than it should be the same as an increase in the ambient temperature. But experimentally, it has been shown that this hypothesis is not valid. The electrical factor has a thermal effect, but also a specific electrical effect due to electrical field or electrical current.

Often, the components have to work in an intermittent regime. In these cases, the phenomenon limiting the lifetime is the thermal fatigue of the metal contact, produced by the synergy between the thermal factor (thermal effect of functioning) and the mechanical factor, modelled with [29]:

$$N = c(\Delta e_p)^{-n} \qquad (3.2)$$

where N is the number of functioning cycles, c is a constant and Δe_p the thermo-mechanical stress, given by:

$$\Delta e_p = L(\Delta\alpha)(\Delta T)/x \qquad (3.3)$$

where L is the minimum dimension of the contact, $\Delta\alpha$ the average of the dilatation coefficients of the two interfaces, ΔT the temperature variation and x the width of the contact. Experiments on intermittent functioning of rectifier bridges (2000 cycles of 20 minutes each at on- and off-state, respectively) emphasise [30] the main FMs: (i) degradation of the contact between silicon and electrodes and (ii) contact interruptions. Therefore, intermittent function induces different FMs than continuous function.

3.1.4.2 Application-Driven Accelerated Testing

In the 1960s, real-life tests strictly simulating operational conditions became useless due to the continuous improvement of electronic component reliability, because the components did not fail during these tests. The solution was to use tests under the same type of stress, but at higher values than those encountered in normal functioning, with the aim of shortening the time necessary to obtain significant results [31]. This solution is more convenient for component manufacturers, offering information about batch reliability sooner. The accelerated test is an ageing deterioration of an item to induce normal failures by operating at stress levels much higher than would be expected in normal use [32].

There are two main reasons for considering FA to be essential for accelerated testing:

- The basic rule of accelerated testing is the following: the FM acting at the higher stress level must be the same as that acting at the normal stress level. So all possible FMs must be identified.
- The model describing the acceleration obtained by a given stress factor is valid only for a population affected by the same FM. So, before statistical data processing, the failed items have to be carefully analysed, in order to separate the group of components failed by specific FMs.

If the first point is not respected, and a real-life FM is missed in accelerated testing, this could result in a failure at operating conditions. In statistical terminology, such accelerated testing carries a risk of confounding. For linear models, it is possible to determine theoretically which models are confounded with others. An analogous theory for a simple kind of confounding model, evanescent processes, in which kinetics is used as the basis of modelling accelerated testing, is proposed in [33]. A heuristic for identifying simple evanescent processes that can give rise to disruptive alternatives (alternative models which reverse the decision that would be made based on modelling to date) is outlined, then activity mapping is developed, a tool for quantitatively identifying the parameter values of that evanescent process which can result in disruptive alternatives. Finally, activity mapping is used to identify experiments which help reveal such disruptive evanescent processes.

Of course, the accelerated testing is used for the two main types of reliability test mentioned before: qualitative life test (which in this case are named accelerated stress tests, ASTs) and QLT (becoming quantitative accelerated life tests, QALTs).

3.1.4.3 Reliability Assessment

Generally, the failure rate of a batch of electronic components is assessed by laboratory tests. For example, one sample of 200 items undergoes a life test at a combination of stresses and at a stress level simulating the operational one. The test is ended when at least 50% of the items have failed. At given moments (e.g. 72, 168, 336, 500, 1000, 2000, 3000, 5000 hours), the test is stopped and, for all components, some electrical parameters (considered relevant to identifying the failed components) are measured. All failed devices are carefully analysed and the population affected by each FM is identified. For each failure mechanism (FM_i) and stress level (S_i), the distribution law (lognormal, Weibull, etc.) is obtained, the distribution parameters (e.g. median time and standard deviation for lognormal distribution, scale and shape parameters for Weibull distribution) being estimated using analytical and/or graphical methods. The estimated parameters of the distribution are checked against the experimental data using concordance tests (Kolmogorov–Smirnov, etc.). Eventually, the failure rate versus time curve is obtained for the given test conditions.

However, for modern devices with small failure rates, long and expensive tests are needed, because one must wait a long time (one year or more) before significant failures (at least 50% of the whole sample) occur. This is why a method for shortening the duration of life testing was proposed, by testing the components at higher stress level than the operational ones [34]. QALT is used to minimise the test duration by employing high stress levels. The curve failure rate versus time is obtained at this higher-than-normal stress level and then, using the acceleration law for this type of stress, the same curve is obtained for the operational conditions. QALT must fulfil a decisive condition: the FMs at the accelerated stress level must be the same as for the operational stress level. This means the

FA is an important component of the QALT, which produces time-to-failure information and can be used to assess the failure rate of a batch of devices. In order to do this, some samples are randomly withdrawn from a batch of devices and tested at various stress levels. The acceleration laws are different for various FMs, so the failed devices must be carefully analysed in order to separate the populations affected by different FMs. Then the statistical processing of data is performed separately for each population. The time dependence of the failure rate is estimated for each stress level used in accelerated testing and the activation energy is calculated, which describes the slope of the acceleration law for the given type of stress. This allows calculation of the time dependence of the failure rate at the operational stress level.

3.2 Failure Analysis during Fabrication Preparation

The failures of electronic components and systems have well-defined causes, whether foreseen or not. Failures caused by design were dealt with in Section 3.1. Many other causes are related to fabrication preparation, the main issue being poor selection of materials or combinations of materials.

3.2.1 Reliability Analysis of Materials

For electronic systems, the materials to be used are analysed at the input control. Generally, FA is applied under the form of microphysical characterisation and only if some doubts about the quality of a material have arisen.

For electronic components, the input control of the materials used is well documented, because the quality and reliability of the materials is extremely important to the quality and reliability of the future component. Generally, the input control consists of the measurement of physical and/or electrical parameters of the material. If some faults are found, the user (or more likely, the manufacturer of the material) performs some reliability tests on the whole batch, in order to analyse further the reliability of the material. These tests (which may be executed by independent testing laboratories) refer to the future use of the material in component construction. FA of the failed items allows identification of the FM. Based on this, the manufacturer of the material may suggest corrective actions to avoid the FM identified.

In Table 3.5, the main materials used for electronic component manufacture are presented, together with their possible usages.

Obviously, when manufacturing reliable components, reliable materials are needed. From the viewpoint of the component manufacturer, it is important to be able to assess accurately the reliability of the materials used. Moreover, when the electronic systems containing the components are to be used in harsh environments, the study of material reliability becomes compulsory.

The need for an accurate assessment of material reliability has led to the development of a large number of independent institutions focused on studying and delivering reports on the materials used in components and systems. Such institutions have the advantage of being used by both the manufacturers and the users of materials. The best example is the National Institute of Standards and Technology (NIST) in the USA; its Materials Reliability Division (http://www.nist.gov/msel/materials_reliability/index.cfm) has established several capabilities for analysing the reliability of small-scale structures, purchased from several manufacturers, such as a field emission scanning electron microscope (SEM) with automated electron backscatter diffraction system (EBSD), a 200 kV transmission electron microscope (TEM), a Brillouin light-scattering system, a tip-enhanced Raman spectroscopy system, a nanoindenter and a scanning thermal microscope. A number of custom instruments have also been developed in-house specifically for size-appropriate testing, such as a micro tensile testing apparatus, a contact-resonance force microscope, a surface acoustic wave spectroscopy system and a probe station equipped for automated AC electrical fatigue testing.

Table 3.5 Materials used in electronic component technology

Family of materials	Examples	Usage
Semiconductors	Silicon (single crystal, polycrystalline, porous)	Principal component of most semiconductor devices
	Polycrystalline SiGe	Semiconductor material in ICs, heterojunction bipolar or CMOS transistors, MEMS
	Gallium arsenide (GaAs)	Manufacture of devices such as microwave frequency ICs, infrared light-emitting diodes, aser diodes, solar cells and optical windows
	Indium gallium arsenide (InGaAS)	High-power and high-frequency electronics, because of its superior electron velocity with respect to the more common semiconductors
	Diamond	Electrical insulator, but extremely efficient thermal conductor
Inorganic dielectrics	Si_3N_4 and SiN_x (PECVD, LPCVD), SiO_2 (thermal, PECVD, sputtered)	Insulators
	USG, PSG, BSG, Pyrex	Glasses
Metals and alloys	Aluminium (Al), gold (Au), copper (Cu), silver (Ag), tungsten (W), platinum (Pt)	Interconnections at chip level; connectors; contacts
	Nickel (Ni), chromium (Cr), tantalum (Ta)	Resistors
	Iron (Fe), silver (Ag); AuSn	Packages of semiconductor devices
Polymers, plastics	Resists, polyimides; BCB; silicone resins; thermoplastics (PEEK, etc.)	Plastic encapsulation of devices; polymer microchips; sensors using conductive polymers

Other groups which study the reliability of materials belong to the companies manufacturing such materials. One example is the Center for Materials Science and Engineering of Sandia National Laboratories, USA (http://www.sandia.gov/materials/science/capabilities/materials-analysis.html), which provides knowledge of material structure, properties and performance and the processes to produce, transform and analyse materials to ensure mission success for its customers and partners, both internal and external to the laboratories. The micro-structural mechanisms that underlie material behaviour must be understood and controlled through processing to ensure that as-fabricated products meet reliability requirements. Moreover, the material ageing mechanisms must be understood and quantified to provide the basis for predicting reliability throughout design lifetime. In the last several years, the center has performed studies on: the degradation of solder materials due to thermo-mechanical fatigue and intermetallic compound growth; material mechanisms controlling the reliability and ageing of MEMS devices in the as-manufactured state and under dormant storage conditions; corrosion effects, which can affect electronic components by removing material (causing an open circuit) or adding corrosion product (causing a short circuit); electrical contacts used in switches and other electromechanical devices, which can degrade due to contamination, corrosion and fretting/wear; stress voiding of aluminum interconnect lines in ICs, which can occur during long-term dormant storage as shown above; polymeric materials, which may perform critical functions at both component and system levels; and glasses and ceramics, which are chemically very stable, but may be sensitive to process-induced defects and long-term ageing effects.

Reliability tests for materials have two main purposes: (i) to identify the possible weaknesses of the product and remove them by specific corrective actions and (ii) to quantify the reliability level of the product in various applications. Because modern materials have a high reliability, in recent years reliability tests became too long and even useless (if no failures arose during testing). The solution was to use tests at stress values higher than those encountered in normal functioning, aiming to shorten the time necessary to obtain significant results [5]. This solution is more convenient for material manufacturers, offering information about batch reliability more quickly. In fact, the accelerated test is an ageing deterioration of an item to induce normal failures by operating at stress levels much higher than would be expected in normal use [32]. The basic rule of accelerated testing is to deeply investigate the FMs, because it is essential that the FM acting at the higher stress level be the same as that acting at the normal stress level. An accelerated test is only useful if, under the accelerated conditions, the item passes through all the same states, in approximately the same order, as are expected in normal use, but in a much shorter period of time. The model describing the acceleration obtained by the given stress factor must also be developed, otherwise extrapolation of the results obtained by the accelerated test to normal functioning is impossible. For materials, the reliability tests are performed on test structures, which are specific for the studied FMs. This is an 'application-driven' process, because the stresses used must simulate those encountered in 'real life' (industrial applications). Specific test structures for reliability testing have been reported by various companies and researchers (e.g. US patents: [35–37]).

A systematic reliability study of the materials used in the manufacture of electronic components must follow a methodology with the following steps:

- **Manufacturing.** Test structures are designed and fabricated for the material under study.
- **Material properties.** Material properties required for the simulations are measured. These include but are not limited to: coefficient of thermal expansion, hardness, Young's modulus, Poisson ratio and grain structure. Many of these properties have to be obtained as a function of temperature.
- **Reliability.** Tests are executed in conditions that simulate those found in the main applications (accelerated tests are used: that is, tests with the same loading conditions but at a higher stress level). Examples include high-temperature ageing, thermal cycling, drop/shock tests, vibration tests, thermal cycling and so on.
- **Failure mechanisms.** Typical FMs (e.g. temperature-induced plastic deformation, corrosion, fracture, fatigue, wear, stiction, electromigration, hillocking and so on) are identified and their dependences on time, stress and material properties are obtained.
- **Modelling.** Models based on structural information about the physical or chemical phenomena involved in degradation and failure and on advanced multi-scale approaches are developed and experimentally validated; the models are used to predict the material's lifetime. For example, one has to study, formulate and implement models able to describe the polycrystalline nature of the material under study, the process of initiation and propagation of micro and macro cracks, long-term behaviour in subcritical conditions and high-cycle loading (i.e. fatigue), the influence of defects of various natures, and thermoelastic and thermoplastic behaviour. The numerical models reproduce the material behaviour at various scales and therefore the use of multi-scale approaches is of paramount importance. In this field, atomistic and continuum descriptions are coupled in order to simulate the material behaviour in a highly realistic way, with computational costs comparible with those of modern computing facilities.
- **Material improvement.** Based on the obtained experimental results and on the proposed models, corrective actions for fabricating the material are proposed and applied by the manufacturer of the material.
- **Reliability check.** By re-executing the same reliability tests on test structures, the effect of the improvements is checked in new batches of the same material.
- **Design guidelines.** The possible implications of the reliability of the components using the material under study are identified by means of the involvement of an industrial user.

The working methodology is synthesised in Figure 3.1.

The final results of the above methodology are: (i) models of the typical FMs (starting with the chemical and physical mechanisms that cause material properties to change during exploitation);

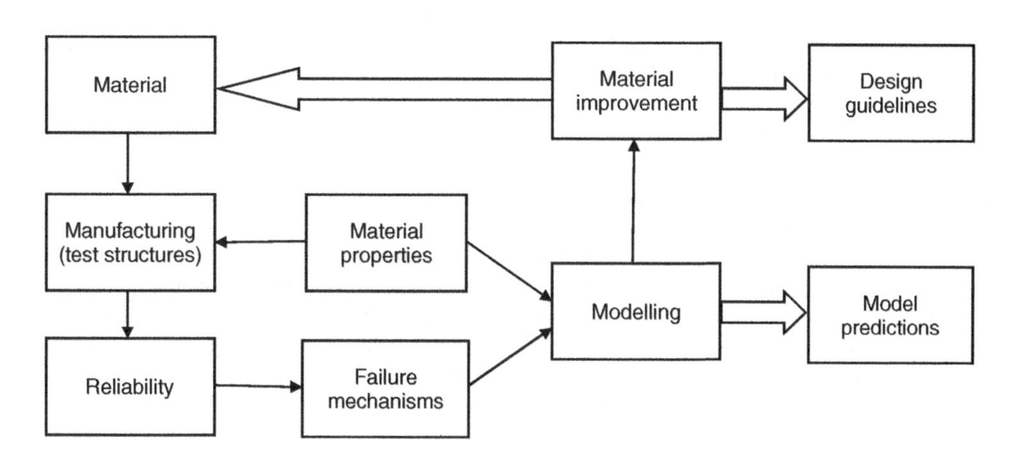

Figure 3.1 Interactions between the steps of the working methodology for developing reliable materials for electronic components

(ii) design rules for achieving reliable materials for the same applications; and (iii) new and more reliable variants of the studied material.

3.2.2 Degradation Phenomena in Polymers Used in Electron Components

The polymers were used intensely in the technology of electron components, especially as photoresists (in lithography) and encapsulating materials. In recent years, new applications have arisen, and the polymers have become a necessary material for microtechnologies[2].

3.2.2.1 Plastic Encapsulation

In the last 15 years, plastic (polymer) encapsulation has become the main choice for the electron components. The reason for this is the significant improvement in the quality of plastic cases, which are sometimes better than metal and ceramic ones. A milestone of this trend is the so-called 'Acquisition Reform', launched in June 1994 by the US Army, which accepted of plastic-encapsulated devices as fulfilling military requirements. Furthermore, the idea of developing purely military components was rejected, because the commercial ones seemed to be good enough for military equipment.

This does not mean that the polymers used in plastic-encapsulated devices will last forever. Degradation phenomena are still identified and some examples are presented below.

Oxidative Degradation

Oxygen degrades the polymers by diminishing their molecular weight, reacting with the free radicals of the polymers to form peroxy and hydro peroxide free radicals [38], producing chain scission. The properties of the polymers are drastically modified. Even a reduction of 5–10% of the molecular weight can produce the failure. To prevent this, it is recommended to avoid contact with oxygen and to use an antioxidant. But processing the polymers used for encapsulation may involve high temperatures, required to reduce melt viscosity. The presence of oxygen, in small amounts, can cause the degradation of the polymer, even if an antioxidant is used. The solution is to optimise the processes, but in order to do this, reliable methods of analysis are needed, allowing rapid verification of the various solutions.

[2] The microtechnologies are defined as technologies which obtain structures, subsystems and technical systems with functional dimensions of the order of micrometres (e.g. microsystems).

To identify the oxidative degradation, differential calorimetry, infrared spectroscopy and measurement of the change in molecular weight are used.

Environmental Stress Cracking

Breakage through the combined action of stress and environment (environmental stress cracking, ESC) is one of the main causes of failure in components encapsulated by plastic injection. ESC is an accelerated fracture of the polymeric material, caused by the combined action of mechanical stress to the case and the environment. This phenomenon was identified from the beginning of polymer use in the encapsulation of electronic components and could not be entirely stopped, although remarkable steps were made in this direction. In order to control the susceptibility of a plastic material to ESD [39], one must verify chemical exposure, mechanical stress, working temperatures, humidity and exposure to ultraviolet radiation. The NPL Materials Centre (UK) produced a programme called CAMPUS (Computer Aided Material Pre-selection by Uniform Standards) to allowing manufacturers to choose the most appropriate polymer type for a required application.

3.2.2.2 Sensors with Polymers

This a very important area in polymer use in microtechnologies. The polymer can be the dielectric of a capacitor, or its conductive properties can be exploited.

1. **Polymer as a dielectric.** Humidity and temperature sensors are accomplished as capacitors, the polymer being the dielectric. In this field, some improved devices have arisen lately, with better time stability of performances. For example, the Swiss company Sensirion AG [40] produced a new generation of temperature and humidity sensors, fully integrated, digital and calibrated, by using the micromachining CMOS technology. The new product, SHT11, is a multisensor one-chip module, for temperature and relative humidity, with a calibrated digital output, easy to integrate in a system. The polymer used as a dielectric may adsorb and release the water, in proportion with the relative humidity, modifying the capacitor capacity. The temperature and humidity sensor are a single device, so the calculation of the dew point is more accurate. The degradation phenomenon for this type of sensor is the ageing of the polymer layer, producing important variations in sensor sensitivity. The process becomes critical because the dimensions of the sensing elements are relatively large: 10–20 mm^2. But design methods are able to minimise the failure risk from polymer ageing.
2. **Conductive polymers.** These are newcomers, but they already have many applications in microtechnologies [41]. A polymer becomes conductive by injecting mobile charge carriers in polymeric chains. This may be done by reaction with an oxidant (to remove the electrons from the polymer) or with a reductant (injecting electrons in the polymer), in an analogous way to n and p doping in semiconductors. For this discovery, Alan J. Heeger, Alan G. MacDiarmid and Hideki Shirakawa (USA) received the 2000 Nobel Prize for Chemistry. There are basically two main methods to synthesise conductive polymers: electro-polymerisation (particularly used to obtain polymer films: the electrodes are inserted in solvent solution containing the monomer and the dopant electrolyte; when biased, a polymer films grows on the working electrode, usually a platinum one) and chemical reaction (the monomers react with an oxidant in excess, dissolved in a solvent; the polymerisation arises spontaneously, and the method is particularly used to obtain volume polymers).

The most important applications of conductive sensors in microtechnologies are in biological and chemical sensors, using conductive polymers as sensitive materials. The advantages include the following:

- There is a great variety of polymers.
- Electrochemical preparation allows mass production and sensor miniaturisation.

- Biomaterials (enzymes, etc.) can be easily incorporated in polymers.
- The oxidation state of the polymer can be easily changed after deposition; hence one may build the sensitive characteristics of the film.

For these reasons, the conductive polymers are used in various types of sensor: pH mode (pH variation), conductometric modes (conductivity variation), amperometric mode (current variation), potentiometric mode (variation of the open circuit potential) and so on. For example, the gas sensors working in a conductometric mode use the conductivity (or resistivity) time variation when the polymer is surrounded by gas molecules. Examples of polymers used for gas sensors include polypirol and polyaniline.

The electrical resistance can change when the polymer is exposed to air for a long time, because the oxygen produces an increase in resistance. To date, optimisation of the manufacturing process to deal with this degradation mechanism has been unsatisfactory.

3.2.2.3 Polymer Micromachining

Polymer micromachining represents a new direction in the field, being used for example in biochip manufacturing. The Swedish company Amic AB reported in 2002 that it employed micromachining technology for polymers in order to achieve a *CD microlab* [42]. The basic idea is to manufacture polymer microchips, achieved by replicating a master on silicon. The microlab contains capillars, reaction chambers, mixers, filters and optical elements. The dimensions of these microfluidic structures are in the range of 10 μm to some mm (lateral dimensions) and 5 μm to hundreds of μm (layer depth). All these are first achieved on silicon, the polymer replica being obtained as CDs. The advantages are obvious: multiplication in polymer is much cheaper and less complicated than mass production on silicon of these microlabs. The CD microlabs do not require tubes, because the microfluids are conducted through centrifugal action, by CD spinning. The micro-optical elements are integrated on the rear of the CD: the optical surface collects the luminescence from CD reaction chambers. Hence, one may avoid using an external collector lens and the micro-optical elements are integrated on a chip.

The company MIC (Denmark) [43] developed, in the frame of the programme μTAS, research on polymer micromachining. Thereafter, in November 2001, a spin-off called POEM (Centre for Polymer Based Microsystems Dedicated to Chemical and Biochemical Analyses) arose, focused on manufacturing integrated chemiluminescence sensors and miniaturised electrochemical sensors, and on fundamental research into polymer micromachining. POEM developed collaboration relations with European companies interested in methods for micromachining micro- and nanostructures on various polymers.

Polymer micromachining is a field with a bright future. Some advantages are listed below:

- Mass production is cheaper than for silicon devices, allowing single-use devices to be manufactured (e.g. for blood analysis).
- Many polymers are transparent in visible light, making them appropriate for systems based on optical detection.
- Polymers can be made resistant to most chemicals used in microsystem manufacturing.

In a polymer exposed to various stresses (heat, light, air, water, radiation, mechanical strain) chemical reactions produce a change in chemical composition and molecular weight, inducing modifications in physical and optical properties of the polymer. In practice, any change of the polymer's properties is called degradation. As a rule, degradations are detrimental to the device in which the polymer is used, but there are also some situations (see Section 3.2.2.5) in which degradation is beneficial to device function.

Polymer degradation may result from the action of an aggressive environment or external agent, unforeseeable by the device's designer [44]. The main FMs of polymers are:

- thermo-oxidation,
- photo-oxidation,
- degradation due to ionising radiation,
- chemical attack,
- environmental stress cracking,
- electrochemical degradation,
- biodegradation.

Oxidation is the most common degradation mechanism in polymers. It is important to note that if oxidation begins (by thermal or photo effects), a chain reaction is produced, accelerating the degradation. Oxidation degradation can be halted by using stabilisers capable of stopping the oxidation cycle.

As noted above, one of the causes of oxidation initiation is exposure to sunlight or to some types of radiation. For example, ultraviolet radiation may affect polymers, by breaking their chemical bounds. This process, called photo degradation, causes breakage, colour change and loss of physical properties. It should be noted that generally polymers do not absorb ultraviolet radiation, but other compounds in the polymer (degradation products, catalyst residuum, etc.) may do so. For sunlight, the photo degradation is accomplished by thermo degradation.

A failure may also originate from contamination with inclusions (nonpolymeric materials). These inclusions may act as stress concentrators and cause premature mechanical failure, far below the mechanical stress limit taken into account in the design.

Another class of failure originates in external sources: an unusual distribution of the additives and modifiers in the polymer. These additives are intended to protect the polymer against oxidative degradation, so an unusual distribution leaves part of the product unprotected during function, producing early failures. These types of defect are an important cause of loss of function.

3.2.2.4 Biodegradation

This is a degradation process involving bio factors, such as bacteria or fungi. In order to study polymer degradation produced by bacteria, it is necessary to use advanced investigation methods. In [45] the results of a study performed on a polyimide coating exposed to a fungal consortium were isolated, identifying the existing species: *Aspergillus versicolor*, *Cladosporium cladosporioides* and a *Chaetomium* species. Actively growing fungi on polyimides yield distinctive electrochemical impedance spectroscopy (EIS) spectra through time, indicative of failure of the polymer integrity compared to the uninoculated controls. First a decrease in coating electrical resistance was noticed, which correlated with a partial ingress of water molecules and ionic species into the polymeric matrices. This was followed by further degradation of the polymer produced by the activity of the fungi. The relationship between the change in impedance spectrum and microbial degradation of the coating was established by SEM and an extensive colonisation of polyimide surfaces by fungi was shown. EIS proved to be a sensitive tool for evaluating the polymer bio-susceptibility to microbial degradation.

In order to detect the source of microbial contamination, research on metallic substrates covered by polymers was performed [46]. Long-term electrochemical tests in aqueous NaCl produced from deionised water were performed for these structures. Microscopic inspection revealed that the surface of the polymer was heavily colonised by micro-organisms, which had developed naturally. The deionised water was considered as a possible source and, to establish the extent of contamination, samples were procured from Italy, the USA and the UK. In all cases, the presence of biological contaminants, both bacterial and fungal, was noted and growth ensued, reflecting the widespread contamination of deionised water systems. As a conclusion, the bacterial biodegradation of polymers seems to be a real and nocive phenomenon. To prevent it, decontamination of deionised water must be accomplished.

3.2.2.5 Beneficial Degradations

Generally, polymer degradation is a destructive phenomenon, which needs to be diminished. But there are situations in which polymer degradation proves to be beneficial, even becoming the basis of the functioning principle of the micromachined device incorporating the polymer. Some example are relevant:

- Chemical sensors based on the degradation of polymer films in the presence of chemical substances.
- Biodegradable sutures, avoiding the operation of string removal.
- Drug delivery at a 'fixed point' in the human body. The idea is simple: the drug is encapsulated in a polymer that is resistant to normal environment but which degrades in the environment specific to the desired area of the body.

Some examples of such applications of polymer degradation are presented below.

At the end of the 1990s, researchers from Sheffield University [47] reported a biosensor based on capacitance modification produced by the dissolution of a polymer catalysed by an enzyme. Thin films of a copolymer of methyl-methacrylate and methacrylic acid were deposited on to gold-coated electrodes, setting up a capacitor. Through the enzymatic action of urease on urea, a local pH increase, triggering the dissolution of polymer films, was noticed. This was accompanied by an increase in capacitance of up to four orders of magnitude – a very accurate sensor for urea in serum – and whole blood was obtained. This degradation is reproducible, the degradation rate being proportional to enzyme concentration. Later, the group from Sheffield developed sensors for other enzymes on the same principle [48].

In March 2003, researchers from California University, San Diego, reported the transfer of the optical properties of a semiconductor crystal to a plastic material [49]. This important achievement opened the possibility of using implantable devices to monitoring the 'delivery' of drugs in the human body and of manufacturing biodegradable suture. Recently the team, led by Prof. Michael J. Sailor, developed sensors on porous silicon capable of detecting biological and chemical agents (useful in case of terrorist attack), and an optical sensor (on porous silicon) which changes colour in the presence of sarin or other toxic gases. This team has now reported a method for transfering the optical properties of such sensors (specific to nanostructured crystalline materials) to some organic polymers. The advantage of this is the biocompatibility and flexibility of new sensors, which is very useful in medical applications. Moreover, plastic materials have a much higher reliability and durability. The method used is similar to the manufacture of plastic toys from a mould. First, a silicon chip containing an array of nanometre-size holes is manufactured, giving the chip the optical properties of a photonic crystal (a crystal with a periodic structure that can precisely control the transmission of light, much as a semiconductor controls the transmission of electrons). Then, a molten or dissolved plastic is injected into the pores of the finished porous silicon photonic chip. The silicon chip mould is dissolved away, leaving behind a flexible, biocompatible 'replica' of the porous silicon chip. Sailor's team is now able to 'tune' these sensors to reflect over a wide range of wavelengths, some of which are not absorbed by human tissue. In this way, polymers can be fabricated which respond to specific wavelengths that penetrate deep within the body. A physician monitoring an implanted joint with such a polymer would be able to see the changes in the reflection spectrum as the joint was stressed at different angles. A physician in need of information about the amount of a drug being delivered by an implanted device could obtain this by seeing how much the reflection spectrum of a biodegradable polymer diminished as it and the drug dissolved into the body; such degradable polymers are used to deliver antiviral drugs, pain and chaemotherapy medications and contraceptives.

To demonstrate that this process would work in a medical drug-delivery simulation, the researchers created a polymer sensor impregnated with caffeine. The sensor was made of polylactic acid, a polymer used in dissolvable sutures and a variety of medically implanted devices. The researchers watched as the polymer dissolved in a solution that mimicked body fluids and found that the absorption spectrum of the polymer decayed in step with the increase of caffeine in the solution. Hence, the drug was released on a time scale comparable to polymer degradation.

3.3 FA during Fabrication

For any technical product, failures are typically attributed to a large range of factors related to its fabrication, such as inappropriate manufacturing and assembly processes, lack of adequate technology and so on. For electronic components, other failure-generating factors may be involved, as some additional fabrication needs refer to the fabrication environment, for example harsh requirements about the quality of fluids and working atmosphere (a small quantity of dust particles, etc.). The involvement of FA in identifying the causes of the FMs initiated during fabrication will be discussed in this section.

3.3.1 Manufacturing History

The first requirement when FA is to be executed in a batch of components or a system is to know the manufacturing history. This means the specialists performing FA must have access to all the information about the quality of the materials, fluids and working environment, the process parameters and the measurements performed at all quality and reliability monitoring points. Otherwise it is virtually impossible to understand why the items failed and which corrective actions are the most appropriate.

3.3.2 Reliability Monitoring

The process of building the reliability of an item is initiated at the design stage, when all the necessary measures are elaborated. Fabrication is then prepared by checking the quality and reliability of the materials. When fabrication begins, it is essential to continuously monitor the reliability, because failure may arise at each step of the process. In this context, it is interesting to synthesise the main types of FMo and FM linked to the process operations for an electronic component (Table 3.6).

There are two main causes of failure risks: human errors (human reliability) and unexpected modifications in process quality (process reliability). Human reliability is the more important, because failures are mainly caused by people: (i) designers failing to take into account all important factors, (ii) suppliers of materials delivering bad products, (iii) process engineers or assemblers failing to observe well-defined procedures, (iv) users misusing items (components or systems) and (v) maintainers not observing the correct maintenance principle. Therefore the achievement of reliability is essentially a management task, ensuring that the right people, skills, teams and other resources are applied to prevent failures. It is not easy for managers to think long-term about reliability, especially when they are not engineers or when their motivations are geared to short-term objectives. As O'Connor said [23]: 'The management dimension is crucial, and without it, reliability engineering can degenerate into ineffective design analysis followed later by panic FA, with minimal impact on future business. Reliability philosophy and methods should always be taught first to top management, and top management must drive the reliability effort.'

The most important elements of the reliability effort during fabrication are the reliability monitors. There are two types of reliability monitor [51]:

- Quick-reaction reliability monitor (QRRM).
- Long-term reliability monitor (LTRM).

Table 3.6 Knowledge matrix for semiconductor failures

Process step	Failure site	Failure mechanism	Failure mode
Diffusion	Substrate	Crystal defect	Decreased breakdown voltage
Junction	Diffused junction isolation	Impurity precipitation, photoresist mask misalignment, surface contamination	Short-circuit, increased leakage current
Oxide film	Gate oxide film, field oxide film	Mobile ion, pinhole, interface state, TDDB, hot carrier	Decreased breakdown voltage, short-circuit, increased leakage current, drift of h_{FE}
Metallisation	Interconnection, contact hole, via hole	Scratch or void damage, mechanical damage, non-ohmic contact, step coverage, weak adhesion strength, improper thickness, corrosion, electromigration, stress migration	Open circuit, short-circuit, increased resistance
Passivation	Surface, protection film, interlayer dielectric film	Pinhole or crack, thickness variation, contamination, surface inversion	Decreased breakdown voltage, short-circuit, increased leakage current, h_{FE} and/or V_{th} drift, noise deterioration
Die bonding	Wire bonding connection, wire lead	Wire bonding deviation, damage under wire bonding contact, disconnection, loose wire, contact between wires	Open circuit, short-circuit, increased resistance
Scaling	Resin, sealing gas	Void, no sealing, water penetration, peeling, surface contamination, insufficient airtightness, impure scaling gas, particles	Open circuit, short-circuit, increased leakage current
Input/output pin	Static electricity, surge, over voltage, over current	Diffusion junction breakdown, oxide film damage, metallisation defect/destruction	Open circuit, short-circuit, increased leakage current
Others	Alpha particles	Electron–hole pair generation, surface inversion	Soft error, increased leakage current

After [50].

QRRM is a monitor programme focused on assembly-related issues, providing quick feedback on reliability assessment of the assembly operation. The tests in the QRRM programme are designed to identify reliability weaknesses associated with wire bonding, die attach, package encapsulation and contamination-related failures. Generally, this is a weekly test of the reliability of all packages from all assembly locations. If at least one failure occurs during QRRM testing, the entire production lot is stopped before shipment. Failures are analysed to determine validity and the root cause of any valid failure. Quite often additional samples are pulled and tested for an extended period of time. Lots with substandard reliability performance are rejected. The data generated from this programme is used to establish a programme for continuous quality improvement of assembly facilities.

LTRM is used for extended-life and end-of-life approximations such as failure-rate calculations. It also serves as a check against short-term reliability estimates. Long-term reliability tests are designed to evaluate design, wafer fabrication and assembly-related weaknesses. Industry-standard reliability tests and highly accelerated stress tests (HASTs) have been incorporated into this programme. The most severe tests for plastic package devices are the temperature and humidity tests, particularly HAST testing, which accelerates the penetration of moisture through the external protective encapsulant or along the interface between the encapsulant and the metallic lead frame. Additionally, the HAST is conducted with the device under bias.

At the company LTC [51], the HAST places the plastic devices in a humid environment of 85% relative humidity, under 45 psi of pressure, and at a temperature between 130 and 40 °C. Under these conditions, 24 hours of HAST at 140 °C is roughly equivalent to 1000 hours of 85 °C/85% RH testing. The employment of HAST has dramatically reduced the length of time required for qualification.

3.3.3 Wafer-Level Reliability

The vehicles for monitoring the reliability at wafer level are the test (diagnostic) structures, which are specifically designed as reliability test patterns and are stepped into all wafers. These structures are tested during fabrication using a parametric analyser. They are then used to investigate and detect potential yield and reliability hazards after assembly.

Basically, the test structures are aimed at various goals for the two main processes used in manufacturing electron components [5], as shown in Table 3.7.

The use of test structures allows any device to be monitored and gives faster unambiguous feedback than is normally achieved by performing reliability testing on an assembled product. Reliability data is generated in less than one week, giving immediate feedback to the production line.

3.3.4 Yield and Reliability

There is a correlation between test yield and reliability, as shown in many papers (e.g. [52]). At wafer level, this means that lower-yielding regions will have proportionately more failures than higher-yielding regions. The explanation is simple: both yield and reliability are linked to manufacturing-process defects and the ability of the test programme to screen these defects. A procedure for improving yield and reliability is given in [53]: applying an ST at wafer sort is as effective in identifying particular early-life fails as production burn-in. This is a so-called 'EVS/DVS programme' (an elevated stress test followed by a dynamic stress test), which can eliminate failing die earlier in the flow, thus saving additional test time/cost. It should also be noted that additional package testing of EVS/DVS-screened material shows improved yields and more uniform current behaviour.

Given the importance of using the yield as an indicator of the future product reliability, efforts have been made in this area by many specialists. A model to estimate field reliability from process was proposed in [54], which is an explicit yield–reliability relationship for various defect densities, such as Erlang, uniform and triangle distributions, using a multinomial distribution to consider a correlation between the number of yield defects and the number of reliability defects. The proposed

Table 3.7 Characteristics and goals of test structures for bipolar and CMOS processes

Process	Characteristics of test structures	Goals of test structures
Bipolar	A three-terminal structure is scribed from a wafer and assembled in either a hermetic or a plastic package. These devices are burned in for a predetermined temperature and time. A limit is defined for the leakage current change during burn-in	Optimised to accelerate, under temperature and bias, the two most common FMs in linear circuits: mobile positive ions and surface charge-induced inversions. The same structures become sensitive to either FM depending upon the bias scheme used during burn-in
CMOS	Allows measurements of thresholds of various sizes and kinds of n-channel and p-channel MOSFET. Body effects, L effective, sheet resistance, Zener breakdown voltage, contact metal resistance and impact ionisation current are measurable with this chip, which is assembled in a 20-lead DIP	Electrical testing is performed on the structure before and after burn-in. After evaluating any sample population shifts or failures, process engineering is appraised of the results of this process monitor

model has advantages over previous models in determining the optimal burn-in time for any defect density distribution. Of course, reliability and yield depend on timely accurate FA, as mentioned in [55], where the necessity of developing new analytical methods of FA is stated, especially while scaling and new materials continue to drive electron component performances.

3.3.5 Packaging Reliability

For electronic components, the package is the origin of many reliability risks. FA methods are therefore widely used to identify possible future failures. This must be done by reliability monitors at all packaging processes. However, the introduction of new automated equipment and techniques in the assembly process significantly improves the reliability, by reducing the handling and subsequent damage of die and wafers. Where die or wafers have to be handled, vacuum wands and vacuum pens have replaced tweezers and thereby decreased damage due to scratches. Automated wire-bonding machines have produced more consistent wire-bonding quality and improved productivity. However, package contamination remains an important problem. It may arise from material sourcing and assembly process excursions, and has a high impact on product reliability. It is thus essential to achieve integrity at all internal interfaces in the flip-chip, wire-bond and novel mixed technology packages, because both organic and low-level ionic contaminants may lead to interface delamination, metal migration, micro-cracking and so on, all of which can cause a premature failure of product. In [56], several examples of FA relating to package contamination in the assembly world are given, together with the effectiveness of time-of-flight secondary ion mass spectroscopy in isolating and understanding failures.

In recent years, improvements in the traditional packaging process have been developed. New technologies, such as system-in-package (SIP) or system-on-package (SOP) have arisen more or less from the traditional board packaging, with discrete components assembled by surface-mounting-technology (SMT) processes. The SOP concept proposes to improve system integration by embedding passive and active components and including such components as filters, switches, antennas and embedded optoelectronics. The first product introduction of this concept was the embedding of RF components [5]. Embedding of optical components in the board by means of waveguides, gratings and couplers has also been achieved [57].

With MEMS devices, a new and major challenge has arisen in the field of packaging reliability. De Wolf identified five main reliability issues linked with MEMS packaging [58]:

- The temperature required to mount/seal the packaging can affect the components.
- Outgassing of materials used in the package or the devices can contaminate the components and affect their functioning and reliability.
- The environment inside the package (pressure, gasses, humidity, particles) and changes in this environment can alter performance and/or reliability.
- Acoustic coupling between the package and the component is difficult to obtain.
- The packaging processes can induce stress and stress variations.

The predictable future trend is towards nano-packaging for nano-systems. The transition is towards nano-chips, with <100 nm features. Some of these chips will have several hundred million transistors, which require I/Os in excess of 10 000 and power in excess of 200 W, providing computing speed in terabits per second and supporting 20–50 GHz digital/RF applications. These nano-chips require nano-packaging at the wafer and system-board levels. Wafer-level packaging provides unprecedented advances in electrical, mechanical and thermal properties as well as the smallest form factor and lowest cost. New challenges to packaging reliability are microscale and nanoscale material characterisation, fatigue modelling and DfR [6], which provide an essential role for FA.

3.3.6 Improving Batch Reliability: Screening and Burn-In

Reliability screening is aimed at removing the weak items from a batch of devices in order to improve batch reliability. This can be done before product delivery, by applying a series of STs to the whole batch of components, in order to induce the failure of weak items. The role of the screening tests is to identify and produce the failure of partially unreliable components, with defects that do not lead immediately to non-operation in real life. After reliability screening, the remaining batch is composed only of the most reliable components and the batch reliability is improved. Generally, the defects result from intrinsic weaknesses of the components. However, there are two important rules to be observed when screening tests are designed: (i) there should be no activation of FMs that would not appear in field operation and (ii) there should be no damage to the remaining batch items, which might be the cause of further early failures. Consequently, FA is an important element in reliability screening.

Reliability screening may be applied in the following situations:

- At component manufacture – before component delivery.
- At system manufacture – as input control.
- At system manufacture – at PCB test level, for manufacturing highly reliable systems.

Generally, the selection is a 100% test (or a combination of 100% tests), the stress factors being temperature, voltage and so on, followed by a parametric electrical or functional control (performed 100%), with the aim of eliminating the defective and marginal items and those items that will probably have early failures (potentially unreliable items) [5].

Burn-in is the most frequently used method of reliability screening. This is method 1015.2, the first group of methods from Mil-Std-883D (the standard for screening tests), which is aimed at detecting latent flaws or defects that have a high probability of coming out as infant mortality failures under field conditions. Although the major defects may be found and eliminated in the quality and reliability assurance department of the manufacturer, some defects remain latent and may develop into infant mortality failures over a reasonably short period of operation time (typically between some days and a few thousand hours). It is not simple to find the optimum load conditions and burn-in duration, in order to eliminate nearly all potential infant mortality items. There must be a substantial difference between the lifetime of the infant mortality population and the lifetime of the main (or long-term) wear-out population under the operating and environmental conditions applied in burn-in [59]. The trend is towards monitored burn-in [60]. The temperature should be high, without exceeding +150 °C, which is high enough for the semiconductor crystal.

A clear distinction must be made between burn-in as a test and as a treatment. A test is a sequence of operations for determining the manner in which a component is functioning, and also a trial with previously formulated questions, without expectation of a detailed response. That is why the test time is short and the processing of the results is immediate. It is an attributive trial, which gives us information about the type, whether good or bad. As a treatment, the burn-in must eliminate the early failures, delivering to the client the rest of the bathtub failure curve.

Basically, burn-in is a procedure to remove weak components from a mixture of strong and weak populations. Recently, a method called receiver operating characteristic (ROC) was proposed for optimising the burn-in [61], based on an optimisation criterion for minimising the burn-in error and on an algorithm for finding the optimal burn-in time.

The basic questions in designing a burn-in for a system are: (i) whether or not to perform the system burn-in and (ii) how long the burn-in period should be. The scientific way to answer is to use a probabilistic model relating component burn-in information and assembly quality to the system lifetime, assuming that the assembly defects introduced in various locations are capable of connection failures represented by an exponential distribution. A relatively recent paper [62] proposed an extension of the exponential-based results to a general distribution so as to study the dependence of system burn-in on the defect occurrence distribution. A method of determining an optimal burn-in period that maximises system reliability was also developed, based on the system lifetime model, which assumes that systems are repaired at burn-in failures.

A direct link between burn-in and FA was proposed by Cha [63], in a paper about burn-in procedures for a general failure model. Two types of failure are considered in the general failure model:

- Type I failure (minor failure), which can be removed by either a minimal or a complete repair.
- Type II failure (catastrophic failure), which can be removed only by a complete repair.

During the burn-in process, two types of burn-in procedure are considered:

- **Procedure I.** The failed component is repaired completely regardless of the type of failure.
- **Procedure II.** Only minimal repair is done for the Type I failure while a complete repair is performed for the Type II failure.

The two burn-in procedures are compared in cases in which both procedures are applicable. In each case, optimal burn-in time and minimum cost rate are calculated.

3.4 FA after Fabrication

The finished product has to be introduced into the marketplace, so there must be an assurance that the product functions are fulfilled. This can be obtained by performing reliability testing after fabrication in order to qualify the product. For electronic components and systems, this issue has become crucial in recent years, with proposals for several new technologies and materials, such as:

- Plastic semiconductor packages, replacing the ceramic ones for high-reliability components.
- Copper replacing aluminium for the circuitry inside semiconductor devices.
- Lead-free products replacing the traditional packaging technologies of electronic components.

All these and many others represent new challenges for the reliability of components and systems. Long-term reliability tests are needed to certify the newcomers, as there is little or no evidence for the reliability of these new materials and technologies in various harsh operational environments.

Meanwhile, in the field of reliability testing there are two basic philosophies [64]:

- *Standard-based testing* recommends the use of a list of standardised testing methods, in which stresses and their time and temperature durations are outlined. The results can be compared to a standard set of data to determine whether the product has passed the tests or not.
- *Knowledge-based testing* can be defined as a concerted effort to understand the expected stresses and demands placed on a product in its end application and to derive a set of tests that simulate these conditions.

3.4.1 Standard-Based Testing

Standard-based testing (also called the 'cookbook approach' or 'stress-based testing') employs standardised reliability tests, the best example being Mil-Std-883 (promoted by US military organizations). Real-life conditions are accelerated and products have to survive for a prescribed duration in order to demonstrate that they will fulfil the reliability requirements for field applications (generally, military ones). Complex FA is used to identify the FMs.

Unfortunately, these standards act like 'black boxes', difficult (if not impossible) to adapt to other components or applications than those they were designed for. For example, if you wish to test the hermeticity of a small-volume MEMS package, you will find method 1014.9 of Mil-Std-883 D leaves an undefined gap between the gross leak test and the fine leak test [65]. Or when using the pressure cooker test for MEMS, you will find that the FM is the corrosion of the aluminum connections, which is normal for the given testing conditions (121 °C, 100% humidity, 1 hour), but nothing is said about the quality of the package, which may allow the humidity to enter the chip [6]. Obviously, the test is not the most appropriate for the given device and FM.

In conclusion, standard-based testing has a series of weak points, which may limit the usefulness of this approach [6]:

- The acceleration factors for various stresses are not measured.
- The stress conditions used are not necessarily the correct ones for every certain application.
- You cannot be sure that the important FMs have been triggered.
- Many tests are superfluous, expensive and give limited (if any) practical results.

3.4.2 Knowledge-Based Testing

Knowledge-based testing aims to investigate/predict the effects of design, processing, packaging and usage in different environments and conditions on the functioning and lifetime of devices and systems and to define corrective actions [6]. This is a concerted effort to understand the expected stresses and demands placed on a product in its end application and to derive a set of tests that simulate these conditions [64]. These demands can be defined by various means, including actual observation, customer specifications or by referal to various standards agencies that have already established preset expectations of product [66–68]. The basic element is to identify all encountered FMs, to establish their impact on the final product and to form a database to generate acceleration models for future products.

The stress programme elaborated by the knowledge-based testing approach is based on various elements, including yield, manufacturability, use conditions, expected reliability performance and data gathered during the whole development cycle [62]. The types of stress to be used have to be chosen from among environmental tests (thermal, mechanical, solderability, terminal strength, etc.) or endurance tests (high-temperature storage, high-temperature bias, temperature–humidity bias, pressure–temperature bias) [64].

A procedure recommended for the elaboration of the accelerated reliability stress programme for knowledge-based testing is indicated in Table 3.8 [64]. As can be seen, in this complex process of elaborating the accelerated reliability stress programme for knowledge-based testing, FA has a well-defined and important role.

Some examples of the use of knowledge-based testing to qualify products and technologies are given below.

The European project MEDEA + A407 FdQ [69] developed a methodology for using FMs, their effects and interactions to provide a more detailed and more efficient reliability forecast. The presence of suitable FA procedures is a basic requirement for a failure-driven qualification methodology and the compilation of a knowledge matrix. New technologies and the increasing integration of semiconductor devices mean that additional research is required in the area of process technologies for failure localisation, failure preparation and FA. The MEDEA + FdQ project was focused on the products of the automotive industry, the idea being to establish and standardise a method on a broad European basis.

The methodology involved well-known FMs, and an efficient preventive qualification strategy for the avoidance of failures had to be developed, standardised and implemented for the benefit of automotive industry customers and the semiconductor industry as a whole. A premise for developing such a methodology was to provide additional detailed information, not only about FMs, but about

Table 3.8 Procedure for elaborating the accelerated reliability stress programme for knowledge-based testing

Steps of the stress programme	Details
Define the technology envelope	Requirements about business and marketing, technology and materials, quality and reliability are gathered. Preferably, the final product will be sold into more than one market, necessitating a testing envelope that encompasses all the relevant markets
Identify the areas of impact of product reliability	These are usually areas of storage, shipping and handling impact, system assembly or other further processes downstream, and the final end user. For each area of impact, the relevant environmental boundary conditions are defined
Identify the potential product FMs and modes	Knowledge from previous FA activities is used, such as: risk-assessment process (identifying the FMs that are of highest concern in the particular application) or failure modes effects analysis (FMEA) (detailed in Section 2.2.1). Literature searches for technologies that have not been evaluated before are also helpful, starting with similar existing technologies as a basis to begin the search
Understand the appropriate STs	The most appropriate test for activating or accelerating the specific FM of concern is identified. Based on the information gained at the previous step, an understanding of the ST material limitations and selection of the right stress and acceleration models for the test duration is needed. This information is required so that no stress conditions are set beyond the physical capability of the assembly material. The purpose is to not activate the FMs or FMos that have no linkage to the use conditions
Elaborate acceleration models	These models link use conditions and STs. As a starting point, standard reliability models with the acceleration factors, environmental conditions, material property data and lifetime requirements are used. These speculative models may help define the initial reliability stress conditions until adequate data is available to validate the models or previous data indicates an acceptable substitution
Elaborate stress models	This allows some preliminary indication, based on material properties, of the reliability and performance of the product in the end application to be obtained. This is usually used to narrow down the possible permutations needed to validate a product
Elaborate validation guidelines	The main purposes of validation testing are: (i) to avoid misunderstanding of the FMs of the new technologies using unrealistic stress conditions that are not experienced in the field and (ii) to update FMEA and test plans in order to reflect the correct priority of FMs
Run STs to failure	If possible and where time permits, the STs are run to failure. Based on the results obtained, the discrepancies between models and data can be analysed, the acceleration factors for various stress conditions recalculated and the reliability plan revised
Document the test plan requirements for execution or implementation	It is recommended that a company-standardised format is used, so that everyone dealing with the product understands what they are reading, thus avoiding confusion and time-consuming explanations
	Finally, the data summary is presented to the project team and management. Based on the goals previously established in the use-condition criteria, it will be obvious whether the product passed the requirements or not. This information is then fully archived for future reference

After [64].

the interaction between the various types of FM. One of the goals of the project was to investigate typical FMs and create a knowledge matrix. Such a matrix became a necessary tool for comparing requirements at an early stage of technology development, for analysing possible failure mechanisms at a preliminary stage in parallel with product development, and for defining measures for qualification and for avoiding future reliability problems.

In [70], a systematic accelerated product qualification approach based on PoF principles is presented, aimed at determining a durability assessment of an avionics electronic module. The results obtained demonstrate that combined accelerated wear-out tests may produce complex synergies that can only be controlled through precise PoF assessment. A set of generic guideline for designing, planning, conducting and implementing a PoF-based accelerated product qualification process is elaborated. Eventually, the acceleration factors are estimated. When multiplied by the accelerated wear-out test data, these acceleration factors may results in life estimation under life-cycle conditions.

Another European project, called PROCURE, was focused on elaborating a programme for the development of passive devices used in rough environments [71]. Technologies and prototypes of generic passive components were obtained, which are necessary for the realisation of the next generation of automotive electronic control units for harsh environments. These devices are suited for lead-free surface-mount technology. The first step was to identify the dominant failure modes in order to optimise device construction. The following component types were studied: multilayer ceramic capacitors (MLCCs), metallised plastic film capacitors, tantalum chip capacitors, aluminium electrolytic capacitors, SMD inductors, quartz resonators and thin-film resistors. Harsh test conditions ranging up to 175 °C, intended to resemble 17 years of operation in different test sequences, were applied. After each test step, a number of devices were taken out of the sequences and characterised in order to investigate the individual contributions to eventual parametric changes.

3.5 FA during Operation

All the work done during product development and manufacture is validated in real life, when the product fulfils its required function; that is, during operation. In principle, gathering data from device operation seems to be the best choice for quantitatively calculating the reliability level. For example, a statistical model for estimating the failure probability in the field given the known numbers of customers and reported failures is proposed in [72]; two cases are studied, in which the probabilities of fulfiling the specification and reporting the failure are (i) known and (ii) unknown, but the Bayes method is applied to incorporate engineering beliefs. In both cases, the proposed model, based on FA, yields good results. The results obtained depend on the initial choice of the probabilities used to fulfil the specifications and report the failures.

Unfortunately, field data is rarely accurate enough to offer a proper basis for good reliability predictions. There are many explanations for this. First, field data is gathered by a heterogeneous group of specialists, concerned mainly with the reliability of the systems and not with component reliability [34]. Also, the data gives only the FMo (in the best case) and it is almost impossible to identify the FM in each case. Hence, the usual approach is to use an exponential distribution to describe the failure, which is not at all accurate for electronic components (except for the 'useful life' period, but this is in fact an ideal approach). So, the best approach we have is to use laboratory tests to assess batch reliability. However, a method must be found to compare the laboratory reliability with field (operational) reliability. Basically, the values established by the component manufacturer depend on test conditions, and the operational values depend on the incoming control conditions of the components used in electronic systems. In practice, the differences between the two results can reach one or two orders of magnitude. Surely only the operational reliability can include all the stresses that demonstrate a sufficient reliability of the component. As long as exact operational reliability knowledge is not available, the ratio between the reliability measured in the laboratory and the operational reliability will remain a point of discussion between manufacturer and user.

3.5.1 *Failure Types during Operation*

There is a general agreement that the instantaneous failure rate of an electronic component during operation takes the well-known shape of a bathtub. Since this is already reproduced in almost every book about reliability, we did not feel it necessary to show it here. Of course, the shape of the curve depends on how the scales for the two axes (ox and oy) were chosen, but there are basically three regions (or periods, because the time factor is involved):

- The 'infant mortality period', when the failure rate decreases with time.
- The 'useful operating life', when the failure rate is constant with time.
- The 'wear-out period', when the failure rate abruptly increases with time.

More important than the exact shape of the failure rate in each of the three regions is the fact that the types of possible failure are different from one region to another, as shown in Table 3.9.

Obviously, the major task of reliability engineering is to minimise (if not eliminate) the infant mortality and random failures, and to maintain the wear-out period beyond practical duration. A burn-in period can be used as a screening method to determine the failure of the components with defects, by operating the whole lot under severe conditions. The failed items are rejected during an acceptance test [73].

The electronic components fail eventually, after being used for a long time (generally from several tens to several hundreds of years), due to the expiration of their substantial lifetime. During useful operating life, the failure rate has an exponential distribution with a random failure; that is, a constant failure rate. If only the package is discussed, this constant failure rate describes a majority of non-solder joint-related package failures in field applications. The statistical characteristics of the number of failures arising in tests or applications are discussed in [74]. Methods for estimating the acceleration factors (for temperature and voltage, temperature and relative humidity, temperature cycling), based on FA, are described and analysed.

In recent years, a new phenomenon called 'no-fault-found (NFF)' has been studied. NFF describe the situation where a failure is reported to occur during operation but where no failure (or fault) can be found in FA. A common example is the computer that 'hangs up' and, when rebooted, works again. A detailed study has been carried out on NFF [75] (also called 'trouble-not-identified', TNI or 'no-trouble-found', NTF). It seems NFF is reported in automotive, avionics (up to 70% of the total failures in 1990 [76]), telecommunications, computers and consumer industries. In Table 3.10, the possible causes of NFF in electronic products, as identified by the cause and effect diagram [76]

Table 3.9 Types of possible failure in the three periods of component life

Period	Characteristic	Types of failures cause[a]
Infant mortality	Manufacturing defects or faulty design (early failures)	Contamination or manipulation (particles, scratches); defects in isolator (gate oxide defects, pinholes in isolating dielectrics); popcorn damage; encapsulation defects (loose bond wires, shorts, near opens, etc.)
Useful operating life	Extrinsic degradation mechanisms (random failure)	Electrical overstress; latent ESD damage; latch-up; safe operating area violations; load dump spikes; extended early failures
Wear-out	Intrinsic degradation mechanisms (age-related wear and fatigue)	Electromigration; oxide film destruction (TDDB); hot carrier degradation; mobile ion contamination; thermo-migration; stress voiding; surface charges; corrosion

[a] Details about the typical failure mechanisms for electronic components will be given in Chapter 5.

Table 3.10 Possible causes for no-fault-found (NFF) conditions in electronic products

General causes	Subcategories	Examples of particular causes
People (human)	Communications	Field technician, field log, user, service report
	Skills and behaviour	Field maintenance technician, test operators, designer, user
Machines	Measurement tools	Detection limit, calibration, software
	Test equipment	Software, calibration, load (type, level), EMI, parasitics, noise, built-in test, test procedures
Methods	Tests	Load (type, level), procedure, device (analogue, digital)
	Handling	Shipping, handling procedure, frequency, ESD
	Failure analysis	FMEA, nondestructive, destructive
Intermittent failures	Components	Creep corrosion, moisture (Au-Al corrosion), metallisation, soft errors, whiskers, defects in via
	PCB	Conductive filament, pad/trace corrosion, moisture, layer delamination, trace/via crack
	Software	Signal, global variable
	Interconnects	Wiring (ageing), non-wets, stresses, voids/cracks
	Connectors	Broken wire, degradation, fretting, whiskers

After [75].

(also called the Ishikawa diagram or fishbone diagram), are given. FA plays a role in this table, at 'Methods', and has an indirect role in identifying the intermittent failures.

In Table 3.9, one may notice that an important distinction is made between *extrinsic* and *intrinsic* failures:

- **Extrinsic failures** are those related to static or dynamic overload events (electrical, thermal, mechanical and radiative) during the component life cycle or due to misapplication (wrong component for the job).
- **Intrinsic failures** are those occurring mainly (but not exclusively) during the wear-out period, related to component design, materials, processing, assembly, packaging and manufacturing.

The distinction between extrinsic and intrinsic failures is important, because it points to possible causes of failure, and therefore indicates the direction of corrective actions.

Another significant distinction, proposed by O'Connor [24], can be made between intrinsically reliable and unreliable components:

- **Intrinsically reliable components** are those that have high margins between their strength and the stresses that can cause failure, and which do not wear out within their practicable lifetimes. Such items include nearly all electronic components (if properly applied), nearly all mechanical non-moving components, and all correct software.
- **Intrinsically unreliable components** are those with low design margins or which wear out, such as badly applied components, light bulbs, turbine blades, and parts that move in contact with others, such as gears, bearings, power drive belts and so on.

During component operation, there are two possible influences on the lifetime:

- **External influences** operate through direct inductive, capacitive or chemical effects or occur on very sensitive structures through component operation, such as during commutation, running of electrical current or connection to the electrical mass. Other possible effects are produced by UV and X (for example REPROM) irradiation, by radio-wave irradiation and under special conditions [77] through electromagnetic pulse (EMP). The EMP effect is complex [78] and its action mode is multiple, and the protection measures against EMP are not simple. Up to a certain point, they are similar to the

protection measures against lightning. Taking into account the differences between the rise times, frequencies, field intensities and energies, and taking into account that the domain involved can cover some millions of square kilometres, it may be noticed that anti-EMP protection has greater requirements and its realisation is more expensive.
- **Internal influences** act inside an electronic constructive element, or for a semiconductor structure, through the introduction of inductive currents, short-circuit and inductive and particularly capacitive influences.

3.5.2 Preventive Maintenance of Electronic Systems

Preventive maintenance (PM) is effective in improving the reliability of a system. The traditional approach is 'time-based maintenance': various components are replaced at foreseen time periods in order to keep the system reliability above a given level. However, another approach, 'condition-based maintenance', can be better and more cost-effective: the components are replaced based on their condition (i.e. failure risk). This approach is linked to FA. A failure prediction method for PM by state estimation using the Kalman filter is shown in [79]. To improve PM, this study uses a hybrid Petri-net modelling method coupled with fault-tree analysis and Kalman filtering to perform failure prediction and processing. A Petri net arrangement, named the early failure detection and isolation arrangement (EFDIA), is used. The condition-monitoring system of a thermal power plant is used as an example to demonstrate the proposed scheme. The proposed Petri net approach not only achieves early failure detection and isolation for fault diagnosis but facilitates event count, system-state description and automatic shutdown or regulation. These capabilities are very useful in health monitoring and the PM of a system.

A new method which enables monitoring of the state of reliability of a product in real time and provides advance warning about failures is so-called 'prognostics'. Detailed in [80], the purpose of prognosis is to identify in advance the potential failures, as a basis for risk management. The concept might be implemented in the following ways:

- Using expendable prognostic cells ('canaries' or fuses), which fail earlier, providing an advance warning of failure.
- Monitoring shifts in performance parameters, which are precursors of failures.
- Modelling stress and damage in electronics using operational conditions (usage, temperature, vibration, radiation, humidity, etc.) coupled with PoF models to compute accumulated damage and assess remaining life.

Obviously, the third method uses FA as its main tool.

Recently, IEEE Transactions on Reliability dedicated a special section to a new concept, derived from 'prognostics', named 'prognostic and health management' (PHM) [81]. Traditional reliability prediction methods for electronic systems were based on a constant failure rate, which proved to be totally inaccurate, so this new concept was proposed. PHM monitors the health of a product and predicts its remaining useful life by assessing the deviation or degradation of the product from its expected state of health. PoF is the basic approach for PHM. Applicable in several phases of the product life cycle, PHM can provide more effective product qualification, improved next-generation design and increased reliability.

References

1. De Vale, J. (1998) Traditional Reliability, Carnegie Mellon University Report, Spring.
2. Hazelrigg, G. (2009) The Cheshire Cat on engineering design. *Qual. Reliab. Int.*, **25**, 759–769.
3. Visser, I.M. and van den Bogaard, J.A. (2006) The risks of applying qualitative reliability prediction methods: a case study. Proceedings of Annual Reliability and Maintainability Symposium ARMS, 2006, pp. 532–538.

4. Lu, Y. (2002) *Analysing Reliability Problems in Concurrent Fast Product Development Processes*, Technische Universiteit Eindhoven, Eindhoven.
5. Băjenescu, T. and Bâzu, M. (2010) *Component Reliability for Electronic Systems*, Artech House, Boston and London.
6. De Wolf, I. (2006) Reliability issues in MEMS: physics of failure and design for reliability. Patent DfMM & Memunity Workshop, Milan, Italy, November 27.
7. Stock, M. *et al.* (2003) Going back in time to improve design: the function-failure design method. Proceedings of the 2003 ASME Design Engineering Technical Conference, Design Theory and Methodology Conference, Chicago, DETC2003/DTM-48638.
8. Strawbridge, Z. *et al.* (2002) A computational approach to conceptual design. Proceedings of the 2002 ASME Design Engineering Technical Conference, Design Theory and Methodology Conference, Montreal, Canada, DETC02/DTM-34001.
9. Stone, R. *et al.* (2005) Linking product functionality to historic failures to improve failure analysis in design. *Res. Eng. Des.*, **16**, 96–108.
10. Nasser L. *et al.* (2006) Integrated approach to reliability-based design of future electronics systems. Proceedings of the Annual Reliability and Maintainability Symposium, January 23–26, 2006, pp. 122–126.
11. Alam, M. (2008) Reliability- and process-variation aware design of integrated circuits. *Microelectron. Reliab.*, **48**, 1114–1122.
12. Loll, V. (2008) Optimizing the number of failure modes for design analysis based on physics of failure Proceedings of the Annual Reliability and Maintainability Symposium, January 22–25, 2008, pp. 166–170.
13. Lakshminarayanan, V. (2001) Failure analysis techniques for semiconductors and other devices. *Mob. Dev. Des.*, 34–46, http://mobiledevdesign.com/software_news/radio_failure_analysis_techniques/.
14. Mannhart A. *et al.* (2007) Modeling expert judgment for reliability prediction – comparison of methods. Proceedings of the Annual Reliability and Maintainability Symposium, January 22–25, 2007, pp. 1–6.
15. Nicholls, D. (2007) What is 217PlusTM methodology and where did it came from? Proceedings of the Annual Reliability and Maintainability Symposium, January 22–25, 2007, pp. 22–27.
16. Marin, J.J. and Pollard, R.W. (2005) Experience report on the FIDES reliability prediction method. Proceedings of Annual Reliability and Maintainability Symposium, pp. 8–13.
17. Parry, J. *et al.* (2002) Enhanced electronic system reliability – challenges for temperature prediction. *IEEE Trans. Compon. Packag. Technol.*, **25** (4), 533–553.
18. Dasgupta, A. and Pecht, M. (1991) Material failure mechanisms and damage models. *IEEE Trans. Reliab.*, **40** (5), 531–536.
19. Fayssal, S. and Weldon, D. (2007) Design for reliability and safety: approach for the new NASA launch vehicle. 2nd IAASS Conference: Space Safety in a Global World, Chicago, May 14–17, 2007.
20. Bailey, C. (2003) Exploiting virtual prototyping for reliability assessment. Proceedings of the International IEEE Conference on Business of Electronic Product Reliability and Liability, Hong Kong, January 13–14, 2003.
21. Bâzu, M. *et al.* (2004) Reliability assessment by virtual prototyping of MEMS tunable fabry-perot optical cavity. Proceedings of International Semiconductor Conference CAS 2004, 27th edition, Sinaia, Romania, October 4–6, 2004, pp. 249–253.
22. Varvani-Farahani, A. and Mirani, A.S. (2003) Derivation of closure-free crack growth rate under biaxial fatigue loading conditions. *Scr. Mater.*, **48** (3), 241–247.
23. Muhlstein, C.L. and Ritchie, R.O. (2003) High-cycle fatigue of micron-scale polycrystalline silicon films: fracture mechanics analyses of the role of the silica/silicon interface. *Int. J. Fract.*, **119–120**, 449–474.
24. O'Connor, P.D.T. (2000) Reliability past, present and future. *IEEE Trans. Reliab.*, **49** (4), 335–341.
25. Hobbs, G. (1999) *Accelerated Reliability Engineering: HALT and HASS*, John Wiley & Sons, Ltd.
26. Bâzu, M. (1995) A combined fuzzy logic and physics-of-failure approach to reliability prediction. *IEEE Trans. Reliab.*, **44** (2), 237–242.
27. Livesay, B.R. (1978) The reliability of electronic devices in storage environment. *Solid State Technol.*, **21**, 63–68.
28. Calatayud, R. and Szymkowiak, E. (1992) Temperature and vibration results from captive-store flight–tests provide a reliability improvement tool. Proceedings of the Annual Reliability and Maintainability Symposium, Las Vegas, January 21–23, 1992, pp. 266–271.
29. Ciappa, M. (2002) Selected failure mechanisms of modern power modules. *Microelectron. Reliab.*, **42**, 653–667.

30. Dupont, R.L. (2004) Banc de cyclage actif pour l'analyse de la fatigue thermique des brasures de composants IGBTs: les composants de la puissance. *Rev. Électr. Électron.*, (2), 45–51.
31. Peck, D.S. (1961) Semiconductor reliability predictions from life distribution data, *Semiconductor Reliability*, Reinhold Publishers, pp. 51–63.
32. Bâzu, M., Gălăţeanu, L. and Ilian, V. (2006) Basic elements of accelerated testing. Proceeding of 10th International Conference on Quality and Dependability CCF, September 27–29, 2006, pp. 139–146.
33. LuValle, M.J. (2007) Identifying mechanisms that highly accelerate test miss. *IEEE Trans. Reliab.*, **56** (2), 349–359.
34. Jensen, F. (2000) *Electronic Component Reliability*, John Wiley & Sons, Ltd, Chichester.
35. Wieczorek, K. *et al.* (2006) Integrated semiconductor structure for reliability test of dielectrics. US Patent 6, 995, 027 B2, Feb. 7, 2006.
36. Yoshida N. *et al.* (2006) Test structure design for reliability test. US Patent 7, 119, 571 B2, Oct. 10, 2006.
37. Zhai, J. *et al.* (2008) Method and semiconductor structure for reliability characterization. US Patent 0, 102, 637 A1, May 1, 2008.
38. Etzin, M. *et al.* Plastic failure due to oxidative degradation in processing and service. Report no. 07928_01.
39. Maxwell, A.S. (2001) Practical Guide for Designers and Manufacturers of Mouldings to Reduce the Risk of Environment Stress Cracking. NPL Report MATC (a) 05. NPL Materials Centre, Middlesex, March 2001.
40. Anon. CMOS Humidity Sensors, Sensirion Application Note, http://www.sensorland.com/HowPage047.html. (Accessed 2010).
41. Chiu, W. *et al.* (2002) Conducting polymer research and a New Zealand Nobel Prize. Report of the Association of Pacific Rim Universities, June 2002.
42. Bjorkman, H. (2002) Polymer BioChips, Internal Report Amic AB.
43. Anon. POEM – Short Presentation, http://www.mic.dtu.dk/research/POEM/POEM.htm. (Accessed 2010).
44. Wright, D.C. (2001) *Failure of Plastics and Rubber Products. Causes, Effects and Case Studies Involving Degradation*, Springer-Verlag.
45. Gu, J.-D. *et al.* (1998) Microbial degradation of polymeric coatings measured by electrochemical impedance spectroscopy. *Biodegradation*, **9** (1), 39–45.
46. Mitton, D.B. *et al.* (1997) The potential for unanticipated biodegradation during EIS analysis of polymer-coated metallic substrates. *Electrochim. Acta*, **42** (12), 1859–1867.
47. Sumner, C. *et al.* (2000) A transducer based on enzyme-induced degradation of thin polymer films monitored by plasmon resonance. *Anal. Chem.*, **72**, 5225–5232.
48. Sumner, C. *et al.* (2001) Biosensor based on enzyme-catalysed degradation of thin polymer films. *Biosens. Bioelectron.*, **16**, 709–714.
49. Sailor, M. (2003) Flexible, biocompatible polymers with optical properties of hard crystalline sensors. *Science.*, **619**, 253–8188 http://ucsdnews.ucsd.edu/newsrel/science/mcpolymer.htm.
50. Anon. Knowledge Matrix. SAE Technical Report, http://links.sae.org/J1879. (Accessed 2010).
51. Anon. Reliability Assurance Program, Linear Technology Corporation, http://www.linear.com/designtools/quality/Rel_Assurance_Prog.pdf. (Accessed 2010).
52. Riordan, W.C. *et al.* Microprocessor reliability performance as a function of die location for a 0.25 μm five layer metal CMOS logic process. Proceedings of the Annual International Reliability Physics Symposium, pp. 1–11.
53. Flynn, A. and Millar, S. (2005) Investigation into the correlation of wafer sort and reliability yield using electrical stress testing. IEEE 43rd Annual International Reliability Physics Symposium, San Jose, 2005, pp. 674–675.
54. Kim, K.O. (2006) Relating integrated circuit yield and time-dependent reliability for various defect density distributions. *IEEE Trans. Reliab.*, **55** (2), 307–313.
55. Valett, D.P. (2007) Why waste time on roadmaps when we don't have cars. *IEEE Trans. Device Mater. Reliab.*, **7** (1), 5–10.
56. Pathangey, B. *et al.* (2007) Application of TOFSIMS for contamination issues in the assembly world. *IEEE Trans. Device Mater. Reliab.*, **7** (1), 11–18.
57. Anon. IEEE Components, Packaging and Manufacturing Technology (CPMT) Society, http://www.cpmt.org/. (Accessed 2010).
58. De Wolf, I. (2007) MEMS reliability testing. Patent-DfMM Meeting, Lancaster, October 1–4, 2007.
59. Jansen, F. and Petersen, N. (1982) *Burn-In – An Engineering Approach to Design and Analysis of Burn-in Procedures*, John Wiley & Sons, Ltd, Chichester.

60. Robineau, J. *et al.* (1992) Reliability approach in automotive electronics. Proceedings of Third European Symposium on Reliability of Electron Devices, Failure Physics and Analysis ESREF '92, Schwäbisch Gmünd, Germany, October 5–8, 1992, pp. 133–140.
61. Shaomin, W. and Xie, M. (2007) Classifying weak, and strong components using ROC analysis with application to burn-in. *IEEE Trans. Reliab.*, **56** (3), 552–561.
62. Kim, K.O. and Kuo, W. (2005) Some considerations on system burn-in. *IEEE Trans. Reliab.*, **54** (2), 207–214.
63. Cha, J.H. (2005) On optimal burn-in procedures – a generalized model. *IEEE Trans. Reliab.*, **54** (2), 198–206.
64. Greathouse, S.W. (2006) Application-specific reliability in new product applications. *Future Fab Int.*, **20**, www.future-fab.com/documents.asp?grID=217&d_ID=3713.
65. Jourdain, A. *et al.* (2002) Investigation of the hermeticity of RCR- sealed cavities for housing (RF) MEMS devices. Proceedings of MEMS Conference, Las Vegas, 2002, pp. 677–680.
66. Anon. (2003) Failure Mechanisms and Models for Semiconductor Devices, JEDEC spec # JEP122B, www.jedec.org. (Accessed 2010).
67. Anon. (2005) Stress-Test-Driven Qualification of and Failure Mechanisms Associated with Assembled Solid State Surface-Mount Components, JEDEC spec #JEP150, www.jedec.org. (Accessed 2010).
68. Anon. (2004) Reliability Qualification of Semiconductor Devices Based on Physics of Failure Risk and Opportunity Assessment, JEDEC spec #JEP148, www.jedec.org. (Accessed 2010).
69. Anon. (20042007) EUREKA project MEDEA+ A407- FDQ, Failure Mechanism Driven Qualification for Reliability and Analysis of Electronic Components. (Accessed 2010).
70. Upadhyayula, K. and Dasgupta, A. (1998) Physics-of-failure guidelines for accelerated qualification of electronic systems. *Qual. Reliab. Eng. Int.*, **14**, 433–447.
71. Post, H.A. *et al.* (2005) Failure mechanism and qualification testing and passive components. *Microelectron. Reliab.*, **45** (9–11), 1626–1632.
72. Ion, R. (2005) Statistical model for estimating the failure probability in the field. Proceedings of Annual Reliability and Maintainability Symposium RAMS, 2005, pp. 256–260.
73. Anon. (2000) Guideline for accelerated endurance testing of semiconductor devices. EIAJ EDR-4704 Technical Report of Electronic Industries Association of Japan, May, 2000.
74. Yang, L. and Bernstein, J.B. (2009) Failure rate estimation of known failure mechanisms of electronic packages. *Microelectron. Reliab.*, **49** (12), 1563–1572.
75. Qi, H. *et al.* (2008) No-fault-found and intermittent failures in electronic products. *Microelectron. Reliab.*, **48**, 662–674.
76. Pecht, M. and Ramappan, V. (1992) Are components still the major problem: a review of electronic system and device field failure returns. *IEEE Trans. CHMT*, **15** (6), 1160–1164.
77. Taguchi, G. (1995) Quality engineering (taguchi methods) for the development of electronic–circuit technology. *IEEE Trans. Reliab.*, **44** (2), 225–229.
78. Anon. (1975) EMP Engineering and Design Principles, Bell Telephones.
79. Yang, S.K. (2003) A condition-based failure-prediction and processing-scheme for preventive maintenance. *IEEE Trans. Reliab.*, **52** (3), 373–383.
80. Gu, J. *et al.* (2007) Prognostic implementation methods for electronics. Proceedings of Annual Reliability and Maintainability Symposium ARMS, 2007, pp. 101–106.
81. Dolev, E. (2009) Introduction to the special section on prognostic and health management. *IEEE Trans. Reliab.*, **58** (2), 262–263.

4

Failure Analysis – How?

The tools used in failure analysis (FA) are listed in Figure 1.1 of Chapter 1, where a chronological order of the main FA techniques is given. In this chapter, these FA techniques and others will be described, but first we will recommend procedures for performing FA. It is necessary to have procedures in place, because the use of the FA techniques must be the result of a logical process, the final goal being to identify the failure mechanism (FM) and not to display results obtained with as many sophisticated FA methods as possible.

In the second part of this chapter, each FA method mentioned in these procedures will be detailed. Note that a natural conclusion to this chapter is given by the use of FA techniques in studying the typical FMs for various technologies (detailed in Chapter 5) and for the 12 case studies presented in Chapter 6.

4.1 Procedures for Failure Analysis

As mentioned in Chapter 1, FA aims (i) to identify the FM that is the root cause of a failure and (ii) to propose the most appropriate method (so-called corrective action) for diminishing or avoiding that FM. Only if the root cause is identified can solutions be found to the immediate problem and similar failures be prevented from occurring in the future. However, experience suggests that most FAs fall short of this goal and incorrectly use the term 'root cause', when what they really establish is the primary cause of failure or simple physical cause [1].

A typical FA procedure starts from a set of *input data*, follows a number of steps of the *working procedure* and produces some *output data* as its final result. There are two main reasons why an item (electronic component or electronic system) undergoes an FA process: (i) the item has failed in a real-life situation or (ii) the item failed during some reliability test(s). The FA procedure is similar in each case. The procedures for electronic components will be described below, because almost any FA of an electronic system rapidly becomes the FA of the components making up that system.

Generally, the **input data** for an FA performed on electronic components are:

- **Failure mode (FMo)** – the electrical symptom noticed by the person that examined the failed component. This could be a short circuit, an open circuit or the drift of an electrical parameter leading to component failure (note that there are parameter drifts that do not lead to component failure, where the parameter is not essential for the function fulfilled by the component, or the drift value is not beyond the failure limit).

Failure Analysis: A Practical Guide for Manufacturers of Electronic Components and Systems, First Edition.
Marius I. Bâzu and Titu-Marius I. Băjenescu.
© 2011 John Wiley & Sons, Ltd. Published 2011 by John Wiley & Sons, Ltd.

- **Field data** – details about the function fulfilled by the component, the application type, the environmental and operational conditions, and any previous incidents before failure. If the item was failed during reliability tests, specific information about the situation is required (e.g. the equipment used for testing).
- **Lot history** – all the information about the manufacturing process of the component lot; that is, data on the quality of the materials, the environmental parameters during processing, the process parameters and so on. All this information is necessary to elaborating efficient corrective actions after identifying the responsible FMs.

The **working procedure** contains a succession of steps that must follow a logical order. There is a broad spectrum of procedures for FA, but the following steps are compulsory:

- **Failure validation** – The existing input data is compared to the identified FMo (which has to be confirmed by electrical tests) and with all customer-generated documentation about the failure event, including a detailed description of the device history and usage, characteristics of the failure, and any analytical findings. Databases with previous more-or-less similar events can be used for this purpose. Especially for electronic systems, the site of the failure has to be inspected as soon as possible after the failure occurs. If the item failed during reliability tests, it is important to eliminate any possibility that the failure was induced by the testing equipment.
- **Parallel analysis preparation** – Nondefective items have to be prepared for parallel analyses using FA techniques, for comparison with failed items.
- **FA strategy elaboration** – After failure validation, an initial analysis strategy is formed, based on a review of the FA techniques that can be used by the investigator.
- **Root-cause determination** – This is the essential step of the working procedure and in recent years many methods have been refined for use by investigators [2]. First, the fault-tree diagram and/or a 'why-why' method (a brainstorming based on asking 'why' at each step and arranging the possible causes in proper hierarchical order) are useful for identifying all possible causes. Then the likelihood of each possible root cause is evaluated on a failure-mode assessment (FMA) chart, which ranges the probabilities on the scale 'likely–possible–not likely', with rationale for each rating. This step is followed by the elaboration of a technical plan for resolution (TPR), which can take the form of the testing or analysis of failed components or of similar components to demonstrate that the FMo is possible. A hypothesis of the root cause is formulated and all data should be cross-checked against this hypothesis. Obviously, it is possible for a failure to be caused by more than one root cause. The FA must provide support for a legal case in a product liability suit. FA techniques, which are individual analytical steps performed with the aim of identifying the FM, are used as tools for root-cause determination. Basically, the techniques are used for: (i) physical-fault isolation and (ii) electrical-fault isolation. Each FA technique is designed to provide its own, specialised information that will contribute to the determination of the FM. The use of any FA technique must be required by the analysis of the results obtained with previous FA techniques. In other words, although FA techniques are generally independent of each other, their results must nonetheless be consistent and corroborative in order to arrive at a strong conclusion for the FA cycle [3]. For electronic components, this step contains three main parts:
  - Information is extracted from the packaged item (as used in field operation). Exclusively nondestructive FA techniques, which do not permanently alter the item, are used for two main goals: to provide *electrical analysis* of the item and to verify the *package integrity*. Examples of FA techniques used before package-opening are given in Table 4.1.
  - The package of the electronic component is opened. The main unsealing techniques are detailed in Section 4.2.
  - The die and the internal features of the package are investigated using a mixture of destructive (causing permanent changes in the examined item, of whatever sort) and nondestructive FA techniques in three main steps: (i) *optical inspection*; (ii) *fault isolation*; and (iii) *physical analysis*.

Table 4.1 FA techniques used before unsealing

Technique	Character	Details
External inspection	Nondestructive	Visual check (package deformation or discolouration, attachment of foreign substances, large cracks, etc.) and check with optical microscopy (scratches, colour spots, cracks, etc.).
Electrical measurement	Nondestructive	First, the FMo is confirmed; then, if possible, electrical tests that are significant for item reliability are performed using curve tracer, oscilloscope, CV meter, LCR meter, noise meter and so on.
Verification of package cleanliness	Nondestructive	External cleaning of the package, then remeasurement. Leakage due to impurities on the package surface is identified.
Identification of die inversion layer	Nondestructive	Storage at 200 °C, then remeasurement. Repeated if the leakage has diminished. Leakage caused by an inversion layer at the die surface (Na^+ ions into the oxide) is identified.
Hermeticity testing	Nondestructive	Check for hermetic sealing of the package.
X-ray radiography	Nondestructive	Internal X-ray imaging of the package to identify foreign particles or broken wires.
Scanning acoustic microscopy (SAM)	Nondestructive	Detection of possible delaminations of the internal layers.

After [3].

Note that some nondestructive techniques can become destructive if improperly performed (e.g. electrical methods that are improperly used can lead to overstress). Examples of techniques used after package-opening are given in Table 4.2.

- **Hypothesis verification** – After a hypothesis is elaborated, it must be checked, by simulating the proposed events leading up to failure. Very often this step is neglected by investigators, which is dangerous because if the hypothesis is wrong, subsequent batches of the same item may fail.
- **FA report elaboration** – It is essential to record the solution reached, in order to compile a database of solved cases. By documenting the whole procedure of FA, one may avoid the same errors being repeated in future by other specialists, over and over again. The FA report has to be elaborated carefully, combining technical data with legal implications. Generally, technical jargon must be kept to a minimum, because the report has to be comprehensible to a large variety of specialists.

The **output data**, included in the FA report, are variable, depending on customer requirements, but the following elements must almost always be present:

- The *FM* responsible for the failure (e.g. breaks or cracks of the die, intermetallic compounds, oxide defects, pinholes, contamination, metal migration, short circuit of the oxide or dielectric layer, 'bad' solders, overcharges due to incorrect use, open circuits, misalignments, chemical reactions at the level of the metal/semiconductor contact area, metal corrosion inside the package, etc.).
- *Corrective actions* aimed at diminishing or avoiding the action of the identified FM. Generally, a failure can be produced by design factors, non-design factors or a combination of both. Consequently, corrective actions refer to: modifications of the existing requirements for the quality of materials, material exposure to hazardous chemicals or the working environment, changes in the design of the component being studied, changes in the working procedures or in the limits of various parameters measured during the manufacturing flow, proposals for new monitoring points, modifications of the recommendations for component usage and so on. Obviously, these corrective

Table 4.2 FA techniques used after unsealing

Technique	Character	Details
Optical microscopy	Nondestructive	Visual inspection of the die to identify defects of metallisation, soldering, wire, mask or oxide.
Microprobing	Nondestructive	Direct electrical analysis of the die circuit.
Liquid crystal techniques	Nondestructive	Hot-spot detection (heat-generating defects).
Light (photo) emission microscopy (LEM/PEM)	Nondestructive	Detection of light-emitting defects.
Optical beam-induced current (OBIC)	Nondestructive	Induced current imaging of defects.
Scanning electron microscopy (SEM)	Nondestructive	High magnification imaging to identify defects.
Electron beam-induced current (EBIC)	Nondestructive	Induced current imaging of defects (working mode of SEM).
Energy (wavelength)-dispersive X-ray (EDX/WDX)	Nondestructive	Elemental analysis (coupled with SEM instrument).
Atomic-force microscopy (AFM)	Nondestructive	High-resolution probe imaging.
X-ray photoelectron spectroscopy (XPS) or electron spectroscopy for chemical analysis (ESCA)	Nondestructive	Surface analysis.
Verification of the cleanliness of die surface	Nondestructive	Cleaning of the die surface, then remeasurement to identify possible leakage due to surface contamination.
Identification of oxide contamination	Nondestructive	Cleaning of the oxide, then remeasurement to identify possible oxide contamination.
Holographic interferometry	Nondestructive	Measurement of strain and vibration analysis.
X-ray fluorescence	Nondestructive	Identification of the quality of various materials and the state of bonding wires, die bond, voids in the encapsulation resin.
Fourier transform infrared (FTIR) spectroscopy	Nondestructive	Chemical analysis.
Transverse sectioning	Destructive	Identification of bad solders of the connection wires (or of the die).
Junction revelation	Destructive	Structure sectioning and junction revelation to identify deep-diffusion problems and traces of breakdown of the junction, in depth.
Focused ion beam (FIB)	Destructive	High-resolution die sectioning/imaging.
Transmission electron microscopy (TEM)	Destructive	High-precision analysis based on the electrons transmitted through the (very thin, specially prepared) specimen.
Auger electron spectroscopy	Destructive	Surface/depth analysis.
Secondary-ion mass spectroscopy (SIMS)	Destructive	Compositional analysis in depth.

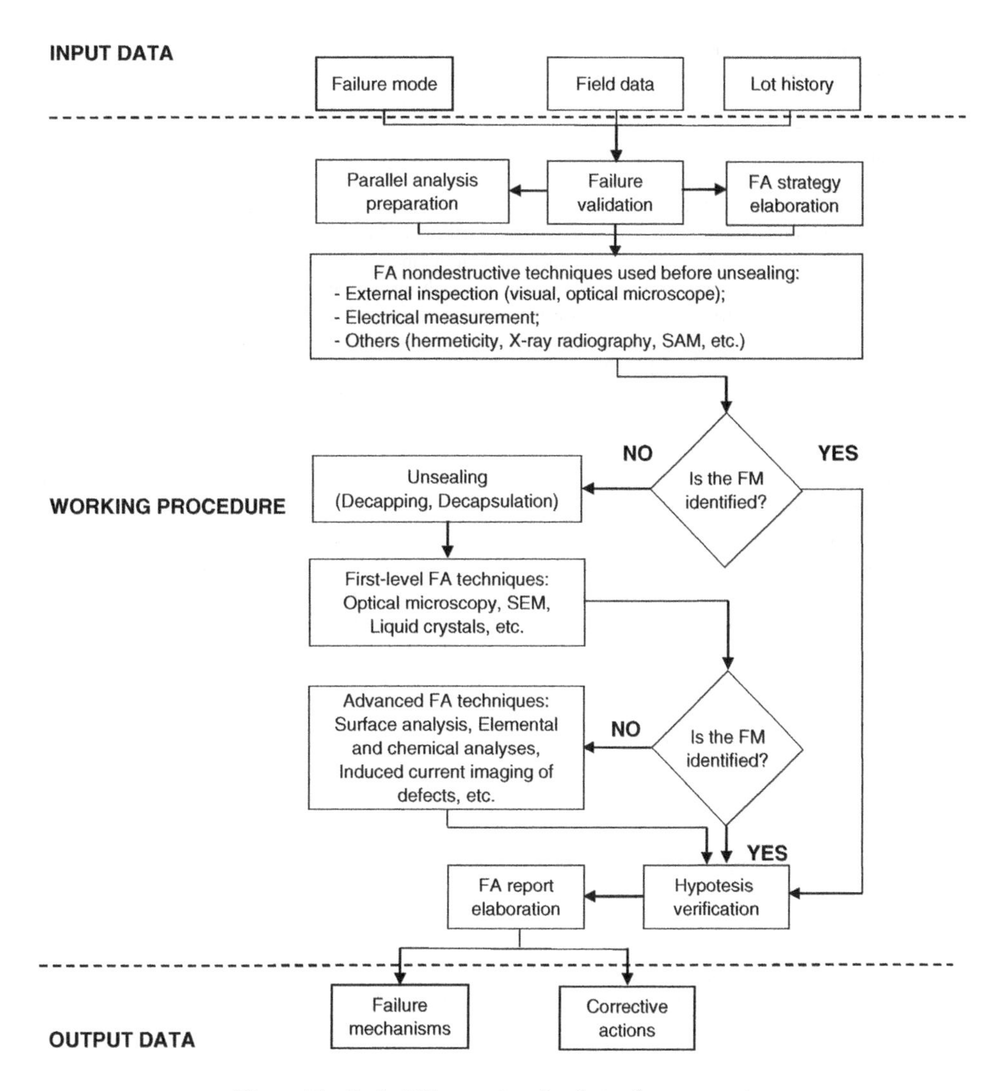

Figure 4.1 Typical FA procedure for electronic components

actions can be elaborated if and only if the failure causes are known and the FM is elucidated. Again, brainstorming can be useful in creating a corrective-action tree and elaborating a corrective-action assessment (CAA) chart.

The above procedure for FA is synthesised in Figure 4.1.

For large-scale integrated (LSI) circuits, it is extremely important to diagnose, localise and physically analyse any faults in a coordinated manner. Fault diagnosis is a technology used to make logical estimations of fault locations based on the internal operations of circuits using design data and inspection results. The main steps are indicated in Table 4.3 [4].

In recent years, the development of nanoelectronics–electronic components at nano-level–has modified the requirements for FA techniques [5]. For physical-fault isolation, with the exception of near-field transistor-level probes (e.g. STM and CT-AFM), current techniques severely lack resolution for nanoelectronic-scale devices. Innovation and development are needed to improve sensors and

Table 4.3 Main steps of FA flow for LSI

Step	Details
Processing input data	Possible input data: circuit information, test patterns, layout information, test results.
Fault diagnosis	Analysing the output of various candidates for faults and fault locations.
Identifying fault location	Narrowing down the fault location to gate levels, using various analysis devices (emission microscopes, etc.) and pinpointing their location; verifying whether the possible fault location and fault phenomena match.
Physical analysis	Detailed observation using analysis devices; clarifying the physical causes of the faults by measuring transistor characteristics and cross-sections observed with SEM, TEM and equipment for performing elemental analysis.

After [4].

detection schemes for far-field imaging of signals in the picoampere to nanoampere range. Wide-area imaging or perturbation of quantum device properties might provide active alternatives for physical-fault isolation. Electrical-fault isolation becomes difficult with the new devices and architectures such as multistate logic. Self-assembly poses altogether new challenges, as the traditional analytical approach is heavily dependent on *a priori* knowledge of circuit layout and electrical response. Finally, reconfigurable and fault-tolerant architectures certainly need new FA techniques.

4.2 Techniques for Decapsulating the Device and for Sample Preparation

Often, the FM is not identified for a packaged component. In that case, it is necessary to unseal the component in order to continue the investigation and discover the FM. Obviously, unsealing the electronic component is an irreversible operation. If the item is packaged in a hermetic (metal or ceramic) package, unsealing can be carried out by a mechanical process known as 'decapping'. The problem is more complicated for plastic packages, in which case the process is called 'decapsulation'.

4.2.1 Decapping Techniques

Decapping is a purely mechanical process involving opening a hermetic package. A low-speed diamond saw can be used to cut the weld around a metal can or to pry the combo lid off a ceramic package. Another solution for ceramic packages is to apply opposite torques to the top and bottom parts of the ceramic DIP to break the seal glass.

4.2.2 Decapsulation Techniques

A list of decapsulation techniques is given in Table 4.4.

4.2.3 Cross-Sectioning

Cross-sectioning (or microsectioning) prepares the plane of interest in a die or package for further analysis or inspection. The working flow is as follows:

- The sample is cleaned, mounted and encapsulated in polyester or epoxy resin. Often, in order to perfectly fit the specimen into the mould, the sample is sawed to reduce its size prior to encapsulation. The positioning of the specimen in the mould must be carefully chosen, in order to minimise the sawing and grinding needed to expose the plane of interest. The resin is usually poured inside a vacuum impregnator to minimise bubbles or air pockets, then the sample is cured at ambient pressure.

Table 4.4 Decapsulation techniques for plastic packages

Technique	Details	Characteristics
Chemical etching	First variant: red fuming nitric acid at 85–140 °C or sulfuric acid at 140 °C are dispensed on the surface of a package to remove the plastic material covering the die. First a cavity is milled on the top surface of the package, then the acid is repeatedly dropped into the cavity to remove the plastic material covering the die. When the die has been exposed adequately, the unit is rinsed with acetone, then with unionised water, before being carefully blow-dried. Second variant (used only for die backside inspection for cracks): the entire package is soaked in a beaker of sulfuric acid heated to about 140 °C. Third variant: the phenolic resin is dissolved in a mixture of 50% sulfuric acid and 50% nitric acid.	The second variant leads to total destruction of the item, leaving behind the silicon die and bits and pieces of undissolved metal parts.
Jet etching	Automated version of chemical decapsulation, using a jet etcher, which automatically squirts heated acid on the area of the package that needs to be removed. During this process, the area to be etched is usually left exposed while the rest of the package topside is covered by a rubber mask.	Jet etching is superior to chemical etching, being sensor-controlled and efficient.
Termomechanical method	The package is heated, then ground, broken and cut to separate the top part of the package from its bottom part. If used carefully, this technique destroys the bond wires but preserves the die.	Avoids the corrosion of metallised areas on a die obtained by chemical methods.
Plasma etching	The plastic of the package reacts with a gas, which can easily be vented out. This requires expensive plasma-etching machines, but is very clean and selective.	The duration is longer than is necessary for the other techniques.

- The encapsulated specimen is sawed, with a diamond wheel cutter, along a plane parallel to the plane of interest. Proper sawing minimises the amount of grinding needed to expose the plane of interest in the specimen.
- The specimen, cut to its optimum size, is grinded, the grinding material being SiC paper, polishing cloth or diamond paste. The grinding is carried out in many steps, separated by rinsing of the specimen. Each step removes all the scratches from the previous step.
- The specimen is polished. This operation is very similar to grinding, except that a Texmet, nylon or silk cloth with diamond or alumina paste or powder on the surface is used instead of SiC paper. All remaining scratches on the cross-section surface should be removed by this step.
- The specimen is stained, ready for optical or electron microscopy.

Note that newer cross-sectioning technologies use specific tools and procedures that eliminate the plastic encapsulation of the specimen.

4.2.4 Focused Ion Beam

Focused ion beam (FIB) is an FA technique used in high-magnification microscopy, die-surface milling or cross-sectioning, and even material deposition. An FIB system works similarly to a scanning electron microscope (SEM), except that it uses a finely focused beam of gallium (Ga^+) ions instead of electrons. The focused primary beam of gallium ions is rastered on the surface of the material to be analysed. On hitting the surface, the beam sputters a small amount of material from it, in the form of secondary ions, atoms and secondary electrons. These ions, atoms and electrons are then collected and analysed to form an image on a screen, allowing high-magnification microscopy. The quantity of sputtered material is directly proportional to the primary beam current. Ion-beam imaging provides the atomic number of the material under study, enabling identification of metal, dielectric and passivation layers. For high-magnification microscopy, only a low-beam operation may be employed.

High-beam operation is used to sputter or remove material from the surface, for example in high-precision milling that produces in situ cross-sections with a high degree of spatial control, providing an effective approach for the analysis of the root cause of a failure.

An FIB system can also be used to bombard an area on the die with various gases as it performs primary beam sputtering. Depending on the mixture used, the gases will react with the primary beam to either etch material from or deposit material on to the surface. Conductors (tungsten or platinum) and dielectric (SiO_2) can be deposited, enabling local modifications in the electrical interconnections of the IC, or the deposition of additional contact areas, which can be used to probe circuit functions and measure voltages on functioning ICs.

For instance, an IC could be prepared for the inspection of voided metal interconnects ('lines') and vias [6]. Conventional FIB milling is combined with a super-enhanced gas-assisted milling process that uses xenon difluoride (XeF_2) to rapidly remove large volumes of bulk silicon. This combined approach allows removal of the TiW underlayer from a large number of lines simultaneously, enabling rapid localisation and plan-view imaging of voids in lines and vias with backscattered electron (BSE) imaging in an SEM. Sequential cross-sections of individual voided vias enable a 3D reconstruction of these voids to be developed. This information clarifies how the voids were formed, allowing identification of the IC process steps that need to be changed.

4.2.5 Other Techniques

Anisotropic removal of dielectric layers (also called 'skeleton etch') allows the anisotropic removal of all dielectric layers, down to the silicon surface. Metal conductors are left sitting on top of pedestals of dielectric material. An anisotropic etch must be used to prevent undercut of the metal lines, or else the stress in the metal will usually cause delamination. When the silicon dioxide etch approaches the polysilicon gate material, a $CF_4 + CHF_3$ gas mix is used to improve selectivity to silicon and decrease erosion of the polysilicon lines [7].

Sequential removal (or **die delayering**) is useful when anisotropic dielectric removal is unable to identify defects or other features of interest that may lie underneath conductors. The metal and dielectric layers have to be removed sequentially and proper etch recipe selection is crucial to preventing inadvertent removal of layers not intended to be etched. It is important that lower-level metals layers are not exposed when the top-level metal is etched, or the lower-level metals will be removed prematurely. Since each layer is chemically and physically different from the others, the delayering steps are different from each other as well. The dielectric etch has to be stopped when a level is reached that is even with the next metal layer to be etched, in order to maintain planarity during sequential delayering of an IC.

Among the die delayering techniques, the most common are: plasma etching (a dry and anisotropic etching process); reactive ion etching (RIE), which involves bombardment of the surface with accelerated reactive ions that hit the surface and react with the substrate material); and wet chemical etching, which involves the application of liquid solutions to the die surface to remove one or more layers of material or to highlight defects. Typically, a die-delayering sequence starts with either plasma etching or RIE to remove the nitride passivation on top of the die surface, followed by a series of wet chemical etching steps to remove the rest of the die layers. Often, the glassivation layer on top of the die surface to be probed (see below) has to be removed by RIE, because the glass layer is hard to penetrate with the probe needle of the microprobing station. Silicon defects may be highlighted with either a Wright etch or a Sirtl etch.

4.3 Techniques for Failure Analysis

It is difficult to choose among the multitude of criteria for grouping the FA techniques. A possible classification is to distinguish between surface and bulk techniques. But the surface techniques, for instance, are very diverse. Some examples are: optical imaging, SEM, energy-dispersive X-ray spectroscopy, Fourier transform infrared spectroscopy (FTIR) and Auger electron spectroscopy. Despite their common goal, these techniques have different basic principles, working modes and so on. Moreover, some are also used for bulk investigations.

We have employed another approach below, grouping the FA techniques according to the type of physical support used to obtain the information, which is generally the natural classification for analytical techniques. This can be considered a hybrid classification, but it seems to be the most intuitive one. Consequently, the main headings in this section will be: (i) electrical techniques (which use the electrical signal as a basic instrument); (ii) optical microscopy (which represents, historically, the first group of FA techniques); (iii) scanning probe microscopy (SPM) (a modern technique developed as the extension of optical techniques); (iv) microthermographical techniques (used to obtain thermal maps of a device during function); (v) electron microscopy (SEM and transmission electron microscope (TEM) are today the basic tools in identifying the FMs of electronic components); (vi) X-ray techniques (which use X-rays as primary beams); (vii) spectroscopic techniques; (viii) acoustic techniques; (ix) laser techniques; (x) holographic interferometry; (xi) emission microscopy; (xii) emission microscopy; (xiii) atom probe; (xiv) neutron radiography; (xv) electromagnetic field measurements; and (xvi) other techniques (a necessary category in any classification).

4.3.1 Electrical Techniques

Generally, electrical characterisation is the first step in FA flow. Any electrical parameter that is significant for the failed item and for the given application has to be measured. The data obtained has to be compared with similar measurements performed on good items and with information received from the client.

The DC characteristics are obtained by using instruments such as a curve tracer, pico-ammeter, C–V meter and so on. These measurements allow identification of open connections, short circuits, drift of DC characteristics, pin damage and so on. For ICs, it is possible to determine the areas with defects on the die, by combining the different characteristics and based on previous experience, possibly with the aid of thermal tests and, if necessary, mechanical ones. By using I–V and C–V curves at various temperatures, relevant information about the failed item can be acquired. An example of the use of C–V measurements to characterise a capacitive radio frequency micro-electro-mechanical system (RF MEMS) switch is given by Lab Microsystem Characterization & Reliability from CEA-Léti, led by Dr Didier Bloch [8]. The device is composed of a platinum bottom electrode, a dielectric layer and an AlSi membrane top electrode, the latter two being separated by an air gap. When a voltage actuation is applied between the electrodes, the membrane collapses, increasing the capacitance of the switch, which behaves as an open circuit for an RF signal. C–V sweep measurements allow the behaviour of the switch to be characterized with the identification of the pull-in/pull-out voltages (corresponding to the actuation voltages required to pull down and then pull up the membrane, respectively), which are good indicators of dielectric charging and are used to compare the behaviours of the switches as a function of the gas environment. C–V sweep measurements were carried out from 0 to ±40 V at wafer level inside a vacuum prober equipped with a thermal chuck in order to manage the gas environment. More than 50 switches were tested per wafer in order to obtain representative results [8]. In the case of SiN_x dielectric material, the results under room environment show C–V cycles that are not symmetrical at all for positive and negative actuations, with a pull-out voltage greater than the pull-in voltage and a stiction in the rest position after actuation. The switch performance under nitrogen environment is significantly improved: the C–V cycle becomes totally symmetrical, with a pull-out voltage less than the pull-in voltage and a more reproducible functional behaviour. The statistical results confirm this behaviour, with a large dispersion of the pull-in/pull-out voltages under room environment and much narrower distributions under nitrogen environment.

Impedance spectroscopy is a general term covering the small-signal measurement of the linear electrical response of a material/device of interest and the subsequent analysis of the response to provide useful information about the physico-chemical properties of the system. Analysis is performed in the frequency domain. Further analysis is performed in the time domain and then Fourier transformed to the frequency domain.

Some specific electrical techniques are detailed below.

4.3.1.1 Microprobing

Microprobing is an FA technique used to achieve electrical contact with or access to a point in the active circuitry of the die. Specific equipment called a 'microprobing (probe) station' is used. Electrical contact is made with fine-tipped probe needles. A micromanipulator allows the needle to be inserted directly on the point of interest, or on an area to which the point of interest is connected. Microprobing is used to access the critical nodes on the microscopic die circuit while analysing the behaviour of the various parts of the circuit, in a process known as 'failure isolation'.

The voltage and current are measured by electrical instruments attached to the probe needle through the micromanipulator, such as voltmeters, curve tracers, oscilloscopes and so on. Circuit excitation from voltage supplies, waveform generators and the like may also be provided to the die circuit in the same manner.

In recent years, a new technique, called 'nanoprobing', has been developed from microprobing. In-depth electrical characterisations using nanoprobing can assist in visualising some nanoscopic defects, which would sometimes be mistakenly classified as 'nonvisible' defects [9]. Nanoprobing can also be used to localise some subtle defects affecting the yield of ICs: the exact failing transistors are isolated and characterised as malfunctioned devices. As a result, the identification process of the FMs is accelerated. The electrical characterisation at the transistor level is an appropriate guide to the subsequent physical analysis that has to be carried out in order to 'visualise' the defects. For instance, marginal failures or degradations relating to the ultrathin gate oxides, variations in the resistance of the implanted layers in the substrate and abnormal passive-voltage-contrast signature were determined [9].

4.3.1.2 IDDQ

IDDQ testing is an electrical technique for production quality and reliability improvement, design validation and FA. It has been used for many years by a few companies and is now receiving wider acceptance as an industry tool.

The procedure for quickly identifying a bad IC chip is to measure the resistance between V_{DD} and GND pins. If there is not a short circuit, a power supply is connected and the power-supply current is measured. If the supply current is more than a few milliamperes, this indicates a bad chip [10]. IDDQ testing represents a similar approach: the bad chips are identified quickly, saving expensive tester time. IDDQ stand for I_{DD} (the supply current) and Q (quiescent). The current of the V_{DD} power supply in the quiescent logic condition (the stable condition between logic-state transitions) is measured. If the circuit being tested contains defects detectable by an IDDQ test vector, the circuit draws excessive quiescent current upon the application of that vector.

In addition to increased defect and fault coverage, IDDQ testing enables rapid identification and physical localisation of many design, layout and fabrication problems (i.e. processing problems of a non-defect nature, such as excessive lateral diffusion). The IDDQ test technique can be applied at wafer level, at packed device level, during incoming inspection, during life test or even during online testing [10].

The main characteristics of IDDQ were identified by O'Connor [11]. IDDQ should be preferred because a small number of test vectors (around 100) are needed and, most importantly, the faults it detects cannot be seen by other testing methods. IDDQ can also be used as a burn-in procedure. The major disadvantage of IDDQ testing is the requirement for special test instruments and extra time to prepare and perform the testing, because very low currents (in the nanoampere range) have to be measured.

In recent years, the range of IDDQ applications was enlarged. For instance, in [12], an IDDQ method based on signature analysis is described and applied, being used to understand defect characteristics. The time-dependent analysis allows specific types of defect to be identified and the footprint analysis is useful for guiding reliability-related test decisions.

4.3.1.3 Reliability Monitors

Very often, electrical techniques are used as reliability monitors, allowing the reliability level of the unfinished item to be estimated during the manufacturing process, and acting against possible failure risks. Obviously, all such techniques (some of them will be discussed below) can be used to identify the FMs.

A novel trap spectroscopy based on *stress-induced leakage current* (SILC) measurements for constant-voltage stress and substrate hot-carrier injection stresses in *n*MOSFET devices is proposed in [13]. When an electrical stress is applied to the gate dielectric layer of a MOSFET, traps form within the oxide, which act as sites for the trap-assisted tunnelling of electrons through the layer, resulting in an increase in leakage current. When the stress is removed and the current is measured at a lower voltage (close to the operating voltage of the device), the trap-assisted component becomes dominant and quite large differences in the gate leakage (up to 2 orders of magnitude) from that of an unstressed device occur. By examining the response of each individual peak to stresses which preferentially degrade the interface of bulk, the spatial locations of the defects can be obtained. Low-carrier-energy stresses preferentially create defects close to the interface, giving rise to SILC at energy positions at which interface traps are conventionally found at 0.2 eV below the Si conduction band and 0.2 eV above the valence band. SILC has been used as a reliability monitor for many years.

Time-domain reflectometry (TDR) can quickly perform nondestructive tests on packaged ICs, being able to isolate open- and short-circuit defects in the three main regions of an IC: the die, the substrate and the interconnects. This is an appropriate FA technique for flip-chip and wire-bonded packages and for detecting failures in packages mounted on PCBs [14]. Currently, a digital sampling oscilloscope (DSO) equipped with a TDR module is used. The TDR module generates a voltage edge with a fast rise time, and the DSO records that edge and the signals reflected back to the TDR. The first voltage step in the waveform corresponds to the signal leaving the TDR, and the second step represents the reflection of the signal from the end of the probe tip.

Charge pumping (CP) was proposed as a technique for Si-SiO_2 interface state analysis of MOS transistors [15], the classical variant being the constant-amplitude technique. Another variant, the *fixed-base-level CP technique*, is described in a recent paper [16] as a very efficient tool which offers the capability of energetic and lateral resolved interface trap analysis in minimal test time. The method was applied for PMOS transistors with a 30 nm SiO_2 gate oxide and a gate length of 6 μm. The advantages of the constant-base-level CP technique over the constant-amplitude technique were emphasised. The greatest fault of the classical technique is that the scanned channel area is not well defined, which may lead to misinterpretation of measurement results. When using the constant-base-level measurement, it is essential to keep the pulse slopes constant in order not to merge local and energetic information, and the density of states in a large range in the upper and lower halves of the bandgap can be determined without taking hundreds of CP curves and extracting the maximum CP currents.

Charge-to-breakdown (QBD) measurement of a MOS device is a destructive test method used to determine the quality of gate oxide. QBD is a measure of the time-dependent gate-oxide breakdown, because the total accumulated charge passing through the dielectric layer is analysed just before failure. Also, QBD can be a useful predictor of product reliability under specified electrical-stress conditions. The most common variants of QBD are: linear voltage ramp, constant-current stress, exponential current ramp and linear current ramp.

4.3.1.4 Noise as a Reliability Indicator

The noise phenomena of the electronic components can be classified in two categories: normal and excess noise. Normal noise includes the thermal and shot noises, while excess noise contains the flicker (or 1/f), the micro-plasma, the generation-recombination and the burst noises. Excess noise (mainly burst noise and flicker noise) can provide some information on the reliability of electronic devices; a time variation of the excess-noise amplitude indicates evolving defects, for instance [10].

The *burst noise* (also called popcorn noise) is generated by current fluctuations in the vicinity of macroscopic crystalline defects or dislocations in the emitter–base junction surface region. From experimental results [17] it seems that the defects induced by ion implantation may cause popcorn noise. A distinct popcorn noise over a large number of transistors from the same batch means a poor quality of semiconductor crystal or oxide layer and consequently a defective fabrication process. Experimentally, it was found that the burst noise depends on recombination–generation centres, on metal atoms from the crystal when precipitating at the junction, and on surface–junction dislocations [18].

Other experiments have shown that the *flicker noise* (or 1/f noise) can be produced by various crystal defects, such as dislocation and precipitates (generated during the diffusion process and dependant on crystal orientation), and by Si-SiO_2 interface states [19]. Consequently, the flicker noise can be used as an indicator for the existence of such failure risks.

4.3.1.5 Deep-Level Transient Spectroscopy

Another current technique in FA, which is in fact a combination of electrical and thermal techniques, is deep-level transient spectroscopy (DLTS),[1] used to characterise the deep level from the forbidden band. From DLTS spectra, data about the traps can be obtained, allowing the mechanism of trapping in semiconductors to be understood, which is a possible FM.

The DLTS technique has a high sensitivity in detecting impurities and defects, reaching concentrations of one part in 10^{12} of the material host atoms. However, there is a disadvantage to DLTS: it cannot be used for insulating materials, because withthese materials it is difficult or impossible to produce a device with a space region whose width can be changed by the external voltage bias and thus the capacitance measurement-based DLTS methods cannot be applied for defect analysis.

4.3.2 Optical Microscopy

Optical microscope is performed with the so-called 'light microscope', which uses visible light and a system of lenses to magnify the image of a sample. Low-power microscopes may magnify a specimen at 5–100× and high-power microscopes at 100–1000×. A new variant, the 'digital microscope', has no lenses, but rather a charge-coupled device (CCD) camera, and displays the image on a computer screen, with a magnification up to 200×.

[1] The DLTS technique was proposed in 1974 by D.V. Lang, from Bell Laboratories.

There are three operation modes in optical microscopy:

- *Brightfield illumination* is the normal mode of viewing, providing the most uniform illumination of the sample. A full cone of light is focused on the sample and the observed results from the various levels of reflectivity are exhibited by details on the sample surface.
- *Darkfield illumination* blocks an inner-circle area of the light cone, such that the sample is only illuminated by light that impinges on its surface at a glancing angle. The method is effective in detecting surface scratches and contamination, because the reflected light is scattered by feature edges, particulates and other irregularities on the sample surface.
- *Interference contrast* (Nomarski) uses polarised light divided by a Wollaston prism into two orthogonal light packets that hit the specimen at two different points and return to the prism through different paths. The differences in the routes of the reflected packets produce interference contrasts in the image when the packets are recombined by the prism upon their return. The method allows identification of surface defects or features such as etch pits and cracks that are difficult to see under brightfield illumination.

Optical microscopy is the basic instrument for nondestructive examination of chips, able to locate many physical defects. The lowest obtainable detail is around 200 nm. For smaller dimensions, other tools are needed. The optical microscope is appropriate for metallographic examination, which requires a small section of material to be cut out, mounted, polished and etched (see Section 4.2). The microstructure of the material can be determined: grain size, inclusion size and distribution of second phases. Crack growth through the microstructure can also be followed.

4.3.3 Scanning Probe Microscopy (SPM)

SPM obtains surface images by scanning the specimen with a sharp tip and detecting the interactions between the tip and the specimen. The first technique from this domain was the STM, followed by many other SPM techniques. Generally, the scanning probe instruments provide an entirely new capability for topographical imaging and analysis of submicron structures, extending spatial resolution well beyond the limits of optical tools. There is a large range of SPM techniques, which are described in a good reference book, *Scanning Probe Microscopy and Spectroscopy* by Roland Wisendanger (Cambridge University Press, 1994).

Details about the main types of SPMs used in FA of electronic components and systems are given below.

4.3.3.1 Atomic-Force Microscopy

Atomic-force microscopy (AFM) measures the local properties of the surface being inspected, such as its height, optical absorption and magnetic properties, by using a tip positioned very close to it. There are two main operation modes of AFM:

- **Contact mode (also called repulsive or static mode** – a soft cantilevered beam that has a sharp tip at its end is positioned very close to the surface, initiating a repulsive interaction between the atoms of the tip and those of the surface. This force exhibits a spring constant (0.001–100 N/m), which is used to nondestructively obtaining information about the surface of the specimen. Contact mode is primarily used to generate images of specimen topography. Cross-section images of polished samples with nanometre resolution are obtained, which are considerably better than those obtained using SEM.
- **Noncontact mode (or attractive/ dynamic mode**) – a small piezo element, mounted under the cantilever, makes the latter oscillate at its resonance frequency. The oscillating cantilever tip is

positioned 10–100 nm away from the sample surface, and longer range interactions, such as van der Waals, electric or magnetic forces, are used to modulate[2] the image contrast. The changes in oscillation are used to generate a map that characterises the surface of the sample. The phase shifts can be used to identify different surface materials. The main issue in noncontact mode is to keep the right tip-to-specimen distance: the maximum distance allowing detection of the inter-atomic forces.

Unlike electron microscopy, AFM generates three-dimensional surface images, does not require specimen preparation (and destruction) and does not require a vacuum environment in order to operate. Also, AFM has a higher resolution than SEM, the high-resolution AFM being comparable to STM or TEM.

The main disadvantages of AFM are related to the small size of the image (maximum height on the order of micrometres and a maximum scanning area around 150 × 150 µm, compared to SEM images which can cover an area on the order of millimetres by millimetres) and the long time required for scanning (several minutes for a typical scan), compared to the real-time images obtained by electron microscope.

AFM is an ideal tool for identifying the small-dimension details (in the nanometre range) on a surface. In Figure 4.2, an AFM topographical image of InGaAs/GaAs quantum dots is shown. The scanning area is 500 × 500 nm and the height of the dots is about 5 nm. In the AFM image of step-patterned Si presented in Figure 4.3, each step has a height of about 15 nm and the monatomic height terraces are clearly visible.

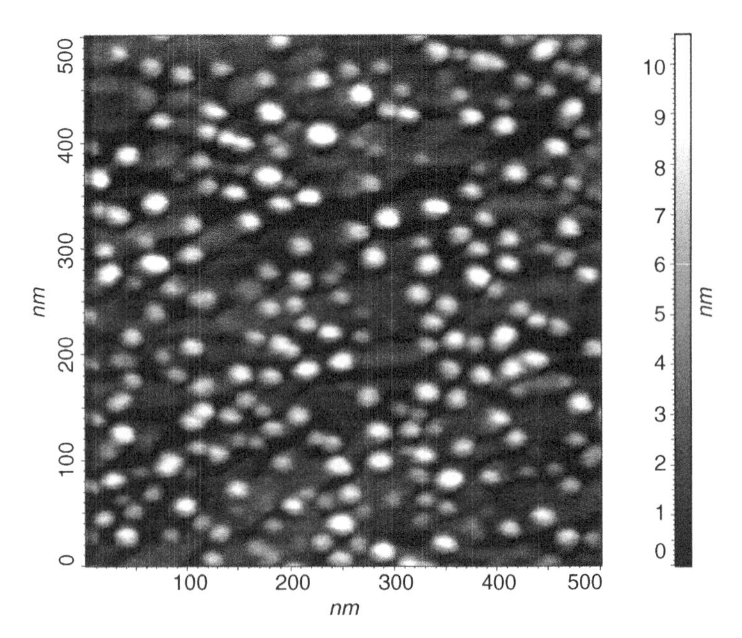

Figure 4.2 AFM image (topography) of InGaAs/GaAS quantum dots. Scan size: 500 nm × 500 nm. The height of the dots is about 5 nm Sample courtesy Dr Emil Pavelescu. Reproduced by permission of Raluca Gavrila (IMT-Bucharest, Romania)

[2] Either frequency modulation (changes in the oscillation frequency provide information about tip–sample interactions) or amplitude modulation (changes in the phase of oscillation are used to discriminate between different types of material on the surface).

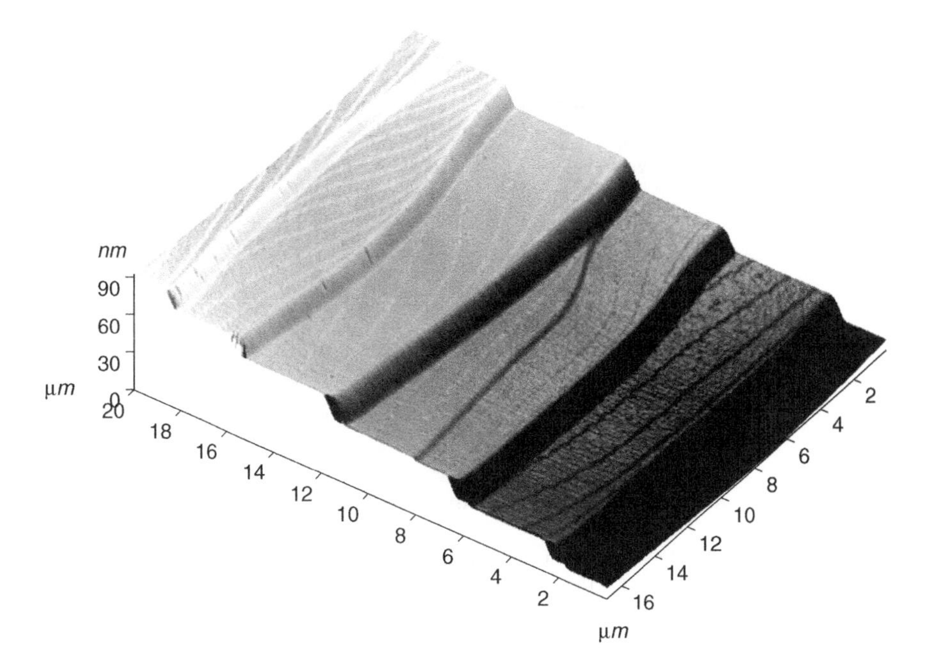

Figure 4.3 AFM image of step-patterned Si. Each step is about 15 nm in height. Monoatomic height terraces are clearly visible The Si STEP test sample was provided by NT-MDT Co. Reproduced by permission of Raluca Gavrila (IMT-Bucharest, Romania)

4.3.3.2 Scanning Tunnelling Microscopy

STM is used for imaging surfaces at atomic level. Invented in 1981 by Gerd Binnig and Heinrich Rohrer (IBM Zurich), STM has a 0.1 nm lateral resolution and 0.01 nm depth resolution, enough to image and manipulate individual atoms. STM functions based on the concept of quantum tunnelling: a voltage difference is applied between the conducting tip and the surface to be investigated, allowing the electrons to tunnel. The tunnelling current depends on the applied voltage, tip position and local density of states of the specimen, and the information acquired by scanning the whole surface is displayed in image form. STM is used in topographic imaging, tunnelling spectroscopy and so on, with specials applications in nanotechnology.

4.3.3.3 Conductive Atomic-Force Microscopy

Conductive atomic-force microscopy (C-AFM) is a combination of AFM and STM. It uses electrical current to obtain the surface profile of the specimen being studied. The current flows through the metal-coated tip of the microscope and the conducting sample. A vibrating tip produces the AFM topography, which is acquired simultaneously with the current (as for STM). A C-AFM microscope uses conventional silicon tips coated with a metal or metallic alloy, such as Pt-Ir alloy. C-AFM can be operated in both imaging and spectroscopic mode.

4.3.3.4 Scanning Near-Field Optical Microscopy

Scanning near-field optical microscopy (SNOM, also called NSOM) has a much higher resolution limit than the 'far-field optical microscopy', because it exploits the properties of evanescent waves. The

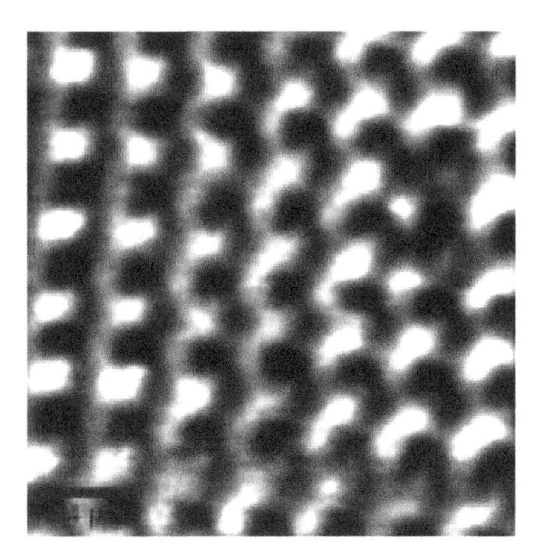

Figure 4.4 Collection-mode SNOM image of metallic nanostructures consisting of arrays of gold disks with a diameter of 200 nm and a height of 50 nm on glass substrate. Reproduced by permission of Cristian Kusko (IMT-Bucharest, Romania)

detector is placed very close (at a distance much smaller than the wavelength) to the specimen surface, allowing the use of high spatial, spectral and temporal resolving power, which limits the resolution of the image by the size of the detector aperture and not by the wavelength of the illuminating light. SNOM is currently used for nanostructure investigation, because the lateral resolution is about 20 nm and vertical resolutions of 2–5 nm have been obtained. A good reference book on SNOM is given in reference [20].

The SNOM equipment contains a light source (usually a laser beam), a feedback mechanism, a scanning tip (which could be a standard AFM cantilever with a hole in the centre of the pyramidal tip or a pulled or stretched optical fibre coated with metal, except at the tip), a detector and a piezoelectric sample stage. There are two main types of operation: aperture operation (the most popular, covering five operation modes) and a non-aperture mode (which needs very sharp tips without metal coatings).

SNOM is an appropriate tool for analysing structures with very small dimensions. For example, Figure 4.4 shows a collection-mode SNOM image of metallic nanostructures consisting of arrays of gold disks with diameters of 200 nm and heights of 50 nm on glass substrate. Figure 4.5, obtained by a group from IMT-Bucharest (Romania), compares an SNOM image in the photon-scanning tunnelling-microscopy mode of a multimode waveguide with the AFM contact-mode image of the same structure.

4.3.3.5 Field Ion Microscopy

Field ion microscopy (FIM) uses a metal tip to scan the specimen surface in an ultrahigh vacuum chamber, filled with an inert gas (e.g. helium or neon). A positive voltage of around 10 kV is applied on the tip, which is sharp (the radius is smaller than 50 nm) and cooled at cryogenic temperatures (below 100 K). The gas atoms adsorbed on the tip are ionised by the high electric field near it, become positively charged, are repelled from the tip in a direction perpendicular to the surface and are collected by a detector. The image formed by the collected ions describes the individual atoms of the tip surface. FIM is a projection-type microscope, so the resolution is not limited by the wavelength of the particle and the magnification is very large.

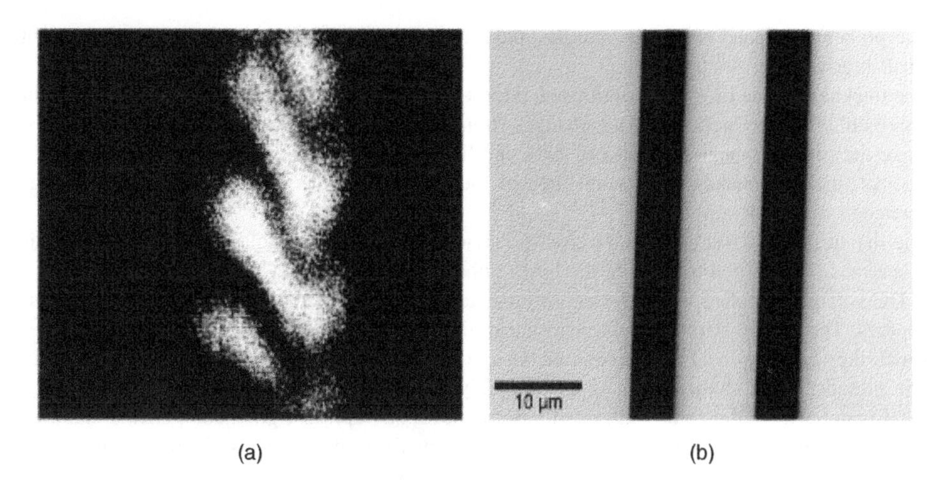

Figure 4.5 A multimode waveguide: (a) SNOM image in the photon-scanning tunnelling-microscopy mode; (b) AFM contact-mode image of the same structure. Reproduced by permission of Cristian Kusko (IMT-Bucharest, Romania)

4.3.4 Microthermographical Techniques

The term 'microthermography' designates a group of FA techniques which aim to locate areas on the die surface that exhibit excessive heating, the so-called 'hot spots'. Excessive heating is an indicator of a high current flow, which may be produced by metallisation shorts, leaky junctions or other die defects.

4.3.4.1 Liquid-Crystal Microthermography

The cheapest microthermographical technique uses liquid crystals (LCs), which have the property (called thermochromism) of changing their visual characteristics according to the environment temperature. As the temperature increases, LCs undergo several phase changes (thermotropic phases): from the crystalline phase (at very low temperatures), the LCs become liquid, though they remains crystal (with smectic, cholesteric and nematic possible phases, as the temperature increases), because their elongated molecules tend to align parallel to each other; at even higher temperatures, the LCs enter the isotropic phase, with their molecules randomly oriented. In hot-spot detection, only the nematic, cholesteric and isotropic phases are exploited.

By viewing with an optical microscope equipped with a polarising filter in the illumination path and a cross-polarised filter (analyser) in the viewing path, thin films of LCs in isotropic and nematic phases appear very different from each other: the nematic films appear to be rainbow-coloured and the isotropic films appear black, since the cross-polarised light is blocked by the analyser. If a film of nematic LCs is deposited on the surface of an electrically biased die, the whole area appears rainbow-coloured, except for the hot spot, which raises the temperature of the nearby LCs, making them appear as a black spot.

The procedure for using the LC technique is as follows:

- The device is electrically characterised and the bias value is chosen carefully, providing enough current from the defect sites to heat the LCs at the hot spot, but not enough to heat the entire LC film, which would darken the entire die surface. The maximum voltage must not exceed the breakpoint as this would introduce an unwanted current path.

- The probes are placed on the die surface, and a thin film of LCs is spread on to the surface with a small brush.
- The thickness of the LC film is optimised: a too-thick film would appear uniformly black, making it impossible to detect nematic phase changes from temperature increases, and a too-thin film would make the surface appear as streaks of dark and light greys, also making hot spots less visible. The optimal thickness makes the general area of the die surface look rainbow-coloured, offering the best contrast to a hot spot.
- The die surface is visualised under cross-polarised conditions. The polarising filters of the microscope must be adjusted to provide the best rainbow colour for the LC film.
- A DC voltage is applied to the device, in order to generate an increased heat, until dark (hot) spots appears. The lack of dark spots does not mean that there are no hot spots. If the hot spots are too small, they heat the local area below the detection threshold.
- The presence of an abnormal hot spot very likely means that something is wrong, but the hot spot itself is not always the actual failure site. Some hot spots represent good components that are just forced to conduct high currents because of an anomaly somewhere else in the circuit. The LC technique must therefore be complemented by other FA techniques.

The LC detects the thermal radiation regardless of the source of heat, which is an advantage over other FA techniques, such as emission microscopy, which detects radiation emitted only from junctions or gates. So, with LC, the heat generated by polysilicon–polysilicon shorts can easily be detected [21].

Fluorescent microthermal imaging (FMI) is an alternative method to LC, which uses the temperature dependence of the fluorescent quantum yield of a rare earth chelate (e.g. europium) to provide a direct, quantitative conversion of surface temperature into detectable photons. A thin film containing the rare earth chelate is applied to the die surface and the surface temperature is determined by imaging the intensity of the visible-light fluorescence from the film when it is pumped with an ultraviolet light source [22]. FMI is harder to use than LC, but is able to provide accurate quantitative results.

4.3.4.2 Infrared Microthermography

All objects emit electromagnetic radiation, its wavelength depending on the temperature of the object. The frequency of the radiation is inversely proportional to the temperature. In infrared (IR) thermography, the radiation is detected and measured by infrared imagers (radiometers), which convert the emitting radiation into electrical signals that are displayed on a colour or black-and-white computer display monitor.

IR microthermography provides a temperature map of the die surface, allowing hot spots (with temperatures 40–150 °C above ambient) to be identified. The system contains an IR microscope, coupled with computer-aided emissivity correction. The spatial resolution is relatively poor, depending on the IR detector employed, while temperature resolution is quite good: temperature differences as small as a few hundredths of a degree Celsius can be identified. After processing the thermal data, the hot spots are displayed on a monitor in multiple shades of greyscale or colour. After thermal analysis, the thin surface film can easily be removed, the method being a nondestructive one.

4.3.5 Electron Microscopy

The smaller size of modern electronic components requires appropriate tools to identify defects. Optical microscopes have well-defined physical limits, linked to the wavelength of the light: details smaller than half a micrometre are no longer visible. The solution proposed to surpass this difficulty was to use a beam of electrons instead of visible light, because the smaller wavelength of electrons allows a much higher magnification to be obtained. The first attempt was reported in 1931, by the German engineers Ernst Ruska and Max Knoll, who succeeded in magnifying an electron image. In 1933, the first prototype of a TEM was built by Ruska (Nobel Prize in 1986) and was capable of resolving to 50 nm. The main modern techniques of electron microscopy are detailed below.

4.3.5.1 Scanning Electron Microscopy

Without doubt, the SEM is the most used tool in FA of electronic components. It is almost compulsory to have an SEM picture in a FA report, sometimes even if such a picture offers no additional information. However, that is a rare event, because very often SEM is the necessary tool for identifying the FM.

SEM magnifications can go to more than 300 000, but most semiconductor manufacturing applications require magnifications of less than 3000, allowing SEM usage in the analysis of die/package cracks and fracture surfaces, bond failures and physical defects on the die or package surface.

The working principle of SEM is as follows. A beam of (primary) electrons is focused on a spot volume of the specimen, transferring the energy to the spot and dislodging electrons from the specimen itself. The dislodged (secondary) electrons are attracted and collected by a positively biased grid or detector, and then translated into a signal. To produce the SEM image, the electron beam is scanned across the area being inspected, producing many such signals. These signals are then amplified, analysed and translated into images of the topography. Finally, the image is shown on a cathode ray tube (CRT).

Several possible working modes of SEM are detailed below.

Secondary Electrons

The quantity of secondary electrons (with energy lower than 50 eV) collected during inspection depends on the energy of the primary electrons (which are emitted by a heated filament). Due to their low energy, these electrons originate within a few nanometres of the sample surface. Very high-resolution images of a sample surface are produced, revealing details about 5 nm in size, with a very large depth of field, yielding a characteristic three-dimensional appearance useful for understanding the surface structure of a sample. The range of magnification goes from 25 to about 250 000, much more than the best light microscope. Basically, the secondary-electron analysis is used to study the topography of the specimen surface. Cross-sections though IC structures can also be investigated. For dielectric surfaces, the sample charging from absorbed electrons (which may reduce the spatial resolution) can be avoided by coating the sample with a thin conductive film or by careful selection of the primary-electron beam energy.

A new type of SEM is the *field emission scanning electron microscopy* (FESEM), which produces electrons by emission from a sharp tip with a high applied electric field. FESEM has a superior spatial resolution at low beam energies, allowing nondestructive imaging of in-process wafers and high-resolution imaging of insulating layers, without the need for conductive films. The change in secondary-electron emission with material is also pronounced at low beam energies, so different layers in a cross-section can be identified in FESEM without using special wet or dry etches to produce topology differences.

Backscattered Electrons (BSEs)

Another effect of the primary electron beam is the emission of backscattered (or reflected) electrons from the specimen by elastic scattering interactions with specimen atoms. They have more energy than secondary electrons, and because they are emitted in a definite direction, the detector has to be directly in their path of travel. All emissions above 50 eV are considered to be BSEs. The yield of the collected BSEs increases monotonically with the atomic number of the material, so they are appropriate for compositional studies, being able to distinguish one material from another, with atomic number differences of at least three. BSEs can be used to determine the crystallographic structure of the specimen, in the form of an electron backscattered diffraction (EBSD) image.

X-Ray Microanalysis

The compositional analysis of a specimen can be obtained by an *energy-dispersive X-ray* (EDX) system. When the electron beam removes an inner-shell electron from the specimen, a higher-energy electron (from an outer shell) fills the shell, releasing energy, under the form of characteristic X-rays. These are used to identify materials and contaminants, as well as to estimate their relative concentrations on the surface of the specimen. An EDX spectrum is obtained, which is a plot indicating how often an X-ray is received for each energy level. Peaks correspond to the energy levels for which the most X-rays have been received. Each of these peaks is unique to an atom, and therefore corresponds to a single element. The higher a peak in a spectrum, the more concentrated the element in the specimen.

Generally, the EDX system is coupled with SEM, but it could operate independently too. EDX is appropriate for identifying FMs related to inorganic contamination or to elemental composition. A variant of EDX is the so-called *wavelength dispersive X-ray* (WDX). The detector uses a crystal that classifies and counts the impinging X-ray in terms of characteristic wavelength by allowing only the diffraction and counting of the desired wavelengths. Compared to EDX, WDX analysis has a better energy resolution (avoiding the errors linked to peak overlap) and a lower background noise. However, the time duration and cost of the analysis are higher than for EDX.

Electron Beam-Induced Current (EBIC)

If the high-energy electrons of the SEM beam inject charge carriers in a sample containing an internal electric field (e.g. at a *pn* junction), the flow of an electron beam-induced current is initiated, showing the presence of local defects that act like electron–hole recombination centres. This EBIC current (I_{EBIC}) can be calculated with the formula:

$$I_{EBIC} = I_p \times n(E_p/E_{eh}) \tag{4.1}$$

where I_p is the primary beam current absorbed by the sample (in the picoampere range), n is the collection efficiency (can be assumed to be 1), E_p is the primary beam energy of the SEM (several kiloelectronvolts) and E_{eh} is the energy needed to create an electron–hole pair (about 3.6 eV for Si). For an E_p of 20 keV, the I_{EBIC} current is more than 5000 times higher than I_p. Obviously, in areas with local defects, the electron–hole recombination is enhanced, reducing the collected current. The 'defect' areas therefore appear to be darker in the EBIC image than areas with no physical defects.

The range of EBIC applications includes detection of collector 'pipes' (resulting in collector–emitter leakage of bipolar transistors), location of *pn*-junction defects, measurement of the width of the depletion layers of the minority carrier diffusion lifetime, and detection of the defects of the crystal lattice. EBIC is a powerful analysis tool for bipolar devices, but it is not so effective in analysing MOS ones, due to the fact that the gate oxides of MOS transistors tend to trap charges from primary beam-charge injection, resulting in false failures.

Absorbed Electrons

The beam current absorbed by the specimen can also be detected and used to create images of the distribution of specimen current, but this technique is rarely used, because special wiring and fixturing are needed for signal detection and implementation is difficult, especially when detecting signals from various sites on the device. However, a combined technique of absorbed electron image (AEI), EBIC and nanoprobing is proposed in [22], based on the implementation of small precise motors inside a vacuum chamber, which allows three-dimensional nanoprobing with fine probes to a semiconductor device under SEM observation. In this new combined instrument, extremely fine tungsten probes are operated three-dimensionally and navigated to random locations to detect signals and measure electrical properties of the device.

Cathodoluminescence

The method is similar to EBIC, the basic phenomenon again being the injection of charge carriers by the electron beam of the SEM, but this time in a direct bandgap material, the result being cathodoluminescence (CL).[3] Light is emitted by atoms excited by the high-energy electrons, which return to their ground state. The CL detector collects the light emitted by the specimen and analyses the spectrum of emitted wavelengths. CL is appropriate for the analysis of the optoelectronic behaviour of semiconductors, particularly when studying nanoscale features and defects.

Voltage Contrast (VC)

This method is based on the idea that local electrical fields can restrict or enhance the emission of secondary electrons from a sample bombarded by an SEM electron beam. SEM images localising the electrical fields are formed as contrast differences: brighter or darker areas, depending on voltage polarity. VC is able to detect open junctions, open conductor lines and reverse-biased junctions. A variant of this technique is *capacitive coupling voltage contrast* (CCVC), which can produce non-destructive images of dynamic voltages beneath passivation layers. The passivation layer is used as a discharging capacitor to generate a dynamic image of changing subsurface voltages. The primary beam energy is low (around 1 keV) and CCVC imaging is performed at fast electron-beam scan rates to increase the time resolution of the dynamic signal.

For both VC and CCVC, typical FMs include: cracks or other defects of die, die attach and package, bonding failures, wire fractures, foreign material on die or package and so on. If an energy spectrometer is used to measure the energy spectrum of the emitted secondary electrons, quantitative voltage measurements can be obtained.

A common problem of SEM analyses is the accumulation of electric charge on the surface of nonmetallic samples, which worsens image resolution. Generally, the method employed to diminish this effect is to cover the nonmetallic or biological materials with carbon or gold. A new type of SEM was proposed by ElectrScan Corp. in 1988 to avoid this problem: *environmental scanning electron microscopy* (ESEM), in which the sample is placed in a chamber with low-pressure gases (typically 1–50 Torr) and well-controlled relative humidity (up to 100%). The detrimental accumulation of negative electrical charge is neutralised by the positively charged ions generated by the beam interactions with the gas. Thus the biological sample and the nonmetallic layers can be investigated without being covered by metallic layers.

The resolution of SEM depends on the size of the electron beam and the interaction volume, which are both much larger than the distances between atoms. This makes it impossible to see individual atoms. In this case the solution is to use a TEM, as shown in the next section. However, SEM has some advantages compared with TEM: the possibility of seeing bulk materials, the comparatively larger area of the specimen, the large range of working modes that may cover various situations (detailed above) and perhaps most importantly, the ease of interpretation compared to TEM. The resolution of SEM is between 1 and 20 nm, but resolutions below 1 nm have been reported in recent years.

If the energy of the SEM beam is greater than 3 keV, the electrons penetrate the surface of the sample and the sample is electrically charged by the difference between the number of electrons that enter the sample and the number that exit as ground current or emitted electrons. This phenomenon can be used to locate opens in metal lines or via chains without external power supplies, since the electron beam biases the sample. As a rule of thumb, the primary beam of an SEM penetrates the materials about 1000 Å per 1000 eV of beam energy. A 10 keV beam will penetrate about 1 μm. An isolated conductor covered with 1 μm of passivation will become negatively charged due to the build-up of electrons trapped from the beam. Grounded conductors remain neutral as electrons flow away.

Two examples demonstrate the practical use of this instrument. In Figure 4.6, SEM is used to identify the process faults: the micrograph of a test fixture for electrical measurements of carbon

[3] The name comes from the fact that CL is encountered in CRTs.

Figure 4.6 SEM micrograph of a test fixture for electrical measurement of CNTs. It clearly shows the unfinished lift-off process. Reproduced by permission of Adrian Dinescu (IMT-Bucharest, Romania)

nanotubes (CNTs) is shown. The unfinished lift-off process may be noted. In Figure 4.7, an ultra high-resolution SEM micrograph of a silicon tip used in AFM analyses is achieved, allowing the contamination of the very end of the tip to be identified.

4.3.5.2 Transmission Electron Microscopy

In the original form of the electron microscope, the TEM, a high-voltage electron beam (between 60 and 350 keV), emitted by a cathode and formed by magnetic lenses, is partially transmitted through the very thin (and so semitransparent for electrons) specimen, and carries information about the structure of the specimen. The 'image' is magnified by a series of magnetic lenses and recorded by hitting a fluorescent screen, photographic plate or light-sensitive sensor such as a CCD camera. The image detected by the CCD may be displayed in real time on a monitor or computer. Eventually, two-dimensional, black-and-white images are produced. As one can see, unlike SEM, which processes the information provided by dislodged or reflected electrons from the specimen, TEM collects the electrons that are transmitted through the specimen. So the basic feature of TEM is the necessity to prepare the sample; that is, to obtain a specimen that is thin enough to allow the electron beam to pass through it.

The resolution is sufficient to produce images of atoms in silicon at 0.078 nm, at magnifications of 50 000 000. TEM became the basic tool for nanotechnologies based on its ability to determine the positions of atoms within materials.

High-resolution transmission electron microscopy (HRTEM) is a new variant of TEM, which does not use amplitudes (like TEM), but rather the contrast arising from the interface in the image plane of the electron wave with itself. The basic idea is that the phase of the electron wave still carries information about the sample and generates contrast in the image. However, this is true only if the sample is thin enough that amplitude variations only slightly affect the image. HRTEM allows imaging of the crystallographic structure of a sample at an atomic scale. Today, the highest obtained resolution is around 0.08 nm; the goal is to push the resolution of HRTEM to 0.5 Å.

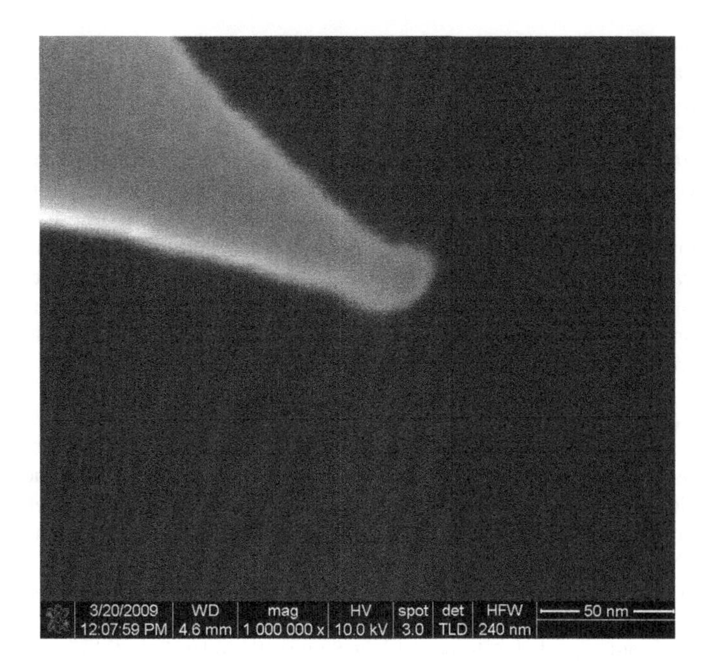

Figure 4.7 Ultra high-resolution SEM micrograph of an AFM silicon tip used in atomic-force microscopy. One can see the contamination of the very end of the tip. Reproduced by permission of Adrian Dinescu (IMT-Bucharest, Romania)

4.3.5.3 Scanning Transmission Electron Microscopy

Scanning transmission electron microscopy (STEM) is a hybrid electron microscope which aims to have the advantages of both SEM and TEM. The working principle is close to that of SEM: the image is obtained by a beam focusing on a small spot which scans over the sample. However, image information is obtained as in TEM, being extracted from electrons that are transmitted through a thin sample. Consequently, STEM conserves many of the advantages of SEM but tries to outperform it by offering stronger materials contrast, longer sample lifetime in the beam and lower effective contamination rates [23]. Moreover, the resolution of STEM can approach that of TEM. STEM is an interesting choice, because it can be adapted to a SEM platform (through the introduction of a transmitted-electron detector) and achieves high-resolution imaging comparable to that of TEM.

The main issue to be solved in STEM analysis is the preparation of thin samples. A method of automated sample-preparation techniques, based on the FIB system, is reported in [23], together with a comparison between the results obtained with SEM, TEM and STEM: STEM (which can be built to an existing field-emission SEM instrument, with a moderate cost) may provide images comparable with those obtained by the very expensive TEM instrument. The automated sample-preparation eliminates much of the time and difficulty associated with conventional thin-sectioning techniques and improves the overall reliability of the sample-preparation process.

4.3.6 X-Ray Techniques

4.3.6.1 X-Ray Radiography

X-ray radiography is a nondestructive FA technique used to examine the interior details of a package of electronics components, before removing the cap. The technique is based on the ability of a material to block X-rays, which increases with its density. The varying densities of the materials included in a packaged component allow different amounts of X-rays to pass through, resulting in varying greyscale levels on the X-ray image. Some materials used in semiconductor assembly, such as aluminum wires, are transparent to X-rays and are therefore invisible in X-ray images. X-ray imaging may be obtained on film, by fluoroscopy or by using image-intensifying video systems.

Generally, X-ray radiography is used to inspect wire-bond problems, die-attach voids, package voids and cracks. Since X-rays are not easy to focus, this method produces low-resolution images, which greatly limits its usefulness. The modern X-ray systems use a microspot source, real-time detection and automated manipulation of the sample to achieve higher resolution and throughput.

Digital radiography is a variant of X-ray imaging, using digital X-ray sensors instead of traditional photographic film. This allows the whole process to be sped up (by removing the chemical processing and the ability to digitally transfer and enhance images) and diminishes the quantity of radiation needed to produce an image of similar contrast to conventional radiography.

4.3.6.2 X-Ray Microscopy

The X-ray microscope uses electromagnetic radiation in the soft X-ray band to produce images. Note that, unlike visible light, X-rays are invisible to the human eye and more difficult to reflect and refract. The working principle of the X-ray microscope is to detect an X-ray passing through a specimen, using a film or CCD camera.

The resolution of X-ray microscopy is higher than that of optical microscopy, but smaller than that of electron microscopy. Resolutions as high as around 30 nm (the best value was 15 nm, obtained by the XM-1 model, manufactured at Advanced Light Source, Berkeley Lab, CA) were obtained using the Fresnel zone plate lens, which forms an image with soft X-rays emitted from a synchrotron (or, more recently, with laser-produced plasma).

One advantage over conventional electron microscopy has to be underlined: X-ray microscopy can view biological samples in their natural state, without any operation needed to prepare the sample as in electron microscopy. However, modern electron microscopes (e.g. the cryo-electron microscope) are able to investigate biological specimens in their hydrated natural state.

An important application of X-ray microscopy is the generation of diffraction patterns, which are used in *X-ray topography*. The internal reflections of a diffraction pattern are analysed by a computer program and the three-dimensional structure of a crystal can be determined down to the placement of individual atoms within its molecules.

A modern technique is the micro-focus X-ray computed tomography (CT), which can be used to observe and evaluate tiny soldering ball joints in the state-of-the-art semiconductor package with spatial resolution as fine as one micron, allowing a 3D configuration of bonding-wire to be obtained [24]. Note that CT is also useful in reverse engineering (see Section 2.4 for details).

In fact, CT has proved to be the best tool for investigating high-density packaging semiconductors of ball-grid arrays (BGAs) and chip-scale packaging (CSP), which cannot be seen in detail by optical inspection of the solder joint. Moreover, the deformation and stress of the cutting associated with optical inspection make true FA extremely difficult.

4.3.6.3 X-Ray Photoelectron Spectroscopy

X-ray photoelectron spectroscopy (XPS, also called electron spectroscopy for chemical analysis, ESCA) uses low-energy (typically 1–2 keV) X-rays to knock out photoelectrons from atoms of the sample using the photoelectric effect. The energy content of the photoelectrons is analysed by a spectrometer and the originating elements are identified.

Basically, XPS can be used in the identification of compounds on the surface of a sample in order to have as narrow as possible X-ray line widths. Light elements, such as aluminium or magnesium, are used as X-ray source material. The X-rays penetrate deep into the specimen, but only the surface electrons can escape with sufficient energy for analysis. So XPS is mainly a surface-analysis technique, used to study organics, polymers and oxides, but also to resolve issues related to oxidation, metal interdiffusion and resin–metal adhesion.

4.3.6.4 X-Ray Fluorescence

X-ray fluorescence (XRF) is an FA technique that employs a primary beam of X-rays to excite a sample into emitting its own X-rays, which are collected and analysed to identify the composition of the sample. XRF has a similar detecting system using EDX and WDX, the basic difference being the primary beam. Moreover, because it uses X-rays and not electrons as primary beams, the advantage of XRF over EDX/WDX analysis is that it can be used to analyse layers that are charged or decomposed by electron bombardment. However, the large diameter of the XRF primary beam limits its spatial resolution.

4.3.6.5 X-Ray Reflectivity

X-ray reflectivity (XRR) is used to characterise surfaces, thin films and multilayers, being complementary with ellipsometry. The basic idea is to reflect a beam of X-rays from a flat surface, and to measure the intensity of X-rays reflected in the specular direction (i.e. the reflected angle is equal to the incident one). If the interface is not perfectly sharp and smooth then the reflected intensity will deviate from the ideal one predicted by the law of Fresnel reflectivity. The deviations are analysed and the density profile of the interface normal to the surface is obtained.

A new in-line metrology tool, which gathers XRR and XRF techniques to monitor film thickness, is presented in [25]. This method is able to cover a wide range of applications, from transparent to metal, ultrathin to thick layers. Comparative results obtained with XRR and TEM analyses for high/low-κ materials (SiOC) and for copper (Cu) barrier layers are detailed:

- For the SIOC film, the XRR spectrum (Figure 4.8) exhibits two critical angles for XRR: θ_{c1} corresponds to the low density of the film, while θ_{c2} is characteristic of the silicon substrate. The SiOC XRR spectrum can be modelled as a three-layer stack, the thin interfacial layer between the SiOC film and the silicon substrate probably being induced by a chemical reaction between the CVD precursors and the silicon-native oxide, and the dense top surface layer being unambiguously induced by the plasma treatment.
- For a Cu layer, if the width is greater than 250 nm, the XRR technique cannot be used due to X-ray absorption by the thick Cu layers. Therefore, XRF has to be used for in-line monitoring of barrier layer thickness. As shown in Figure 4.9a, if it is assumed that the ECD Cu density is constant over the whole Cu film thickness, the K_α Cu XRF intensity increases with the Cu film thickness. In Figure 4.9b, the Cu film thickness is monitored using XRR (2 seconds per point) and XRF (10 seconds per point). In the Cu film-thickness range from 50 to 200 nm, the K_α Cu XRF intensity varies linearly with the Cu film thickness measured using XRR. The K_α Cu XRF intensity exhibits an offset due to the non-negligible contribution of the L_α Ta XRF, which emits in an energy range close to the K_α Cu XRF. Note that XRF can be used to monitor the chemical-mechanical polishing (CMP) of Cu, by measuring the dishing and erosion of the Cu lines. The K_α Cu XRF spectrum allows voids in Cu layers to be detected, in order to tailor the ECD Cu process.

4.3.7 Spectroscopic Techniques

Originally, spectroscopy was used in optics, for the study of the visible light dispersed by a prism according to its wavelength. Later, the area was enlarged to cover any measurement of a quantity as a function of wavelength, frequency or energy. The basic instrument in spectroscopy is the spectrum of emitted or absorbed substances. Today, spectrometry is performed by a spectrometer, the goal being to assess the concentration or amount of a given chemical (atomic, molecular or ionic) species.

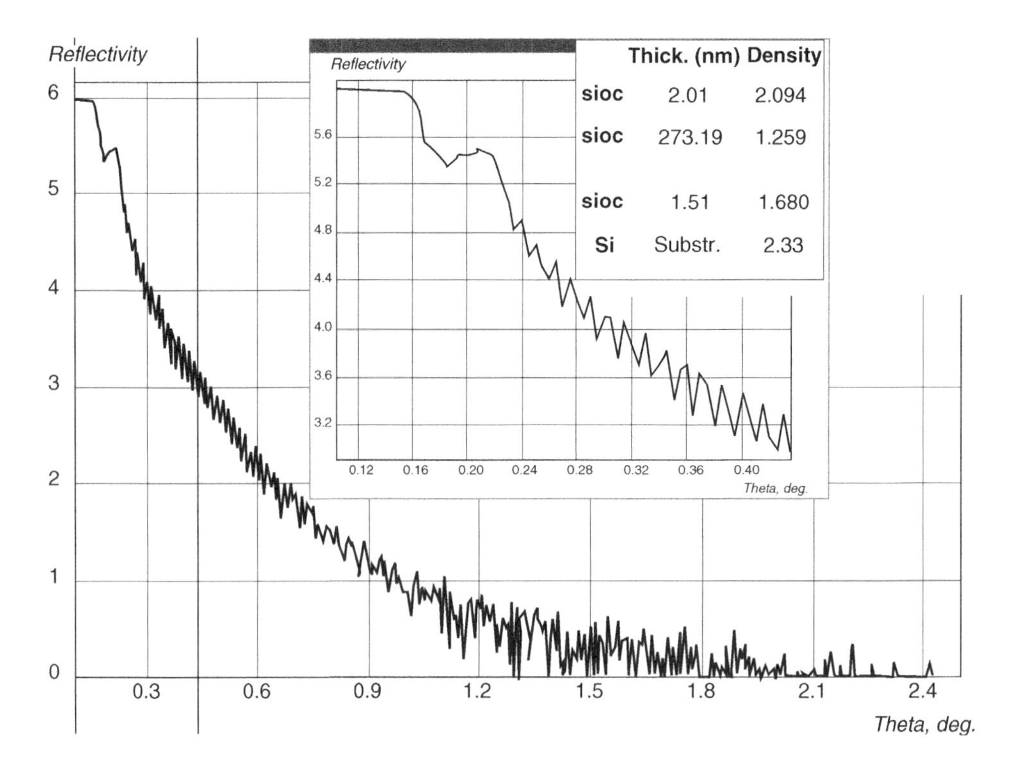

Figure 4.8 Experimental and modelled XRR spectrum for a 275 nm-thick CVD low film deposited on silicon. Reprinted from [25], Figure 4. Copyright 2004, with permission from Elsevier

In this section, the main spectroscopic techniques used in FA are detailed. Some spectroscopic techniques have already been presented, being linked to other groups of techniques: (i) DLTS, which is a spectroscopic technique, was presented as an electrical technique, because the electrical stimulus is essential; (ii) EDX was described in the SEM section, being an accessory instrument of SEM; and (iii) XPS was grouped with the other techniques that use X-rays as primary beams.

4.3.7.1 Auger Spectroscopy

Auger emission spectroscopy (AES) is used to identify the elements on the surface of a sample. Like EDX/WDX, in AES the primary beam is made of electrons. Unlike EDX/WDX, X-rays are not detected, but rather a certain class of electrons known as Auger[4] electrons. They are released by the following mechanism: an electron is ejected by the primary electron beam from its shell, for example the K shell. Another electron from an outer shell of the same atom, for example the L1 level, goes down to the K-shell position vacated by the ejected electron and emits energy in the form of a photon. This photon will either get lost or eject yet another electron from a different level, for example L2. This is called the Auger electron. The generation of an Auger electron requires at least three electrons, which in the example above were the K, L1 and L2 electrons. The emitted Auger electron is referred to as a KLL Auger electron. Obviously, atoms with less than three electrons (e.g. hydrogen and helium) are undetectable by AES.

[4] The Auger effect was discovered in 1922 by the French physicist Pierre Victor Auger. There is a controversy about this subject, because some voices claim that the Austrian-born, later Swedish scientist Lise Meitner was in fact the first person to report this effect.

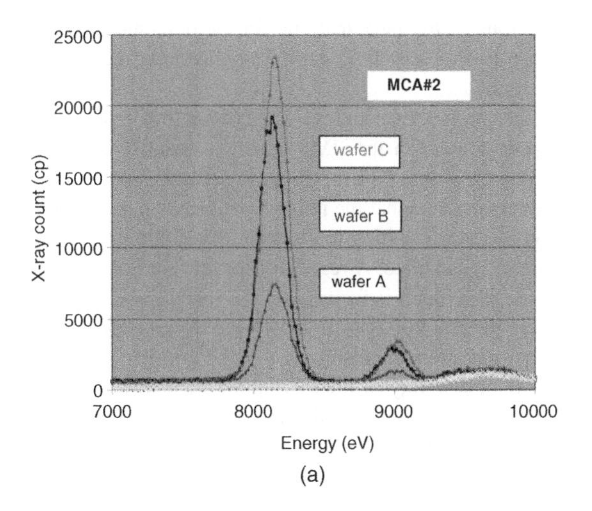

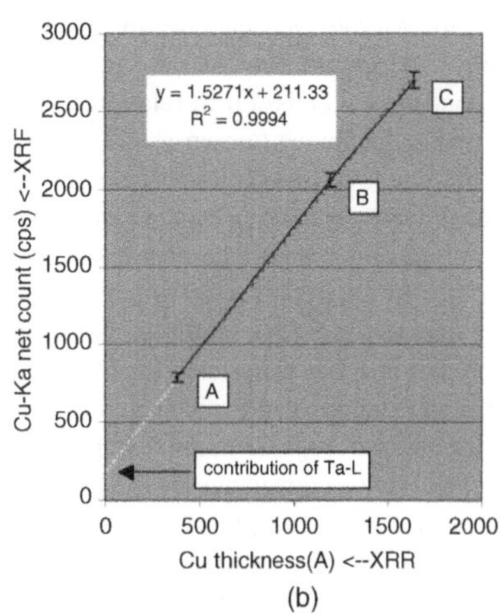

Figure 4.9 Complementary XRF and XRR analysis of a Cu layer with a width greater than 250 nm: (a) evolution of the K_α Cu XRF intensity with an increase in the Cu film thickness; and (b) correlation of K_α Cu XRF intensity with the Cu film thickness measured using XRR. Reprinted from [25], Figure 6. Copyright 2004, with permission from Elsevier

The information contained in the energy of the emitted Auger electrons (which is between 20 and 2000 keV) depends on the specific atom, allowing the element involved to be identified. The Auger electrons originate from the surface, or just beneath the surface (usually less than 50 Å), offering information on the composition in the form of peaks at Auger-electron energy levels, corresponding to the atoms from which the Auger electrons were released. The sensitivity is high, AES being able to detect elements with less than 1% of the atomic composition of the specimen. The main applications are in the detection of surface contaminants, corrosion failures, concentrations of various elements in the dielectric layer and so on. The main limitations are linked to the possible effect of the electron primary beam: charging up of insulator surfaces or damage to certain materials.

AES can also be used for compositional analysis in the depth of the sample, if a crater is milled, by ion-sputtering, on to the sample. Coupled with a raster scanner, the technique is called scanning Auger microscopy (SAM) and can be used to generate three-dimensional maps of the elemental distributions of a volume of the sample.

4.3.7.2 Secondar-Ion Mass Spectroscopy

Secondary-ion mass spectroscopy (SIMS) is used to identify the compositional analysis of a sample. The material is bombarded by a beam of ions with high energy (1–30 keV), resulting in the ejection or sputtering of atoms from the material. A small percentage of these ejected atoms are positively or negatively charged ions, which are referred to as 'secondary ions'. They are collected and analysed by mass-to-charge spectrometry, providing information about the composition of the sample, the elements being identified through their atomic-mass values. Quantitative data can be furnished by counting the number of collected secondary ions. Because the analysed material is removed from the sample by sputtering, SIMS is a locally destructive technique.

The primary ion beam must be carefully chosen, because this influences the detection limit. Oxygen atoms are used to sputter electropositive elements (those with low ionisation potentials, such as Na, B and Al) while caesium atoms are appropriate for sputtering negative ions from electronegative elements (e.g. C, O and As). The sputtered ions originate from shallow depths, so the sputtering of the sample has to be prolonged in order to reach deeper regions of the bulk material. In the dynamic mode of SIMS, depth profiling of the sample composition (up to 10 000 Å) can be obtained by monitoring secondary-ion emission in relation to sputtering time.

In the technique known as 'cluster SIMS', other molecular sources – the cluster ions – are used, leading to a considerable increase in secondary-ion signal. Cluster SIMS is able to identify compositional depth profiling in organic and polymeric systems, without the rapid signal decay that is typically observed under atomic bombardment [39]. The explanation is that the polyatomic beams tend to cause surface-localised damage with rapid sputter removal rates, resulting in a system at equilibrium, where the damage created is rapidly removed before it can accumulate. A review of the current literature on polymer analysis using cluster SIMS is offered in [39], which is focused on the surface and in-depth characterisation of polymer samples with cluster sources, but also discusses the characterisation of other relevant organic materials and basic polymer radiation chemistry.

The obvious advantage of SIMS over AES is the possibility of identifying all elements, including H and He. Elements present in very low concentration levels, such as dopants in semiconductors, can also be identified. The main drawback of SIMS is the high beam diameter (1–20 μm), because the sensitivity decreases if the beam diameter is reduced, as this diminishes the number of ions that are sputtered from the material for analysis.

A new variant of SIMS is the *'time-of-flight' secondary-ion mass spectroscopy* (TOFSIMS). The time-of-flight (TOF) analyser collects the secondary ions, and information about the concentration of a given material and how well it ionises can be obtained from the intensity of a given peak in a TOF spectrum. TOFSIMS is a perfect complement to XPS in analysing the surfaces. TOFSIMS gives elemental and chemical (primarily molecular fragments) information from within a sample depth of about 1 nm with parts-per-billion-level detection limits, whereas XPS gives both elemental and chemical (primarily oxidation state) information from within a sample depth of about 10 nm. The use of TOFSIMS in understanding the role of package-level contamination in the assembly world is detailed in [26]; applications for TOFSIMS in investigating silicon chip-level interface delamination failures due to package contamination and metal migration between the capacitor lands, surface of non-wetting BGA lands and assembly-related residual material analyses due to process excursions are also described.

4.3.7.3 Fourier Transform Infrared Spectroscopy

Fourier transform[5] infrared spectroscopy (FTIR) provides information about the chemical bonding or molecular structure of organic and inorganic materials. In FA, FTIR analysis is complementary to EDX analysis, capable of identifying unknown materials present in a specimen.

The technique is based on the ability of bonds and groups of bonds to vibrate at characteristic frequencies: a molecule exposed to IR rays absorbs infrared energy at frequencies that are characteristic to that molecule. In an FTIR analysis, the specimen is bombarded by a spot that is subjected to a modulated IR beam. The transmittance and reflectance of the specimen to IR rays at different frequencies is translated into an IR absorption plot consisting of reverse peaks. The resulting FTIR spectral pattern is then analysed and matched with known IR absorption plots of identified materials from the FTIR library.

The basic advantage of FTIR spectroscopy comes from the fact that oxygen and nitrogen do not absorb infrared rays, so no vacuum is needed for this analysis, because the air is perfectly transparent to IR rays. Moreover, small quantities of material (solid, liquid or gaseous) are needed for FTIR analysis. Even single fibres or particles are sufficient for material identification.

[5] The Fourier transform is a mathematic operation that transforms a complex-valued function of a real variable into another. It is named after the French mathematician and physicist Jean Baptiste Joseph Fourier (1768–1830).

4.3.7.4 Raman Spectroscopy

Raman[6] spectroscopy uses the inelastic (Raman) scattering of monochromatic light. Usually, the specimen is illuminated by a laser light, which interacts with phonons or other excitations in the system, resulting in the energy of the laser photons being shifted up or down and offering information about the phonon modes in the system: vibrational, rotational and other low-frequency modes.

In semiconductors, Raman spectroscopy is used to identify materials and to obtain information about phonon frequencies, the energies of electron states and electron–phonon interactions, carrier concentrations, impurity contents, compositions, crystal structures, crystal orientations, temperature and mechanical strain. Raman spectroscopy can be used for any semiconductor material: Si, Ge, GaAs, GaN and so on.

An important application of Raman spectroscopy is in stress measurements. In Figure 4.10 [27], a typical Raman spectrum of crystalline silicon is shown. The Gaussian curve indicates the Rayleigh-scattered plasma lines from the argon laser, which can be used for calibration. By monitoring the Raman scattering frequency of Si at different positions on the sample, a 'strain map' can be obtained. Two applications of Raman spectroscopy used for silicon are presented in [27]:

- **Stress measurements** – long silicon nitride (Si_3N_4) stripes with different widths are deposited on a Si substrate, which is under tensile stress because of the deposition process. These stripes compress the Si atoms in the substrate under them, causing compressive stress, and pull at the Si atoms next to the lines, causing tensile stress. In Figure 4.11, the measured shift of the frequency of the Si Raman peak when scanning along a line across the width of such stripes is shown. The phenomenon can be modelled and the full lines in Figure 4.11 show the result of a fit of this model to Raman spectroscopy data, taking into account experimental parameters such as probing-spot diameter and penetration depth. This procedure of fitting theoretical stress models to Raman data can be used for any model describing any device where Raman data can be measured.
- **Packaging** – when information on the variation of stress with depth has to be obtained, it is possible to cleave the sample, and polish if required, and to measure on the cross-section. In Figure 4.12, the results of an analysis by Raman spectroscopy of the stress induced in-chip during the packaging process are shown. A stress map, measured by Raman spectroscopy on the cross-section of an Si chip that was bonded to a copper substrate, is obtained. Stresses are induced in Si because there is

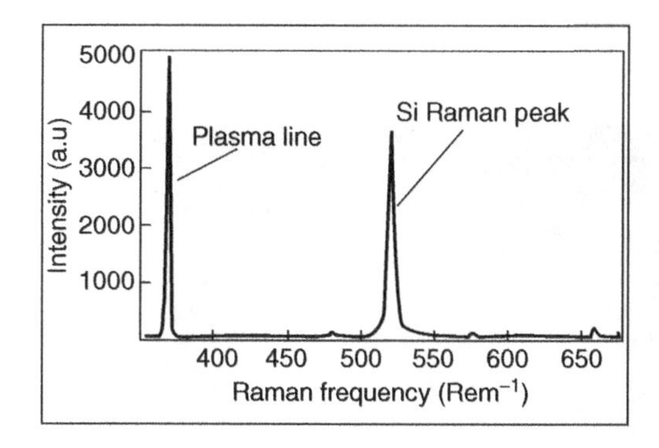

Figure 4.10 Typical Raman spectrum of crystalline silicon, measured using the 457.8 nm line of an argon laser. The Si Raman peak and plasma lines from the laser are clearly visible. Reprinted from [27], Figure 3. Reproduced by permission of John Wiley & Sons Ltd

[6] The Raman effect was discovered in 1928, for sunlight, by the Indian scientist Sir C.V. Raman.

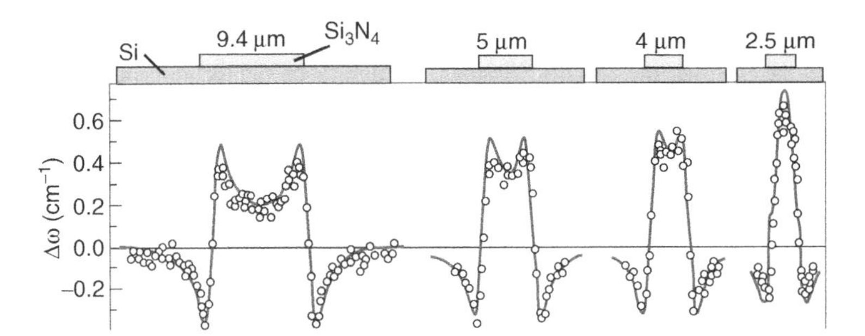

Figure 4.11 Shift of the frequency ($\Delta\omega$) of the Si Raman peak measured on nitride lines with different widths on Si substrate. The rectangles at the top indicate the positions of the lines. Reprinted from [27], Figure 4. Reproduced by permission of John Wiley and Sons Ltd

a difference between the thermal expansion coefficients of Si and Cu. In Figure 4.12, the stresses induced by the Cu substrate and by the solder bump on the top of the chip can be identified.

4.3.7.5 Atomic-Emission Spectroscopy

Atomic-emission spectroscopy is an FA method aimed at determining the quantity of an element in a specimen. The intensity of light emitted at a particular wavelength from a plasma, flame or arc is used. The wavelength of the atomic spectral line allows the element to be identified, while the intensity of the emitted light is proportional to the number of atoms in the element.

4.3.8 Acoustic Techniques

Scanning acoustic microscopy (SAM) uses sound waves to detect defects in encapsulated devices, such as: delaminations between various package interfaces, and faults in bonds, lead frames or die attach materials. SAM is a nondestructive FA technique. Information obtained from the interaction of a sound wave with a package is processed to image the package interior. There are two working possibilities: (i) pulse echo (processing the echoes sent back by the package) and (ii) transmission inspection (processing the sound wave at the other end of the package). The ultrasonic wave frequency used ranges from 5 to 150 MHz.

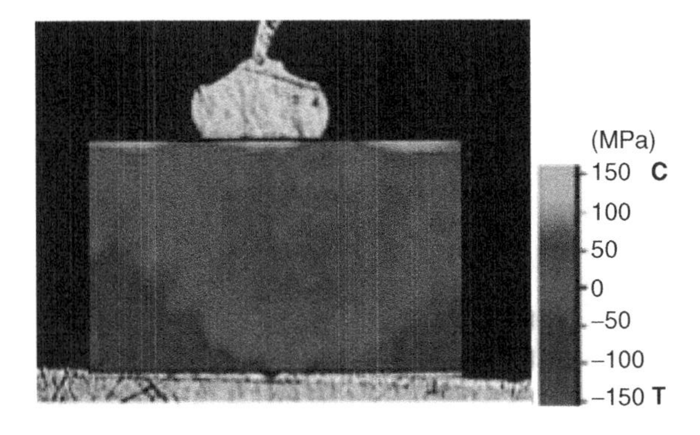

Figure 4.12 Stress in an Si chip bonded to a Cu substrate Reprinted from [27], Figure 7. Reproduced by permission of John Wiley and Sons Ltd

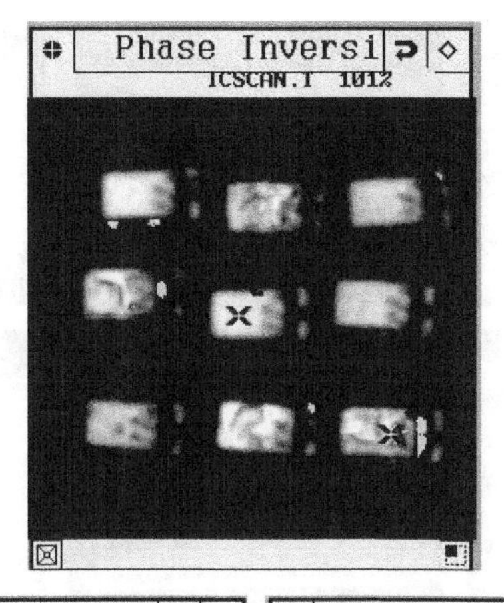

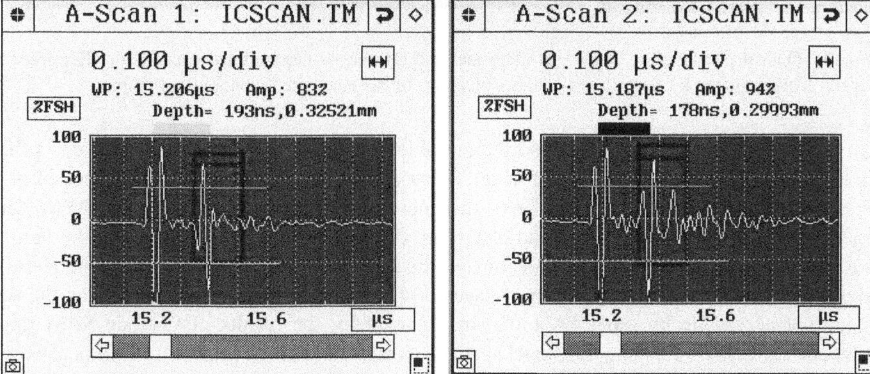

Figure 4.13 SAM images: (a) phase-inversion images of die surface; (b) A-scan image of an abnormal region; and (c) A-scan image of a normal region. Reprinted from [29], Figure 1. Reproduced by permission of the University of Science and Technology of China

SAM has several working modes [28]:

- *A-scan mode* is the real-time oscilloscope waveform of the acoustic signals based on the reflected echoes or collected at a single X–Y portion or point.
- *B-scan mode* furnishes a two-dimensional description along the test line (Y), through the cross-sectional display of the ultrasonic reflection of the various interfaces along the depth of the package or the acoustic data collected along the X–Z plane at depth A.
- *C-scan mode* furnishes a two-dimensional (area) description at a particular depth (Z), through the display of the image of reflected echoes at the focused plane of interest or the acoustical data collected along an X–Y plane at depth Z.

An example of the use of C-mode SAM to detect corrosion in ICs is given in [29]. After a 96-hour autoclave test, a batch of encapsulated ICs was investigated by C-mode SAM at 30 MHz. The basic assumption was that the acoustic impedance between corrosion and the normal area would be different. The presence of inhomogenities and discontinuities along the propagation paths of ultrasonic waves inside the matter caused modifications in the amplitude and polarity of ultrasonic waves. As can be seen in Figure 4.13b,c, the phase amplitude of oscilloscope images using A-Scan showed a great difference between abnormal and normal regions. Some of the ICs under suspicion of corrosion upon

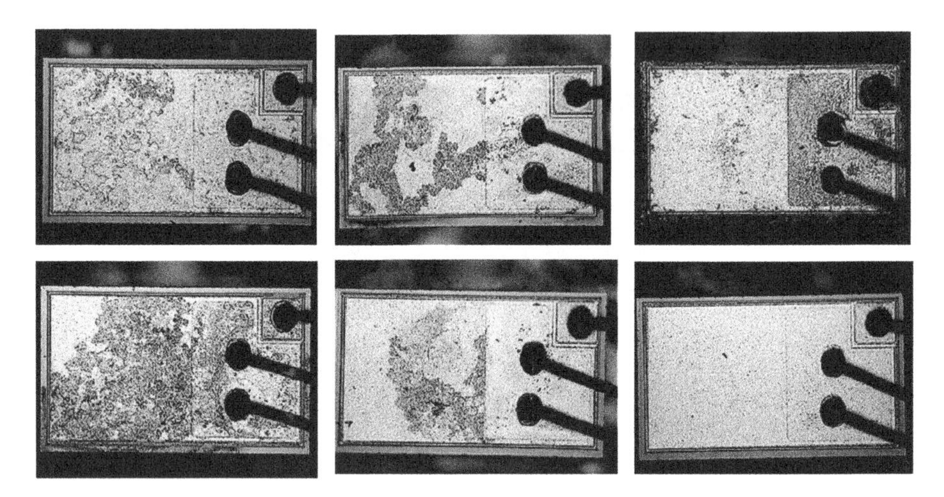

Figure 4.14 Optical photographs of (a)–(e) abnormal and (f) normal units after decapsulation. Reprinted from [29], Figure 2. Reproduced by permission of the University of Science and Technology of China

SAM investigation were decapsulated, and the possible effects of corrosion were observed in the die surface by optical microscope (Figure 4.14a–e), being compared with the good die surface of normal unit by SAM inspection (Figure 4.14f). This experiment confirmed the capability of SAM to identify the incidence of corrosion on the die of an electronic component, before decapsulating the item.

Obviously, SAM is the appropriate tool for investigating the bond process, offering the possibility of analysing inadequate cleaning of wafer surfaces or die attach faults, the key issue being the access to an area not accessible by other FA tools. In Figure 4.15, the results of C-mode SAM through transmission, at 30 MHz, for nine CSP packages are presented [30]. In these very-thin packages, the dice are positioned face-up and the wire bonds run from each die to a substrate attached to solder balls below that die. Moulding compound protects the face of the die and the sides of the package, so that the area of the package is slightly larger than the area of the die. The dark areas in the acoustic images are delaminations between the die-attach material and the die. Probably many of the delaminations are larger than their original sizes because they have expanded laterally as thin cracks.

4.3.9 Laser Techniques

4.3.9.1 Optical Beam-Induced Current

Laser beam-induced current (LBIC) uses a scanning laser beam to induce a current flow within a semiconductor specimen, the current being collected and analysed to generate images that represent the properties of the specimen and allow the detection or location of various defects or anomalies. An ultrafast laser beam scans the surface of the specimen, exciting some electrons into the conduction band through what is known as ‘single-photon absorption’, this absorption involving just a single photon exciting the electron into conduction. In doing this, the single photon has to carry enough energy to overcome the band gap of the semiconductor (1.2 eV for Si). The laser beam scans the specimen and produces current variations that are converted into contrast variations and form the image. An example is given in Figure 4.16.

Note that the use of single-photon absorption in modern ICs, with multiple layers of metal lines, may affect the uniformity in transmitting light through the top surface to the semiconductor layer. A solution to avoid this phenomenon is to perform the LBIC imaging from the IC backside, through the substrate. An even better solution is to use a ‘two-phonon absorption’, which involves two photons

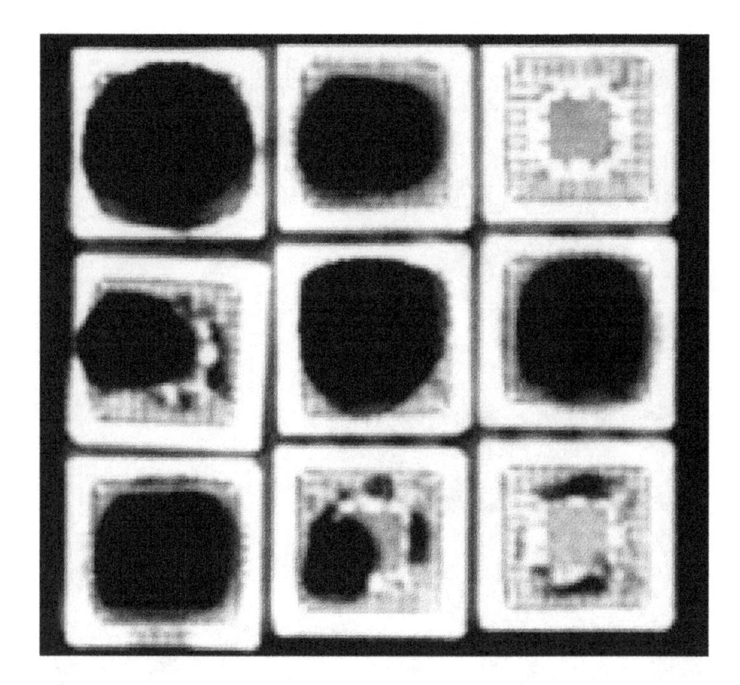

Figure 4.15 Analysis by C-SAM at 30 MHz of a group of nine CSP packages. Reproduced by permission of Sonoscan (Europe) Ltd

(with energies higher than half the band gap) arriving at the specimen at the same time in order to release an electron.

LBIC can be used for a large number of applications, such as: location of weak points of MOS transistors, detection of inter-level shorts and of recombination-generation centres, and detection of latch-up, a common FM of MOS transistors.

4.3.9.2 Confocal Laser Scanning Microscopy

Confocal laser scanning microscopy (CLSM) is used to acquire in-focus images from selected depths, in a process known as 'optical sectioning'. Coherent light emitted by the laser system passes through a pinhole aperture situated in a conjugate plane (confocal) with a scanning point on the specimen and a second pinhole aperture positioned in front of the detector (a photomultiplier tube). The laser beam is reflected by a dichromatic mirror and scanned across the specimen in a defined focal plane. Secondary fluorescence emitted from points on the specimen (in the same focal plane) passes back through the dichromatic mirror and is focused as a confocal point at the detector pinhole aperture. Images are acquired point-by-point and reconstructed by a computer, allowing three-dimensional reconstructions.

For opaque specimens, CLSM can be used for surface profiling, while for non-opaque specimens, a high-quality image of the interior structures can be obtained (better than the image obtained by simple microscopy because image information from multiple depths in the specimen is not superimposed). The principle of CLSM was originally patented by Marvin Minsky, in 1957, but commercial instruments were fabricated only after the development of appropriate lasers. The essential contribution of Colin Sheppard, a theory of image formation [31], must be mentioned.

CLSM has the ability to control depth of field, eliminating or reducing the background information away from the focal plane (which leads to image degradation), and to collect serial optical sections

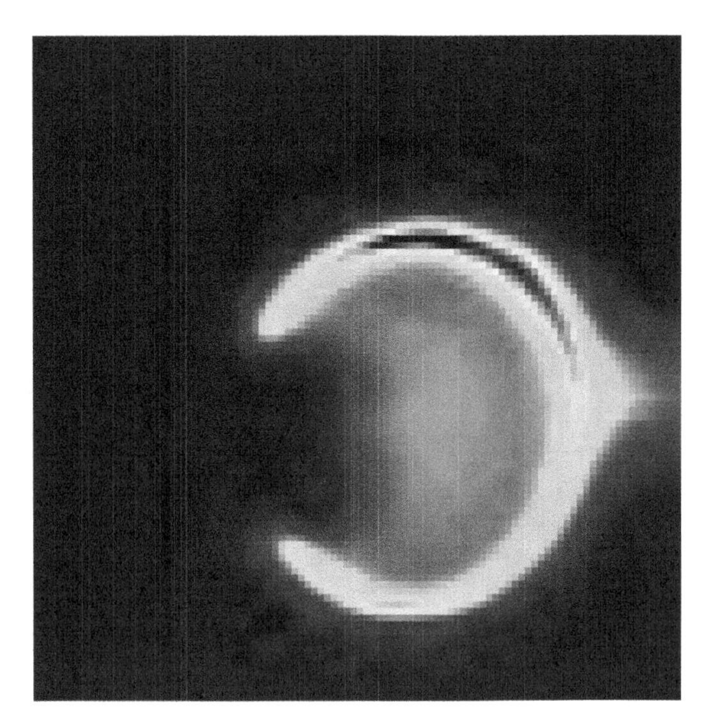

Figure 4.16 Image of the photocurrent in a photonic quantum-ring laser, obtained by laser beam-induced current (LBIC) in laser scanning microscopy. Reproduced by permission of Prof. George Stanciu (Centre for Microscopy, Microanalysis and Information Processing, University Politehnica, Bucharest)

from thick specimens. The basic key to the confocal approach is the use of spatial-filtering techniques to eliminate out-of-focus light or glare in specimens whose thickness exceeds the immediate plane of focus.

4.3.9.3 Laser Terahertz Emission Microscopy

Recently, a new FA technique has been proposed, in [32], which allows a non-contact inspection of defective interconnections in an LSI chip by using a laser terahertz emission microscope (LTEM).

LTEM measures the terahertz (THz)[7] emission images of an LSI chip by scanning it with femtosecond[8] laser pulses, which produces transient photocurrents flowing into interconnections and, eventually, emission of the THz pulse into free space. The characteristics of the THz emissions have been investigated using simple test element group samples, which consist of *pn* junctions connected to metal lines, and it seems the metallic lines connected to photo-excited *pn* junctions work as THz emission antennae, which enhance the emission efficiency of THz pulses near their resonant frequencies corresponding to the line lengths. This means that THz emission signals from *pn* junctions in circuits strongly depend on the structure of the interconnections. In [32], successful application of LTEM for the inspection of defective interconnections in MOSFET devices was reported: by comparing the THz emission images between a normal circuit and a defective one, it was possible to identify the *pn* junctions connected to the defective interconnections without electrical contacts.

[7] 1 THz = 10^{12} Hz.

[8] 1 femtosecond = 10^{-15} seconds.

4.3.10 *Holographic Interferometry*

Holographic interferometry (HI) is an FA technique based on superposing the light field on the 'live field', both scattered from an object. If a small deformation is applied to the object, the relative phase of the two light fields is altered and an interference is noticed. Hence, HI is able to measure strain and vibration analysis, as well as to conduct nondestructive testing, in order to attain optical interferometric precision (fractions of a wavelength of light). The fluid flow can also be visualised, by detecting the optical path length variations in transparent media.

4.3.10.1 Moir Interferometry

Moiré interferometry (MI) is a holographic interferometry FA technique, useful for testing the reliability of many different electronic packages. Historically, MI has been used for monotonic thermo-mechanical loading or steady-state condition-induced strain measurements. Its the application for fatigue testing is shown in [33]. Solder-joint fatigue mechanisms were experimentally observed by means of an MI system. In addition, SEM provided support for documented MI observation. The combination of these two techniques and the robustness of the MI system have been shown to be effective for the determination of the FMs of electronic packaging interfaces.

4.3.11 *Emission Microscopy*

4.3.11.1 Photoemission Microscopy

Photoemission microscopy (PEM), or light emission microscopy (LEM), aims to identify defect sites by detecting the emitted photonic radiation, primarily through carrier recombination mechanisms. During device operation, the defect sites emit light; such photoemissions, being very low-level, are not visible to the naked eye. An image-intensification technology (basically, a CCD camera and a computer) is needed therefore to amplify the light emitted by photo-emitting defect sites. The resulting radiation image is then overlaid with its corresponding die-surface image, such that the emission spot coincides with the precise location of the defect. PEM only identifies the position of the defect sites and other FA techniques are then used to investigate the physical anomaly responsible for the abnormal light emission. Note that the sites emitting light are not always the defect ones, so complementary FA techniques (e.g. microprobing) are needed to identify the FM.

The main applications of PEM are in detecting avalanche luminescence from junction breakdowns, from junction defects, from currents due to saturated MOS transistors and from transistor hot-electron effects. Especially at die level, PEM is the preferred procedure for biasing the device and collecting photons. At wafer level, this task becomes complicated because there are various dice on the reticule field and numerous reticule fields on the wafer. The solution is to perform a die location before wafer-level backside PEM: the wafer is powered by reverse-biased voltage to clamp the current appropriate to the device technology, and then PEM acquisition begins. The die with emission is the target [34].

4.3.11.2 Infrared Emission Microscopy

The infrared emission microscope (IREM) is used to locating hot-carrier-emission and thermal-emission sites in CMOS ICs.

In [35] the use of IREM with cryogenically cooled HgCdTe 256×256 NICMOS3 focal plane array, with a spectral response over the range 0.8–2.5 pm, has been reported for FA. This solution replaces the conventional CCD-based emission microscope (which has a limited spectral response of 0.4–1.1 pm), with the advantage in sensitivity gain due to HgCdTe array being emphasised. In order to enhance IREM detection efficiency and defect localisation resolution, sample-preparation techniques such as substrate global and local thinning, as well as wavelength-optimised anti-reflection coating, are also used.

Note that IREM can be used to localise not just statically excitable defects, but also functional defects excited by specific test vectors. An interesting application of IREM is in locating anomalies in solder bumps of flip-chip packages. In this respect, IREM may be a cheaper alternative to the real-time X-ray system [34].

4.3.11.3 Optical Beam-Induced Resistance Change

The optical beam-induced resistance change (OBIRCH) has the following basic idea: an infrared laser is scanned on the IC backside in order to generate localised thermal gradients in the IC's metallic elements. Such thermal gradients change the metal resistivity, which subsequently modifies the IC power consumption; when the infrared laser heats a metallic line, its resistance increases, since the thermal coefficient of variation of most metals is positive. Detecting and imaging these current or voltage variations at the pins of the ICs allows precise localisation of current-related defects. The heating and the resistance change of thin-metallic-strip lines is favoured over larger ones, so the OBIRCH technique is appropriate for localising microbridges between conductors [36]. OBIRCH can be combined with emission microscopy to perform backside FA. The major advantage is that both techniques can be used on the same tool.

4.3.12 Atom Probe

The atom probe (AP) combines two FA techniques (TOFS and FIM) to provide both one-dimensional compositional maps and three-dimensional maps of metal samples with near-atomic resolution. As in FIM, a sharp silicon tip (prepared by FIB milling) is placed on to the specimen. The atoms at the apex of the tip are ionised by a positive pulsed voltage (typically between 2 and 20 kV) and escape from the tip electrostatically, passing through the probe hole and reaching a detector. The mass-to-charge ratio of the ion is calculated by using a fast timing circuit to measure the time between the pulse and the impact of the ion on the detector, allowing the corresponding element to be identified (similar to the technique in TOFS). By collecting many such ions, the chemical profile of the sample can be created with a relative position accuracy of less than one atom spacing.

A typical weakness of the technique is that the very strong electric field near the apex of the tip induces mechanical forces at the tip, which eventually lead to a mechanical failure there. In [37], the typical FMs of silicon tips are analysed: (i) a deformation of the tip's apex due to an interaction between the oxidised amorphous layer and induced mechanical vibrations; (ii) a rip off of an isolating oxide layer; (iii) the rip off of a cap layer due to insufficient adhesion; and (iv) a failure of the tip in the course of the analysis due to the rising voltage applied there.

4.3.13 Neutron Radiography

Neutron radiography is an FA technique with a working principle that is similar to that of X-ray radiography, the difference being the use of neutrons instead of X-rays. Due to the difference between neutron and X-ray interaction mechanisms, the two techniques produce significantly different and often complementary information: X-ray attenuation is directly dependent on atomic number, and the neutrons are efficiently attenuated by only a few specific elements (e.g. organic materials and water are clearly visible in neutron radiographs because of their high hydrogen content, while many structural materials such as aluminum and steel are nearly transparent).

Neutron radiography is an important nondestructive technique, extensively employed in devices used in space programmes. It is able to detect cracks of 0.1 mm thickness. The ability to detect compounds containing hydrogen atoms is important in inspecting oil levels and insulating organic materials. Neutron radiography also facilitates the checking of adhesive layers in composite materials and surface layers (polymers, varnishes and so on). All types of O-rings[9] and joints containing hydrogen can be observed even through a few centimetres of steel [38].

4.3.14 Electromagnetic Field Measurements

Electromagnetic field (EMF) measurements check surrounding electromagnetic fields using EMF meters (which are antennas with different characteristics). In order to obtain a precise measure, the EMF meters should not perturb the electromagnetic field and must prevent coupling and reflection. Basically, there are *broadband measurements* (performed with a device that senses any signal across a wide range of frequencies) and *frequency-selective measurements* (performed with a field antenna and a frequency-selective receiver or spectrum analyser, allowing monitoring of the frequency range of interest).

4.3.15 Other Techniques

Beside the above 14 groups of techniques, which are mainly used for FA, there are many other characterisation methods that might be considered as additional tools for FA. There are techniques for measuring: the environment of the fabrication process or inside the package (e.g. **residual gas analysis**), the fluid quality (**chromatography**), the mechanical properties of the materials (such as the **fretting test**, **forced motion test**, **Split–Hopkinson pressure bar**, **micro hardness test**, **nano-indentation**, etc.) and others. Generally, these are not used on the flow of FA, but the information from such techniques forms part of the input data for FA. However, in some cases, the information furnished by these 'other techniques' is required by the logic flow of FA and can become important in identifying an FM.

References

1. Bhaumik, S.K. (2009) A view of the general practice in engineering failure analysis. *J. Fail. Anal. Preven.*, **9**, 185–192.
2. Dennies, D.P. (2002) Organisation of a failure analysis, *ASM Handbook*, vol. 11, ASM International, Materials Park, OH, pp. 324–332.
3. Sudolsky, M.D. (1998) The fault recording and reporting method. Poster presented at the Autotestcon'98 IEEE System Readiness Technology Conference, August 24–27, pp. 429–437.
4. Ishiyama, T.S. *et al.* (2006) Failure analysis technology for advanced devices. *NEC Tech. J.*, **1** (5), 47–51.
5. Valett, D. (2002) Failure analysis requirements for nanoelectronics. *IEEE Trans. Nanotechnol.*, **1** (3), 117–121.
6. Antoniuou, N. *et al.* (1999) Die backside FIB preparation for identification and characterisation of metal voids. International Symposium for Testing and Failure Analysis, Santa Clara, November 14–18.
7. Anon. *Failure Analysis of Integrated Circuits*, Trion Technology, http://www.crystec.com/trianae.htm. (Accessed 2010).
8. Souchon, F. (2010) RF MEMS Switches Reliability: Study of the Influence of the Nature of the Gaseous Species Trapped Under the MEMS Packaging, CEA-LETI Internal Report.
9. Toh, S.L. *et al.* (2008) In-depth electrical analysis to reveal the failure mechanisms with nanoprobing. *IEEE Trans. Device Mater. Reliab.*, **8** (2), 387–393.
10. Băjenescu, T. and Bâzu, M. (2010) *Component Reliability for Electronic Systems*, Artech House, Boston and London.
11. O'Connor, P.D.T. (2002) *Practical Reliability Engineering*, John Wiley & Sons, Ltd, Chichester, New York.
12. Nigh, P. and A. Gattiker (2004) Random and systematic defect analysis using IDDQ signature analysis for understanding fails and guiding test decisions. Poster presented at the ITC International Test Conference, Paper 11.3, pp. 309–318.
13. O'Connor, R., *et al.* (2008) SILC defect generation spectroscopy in HfSiON using constant voltage stress and substrate hot electron injection. Poster presented at the 46th IEEE Annual International Reliability Physics Symposium, Phoenix.
14. Odegard, C. and Lambert, C. (2000) Reflectometry techniques aid IC failure analysis. *Test Meas. World*, **1,** www.gigaprobestek.com/images/TDR_tecniques_aid_IC_FA_byTexas_Instruments.pdf. "*Comparative TDR*

[9] An O-ring is a piece of elastomer with a disc-shaped cross-section, used to seal an interface by sitting in a groove and being compressed during assembly between two or more parts.

Analysis as a Packaging FA Tool," by Odegard, C. and C. Lambert, *Proceedings of the 25th International Symposium for Testing and Failure Analysis*, ASM International, Materials Park, OH. November, 1999. pp. 49–55.
15. Ancona, M.G., Saks, N.S. and McCarthy, D. (1998) Lateral distribution of hot carrier-induced interface traps in MOSFETs. *IEEE Electron Device Lett.*, **35** (12), 2221–2228.
16. Aichinger, T. and Nelhiebel, M. (2008) Advanced energetic and lateral sensitive charge pumping profiling methods for MOSFET device characterisation – analytical discussion and case studies. *IEEE Trans. Device Mater. Reliab.*, **8** (3), 509–518.
17. Jaeger, R.C. *et al.* (1968) Record of the 1968 Region III IEEE Convention, pp. 58–191.
18. Roedel, R. and Viswanathan, C.R. (1975) Reduction of popcorn noise in integrated circuits. *IEEE Trans. Electron Devices*, **ED-22**, 962–964.
19. Yamamoto, S. *et al.* (1971) On perfect Crystal Device Technology for Reducing Flicker Noise in Bipolar Transistors, Colloques Internat. du CNRS No. 204, pp. 87–89.
20. Zayats, A. and Richads, D. (2009) *Nano-Optics and Near-Field Optical Microscopy*, Artech House, Norwood, MA.
21. Burgess, D. Failure Analysis Services, http://www.muanalysis.com/services/Failure-Analysis-Microelectronics-Reliability-Testing-Lab.php. (Accessed 2010).
22. Nokuo, T., Eto, Y. and Marek, Z. (2007) A new method for failure analysis with probing system based on scanning electron microscope. Poster presented at the 45th Annual International Reliability Physics Symposium, Phoenix.
23. Tracy, B., Alberi, K. and Tabrez, S. (2007) Adopting Low-voltage STEM and Automated Sample Prep to Perform IC Failure Analysis, Micromagazine.com, http://www.micromagazine.com/archive/04/07/tracy.html. (Accessed 2010).
24. Hirakimoto, A. (2007) *Microfocus X-ray Computed Tomography and it's Industrial Applications*, SA Tech Inc., www.satech8.com. (Accessed 2010).
25. Wyon, C. *et al.* (2004) In-line monitoring of advanced microelectronic processes using combined x-ray techniques. *Thin Solid Films*, **450** (1), 84–89.
26. Pathangey, B. *et al.* (2007) Application of TOFSIMS for contamination issues in the assembly world. *IEEE Trans. Device Mater. Reliab.*, **7** (1), 11–18.
27. De Wolf, I. (2003) Raman spectroscopy: about chips and stress. *Spectrosc. Eur.*, **15** (2), 6–13.
28. Barton, J.J., Compagno, A. and O'Mathuna, C. (2001) Failure analysis of advanced packaging technologies using scanning acoustic microscopy, in *In-Line Characterisation, Yield, Reliability, and Failure Analysis in Microelectronic Manufacturing II*, Proceedings of SPIE, Vol. **4406** (eds G. Kissinger and L.H. Weiland), pp. 31–40.
29. Ma, L.-L. *et al.* (2007) IC surface corrosion inspection by C-mode scanning acoustic microscopy. *J. Electron. Sci. Technol. China*, **5** (4), 336–339.
30. Anon. CSP Die Attach Delamination, Through Transmission, Application Note 657, Sonoscan, http://www.sonoscan.com/resources/appnote-0148.html. (/Accessed 2010).
31. Sheppard, C.J.R. and Wilson, T. (1979) Effect of spherical aberration on the imaging properties of scanning optical microscopes. *Appl. Opt.*, **18**, 1058–1065.
32. Yamashita, M. *et al.* (2009) Laser THz emission microscope as a novel tool for LSI failure analysis. *electron. Reliab.*, **49** (9–11), 1116–1126.
33. Liu, H. *et al.* (2002) Moiré interferometry for microelectronics packaging interface fatigue reliability, *Proceedings of GaAsMANTECH*, GaAsMANTECH Inc., http://www.csmantech.org/Digests/2002/PDF/08f.pdf. (Accessed 2010).
34. Tan, H.S. and Luo, K. (2002) New applications of the infrared emission microscopy to wafer-level backside and flip-chip package analyses. *IEEE Integr. Reliab. Workshop*, **49**, 147–150.
35. Loh Ter Hoe, Y., Mun, W. and Yan, C.Y. (1999) Characterisation and application of highly sensitive infra-red emission microscopy for microprocessor backside failure analysis. Poster presented at the Proceedings of 7th IPFA '99, Singapore, pp. 108–112.
36. Beaudoin, F. *et al.*(2001) Current leakage fault localisation using backside OBIRCH. Poster presented at the Proceedings of 8th IPFA 2001, Singapore, pp. 121–125.
37. Kölling, S. and Vandervorst, W. (2009) Failure mechanisms of silicon-based atom-probe tips. *Ultramicroscopy*, **109**, 486–491.
38. Anon. *Neutron Radiography: Non Destructive Testing*, CEA-LETI, http://www-llb.cea.fr/neutrono/nr1.html. (Accessed 2010).
39. Mahoney, C. M., "Cluster Secondary Ion Mass Spectrometry of Polymers and Related Materials," Wiley InterScience, January 22, 2009.

5

Failure Analysis – What?

The previous chapters have shown *why* failure analysis (FA) is needed, *when* (during the product life cycle) FA must be used and *how* to use FA (main procedures and techniques). Obviously, the last question that needs to be answered is: *what* are the results of using FA? In other words, the failure mechanisms (FMs) identified by FA will be detailed. All the FA methods reported in this chapter were described in Chapter 4.

First we have to remind you (as mentioned before, in Section 2.2.2) of the basic distinction between failure mode (FMo) and FM, because very often these two terms are used interchangeably in the literature, which is wrong.

An FMo is the external symptom of the failed product: for example, a bipolar (Bi) transistor has a short circuit between emitter and collector. Why has this happened? Because a physical process (e.g. a diffusion pipe) led to failure, which is the FM in this case. So, the totality of physical and chemical processes leading to failure composes the FM.

In all cases, an FA must start from an FMo, leading to identification of the FM, because [1]:

- Corrective actions for improving reliability by minimising (or avoiding) this FMo can be elaborated only if the FM is well understood. A reliability growth programme is needed, coupled with management strategy change so that more FMos identified during testing actually receive a corrective action instead of a repair [2].
- FM identification is essential for the reliability of accelerated testing because:
  - FMs produced by high-level stress cannot be different to those observed during actual service conditions;
  - The obtained degradation laws must be extrapolated beyond the time period of the test and the extrapolation must be made separately for each population affected by an FM.

We must also add that some FMs may induce another FM (e.g. corrosion can lead to sticking/binding). Due to the presence of two FMos, the thermal cycle statistical behaviour is described by a bimodal lognormal failure distribution. There are age-related FMs (such as hot-carrier or metal migration) that limit the long-term performance of the semiconductors; all these FMos are far worse at geometries below 0.25 μm. A methodology based on an early detection system has to be developed, correlated with a library of the cells available for semiconductor designers to include on their designs, in order to provide a measure of safety and assurance that the deployed systems will maintain robust performance while in service [3].

Failure Analysis: A Practical Guide for Manufacturers of Electronic Components and Systems, First Edition.
Marius I. Bâzu and Titu-Marius I. Băjenescu.
© 2011 John Wiley & Sons, Ltd. Published 2011 by John Wiley & Sons, Ltd.

Obviously, this distinction between FMo and FM also works at system level, where an FMo is the wrong operation of one circuit in the system and an FM (identified by FA at system level) might be a design fault, or the failure of a component of this circuit. If this FA is continued at component level, a new FMo is noticed, which is the former FM at system level. So, to find the root cause of a failure, one must go deeper, to the component level, and not stop at the system level. Consequently, the 'real' FMs are only those found at component level. This can be seen by the fact that FA at component level provides detailed explanations not only about the structural degradation processes of the device, but also about the degradation issues of the materials used to build the device (degradations during device processing or later, during device operation). So not only the component level but also the material level is covered implicitly. *This is the reason why in the following only FMs at components level are detailed.*

FMs can be grouped in various ways, depending on the purpose of the analysis performed. According to where in the component they develop, Hakim proposed the following classification [4]: bulk, interface, oxide, metallisation and packaging. Bulk material and interface properties usually define the intrinsic reliability characteristics, while defects establish the extrinsic reliability characteristics.

In this chapter, the strategy in presenting the FMs will be the following. First, those FMs that are more or less common to all electronic component technologies (e.g. electromigration (EM), hot electrons, corrosion, etc.) will be detailed, grouped around the process steps where the specific failure risk might be initiated. Then, for each main technology for manufacturing electronic components, the typical FMs will be detailed, which are directly related to the applications of the given family of electronic components.

5.1 Failure Modes and Mechanisms at Various Process Steps

The reliability of a product is built during the manufacturing process. This statement seems obvious today, but not many years ago there was a strong belief that reliability was something that could be 'added' to a final product by a clever combination of tests for reliability selection. In real life, in order to achieve a product with a given reliability level, one has to take reliability issues into account even at the design phase and continue to monitor them during the whole fabrication process, because the FMs are initiated at the process steps. Once the final product has been obtained, the only thing that can be done is to eliminate the weak items from a batch of components by an appropriate reliability selection programme (e.g. burn-in). This will improve the reliability of the whole batch, but not the reliability of each item. The only way to produce a component with a high intrinsic reliability is to avoid the FMs initiated during the manufacturing process.

The manufacturing activity of an electronic component has two main phases: the wafer-level ('front-end') process and the assembly ('back-end') process. The failure risks that may arise at the specific process steps of these two phases are detailed below.

5.1.1 Wafer Level

There are two reasons why wafer is the favourite target for FA: (i) many interesting physical and chemical phenomena are involved in component degradation and failure and (ii) the wafer is a collection of reliability risks, hence almost any process step may induce a new FM or increase the action of FMs induced by previous steps. This is why a huge quantity of literature has been written on this subject and various techniques for wafer-level FA have been developed. For example, optical microscopy, scanning electron microscopy (SEM), transmission electron microscopy (TEM), voltage contrast, conductive atomic force microscopy (C-AFM), energy-dispersive X-ray spectroscopy (EDX) and Auger microscopy are all used in sample inspection, where the samples are prepared by parallel lapping, focused ion beam (FIB) cutting, x-sectioning, wet etching or HDP dry etching [5].

Since the 1970s, test structures[1] have been used for process monitoring. First, the process is qualified using traditional lifetime tests. Then wafer-level reliability (WLR) fast tests can be used to provide a so-called statistical signature of the process reliability. If the WLR measurements on product wafers show a similar distribution to those observed on the qualification lots, it can be assumed that the reliability of the devices on the wafer will be statistically similar to that observed during the qualification tests. However, if this is not the case then some other FM is probably present and the only way of ensuring reliability is to re-test the batch, using traditional reliability measurements [6].

The WLR concept was refined in the WLR Workshop, organised annually since 1982 by the Technology Associates. Tools to investigate the reliability risks at the wafer level and to monitor the process factors affecting reliability were created. Two examples (from a multitude) of extended services in the field: (i) Coventor launched *Coventor Catalyst*, which is a design for manufacturability (DfM) methodology for developing micro electromechanical systems (MEMS) to ensure optimal manufacturing yield [7]; (ii) Dolphin Integration issued a catalogue of test-structure generators, branded RYCS generators, offering a wide range of test-structure kinds [8].

WLR aims to detect potentially unreliable devices and to avoid delivering them to the customer. This requires appropriate measurements done quickly, so that potentially every wafer can be tested, and also that the measurement does not damage the adjacent product devices. Some examples are: hot-carrier measurements, passivation and interconnect dielectrics, gate oxide integrity [9], stress migration, via voiding, current crowding and junction spiking. The four main WLR tests are described in Table 5.1.

The typical FMs induced by various process steps for the fabrication of an electronic component are presented below. They have been grouped around the four most important subjects from the standpoint of reliability: semiconductor-related FMs, oxide-related FMs, metal-related FMs and FMs related to other wafer-level processes (diffusion, implantation, etc.).

5.1.1.1 Semiconductor

A semiconductor is a material that has an electrical conductivity between that of a conductor and that of an insulator, which is generally in the range 10^3–10^{-8} S/cm. It is an appropriate starting material for modern electronic components.

Today, silicon (Si) is still the most important semiconductor support for electronic components. A huge quantity of scientific work has been reported on this material over the last 40 years. However, other semiconductor materials are being used more and more, such as gallium arsenide (for very high-frequency components) and indium gallium arsenide, gallium phosphide and lead sulfide (for optoelectronic components).

As a starting material in most microchip fabrication, single-crystal Si is processed in wafers, with a diameter that has increased steadily from less than 50 mm in the 1970s to 300 mm today and possibly to 450 mm in the near future (2012). Wafer flatness is an important feature, because it directly impacts device line-width capability, process latitude and yield. The polishing process for Si wafers plays a key role in the fabrication of semiconductors, since a globally planar, mirror-like wafer surface is needed to obtain a good device. The surface roughness of the wafer depends on the surface properties of the carrier head unit, together with other machining conditions such as working speed, type of polishing pad, temperature and down force [10]. Cracks or scratches on the surface of the wafer (usually caused by improper handling) may generate open or short circuits in the electronic components. A careful visual inspection is needed to monitor the surface quality during processing and before sealing. The non-uniformity of the resistivity across the wafer is also a failure risk, because the characteristics of the devices may differ depending on the area of the wafer.

The most important characteristic of any semiconductor, as a starting material for a future electronic component, is the number of **crystallographic defects**, which interrupt the regular pattern of atomic

[1] Test structures are chips (or parts of chips) specially designed to allow the identification of some FMs; they are processed under exactly the same conditions as the manufactured product.

Table 5.1 Main WLR tests for general applications

Test	Possible FM	Test description
Contact-integrity test	Aluminum spiking: silicon dissolution into an overlying Al film, followed by Al penetration at contact windows during contact annealing. The dissolved Si precipitates out during cooling, either as Si islands on oxide strips, or as epitaxial Si at contact windows. May result in a total contact occlusion.	Test structures are stressed at high temperature and contact degradation is evaluated electrically at room temperature and in the dark. Contact spiking can be revealed by a severe increase in leakage current, while contact occlusion will reveal an increase in current resistance.
Hot-carrier-injection test	For MOS devices, if large electric fields are applied, carriers can gain sufficient energy (namely hot carriers) to create electron–hole pairs by impact ionisation on the Si atoms. Some of the carriers are injected into the gate oxide and may induce interfacial and/or bulk oxide charges. A degradation of the electrical parameters of the device (mobility, threshold voltage and drain current) occurs.	MOSFET parameters are measured, which are directly related to the capability of the process to generate and trap hot carriers. It is also possible to perform an accelerated stress test under the most severe conditions and to compare the result with the characterisation work performed during qualification.
Metal-integrity test	Electromigration (EM) in thin-film interconnection lines: the electron/ion flux induced in metal tracks by high-current densities. The degradation of the conductor is due to the agglomeration of vacancies, which can result in a void through the metal track or in the local accumulation of metal atoms, which can cause a short circuit between adjacent conductor layers.	Wafer-level test SWEAT (standard wafer-level EM accelerated test), based on the acceleration of EM in the metal line by means of a high current density: the metal line is subjected to a constant current stress defined by a chosen acceleration factor and by the maximum temperature that the metal line can reach without activating other diffusion mechanisms.
Oxide integrity test	Oxide dielectric breakdown occurs when sufficient charge is injected into the oxide by forcing a current through the dielectric or by applying a high electrical field. The damage produces structural changes (traps or interface states), which lead to a low-resistance path through the oxide layer and result in a permanent leakage of the dielectric.	For gate-oxide applications, constant voltage stress (CVS) or linear voltage ramp stress (LRVS) are used. For tunnel-oxide applications, current stresses are used, such as constant current stress (CCS) or exponential ramped current stress (ERCS).

arrangement. Basically, the existence of defects in semiconductors is detrimental to the future devices; there are, however, some beneficial effects of electrically active defects, which may be introduced deliberately with the aim of improving the characteristics of the semiconductor. This is called 'defect engineering'. One example is the intentional introduction of oxygen to reduce the radiation-induced formation of electrically active defects [11], which improves the radiation tolerance of the device. Carbon has an adverse effect, as modelled and discussed in [12].

In Table 5.2 the possible FMs induced by crystallographic defects are shown. Some of these FMs can be avoided by employing a rigorous input control on the semiconductor wafers, using the detection methods indicated in the table. However, some of the defects mentioned are induced in the semiconductor structure by various phases of processing (oxidation, implantation, diffusion, etc.).

Table 5.2 FMs induced by crystallographic defects

Class of defect	Defect type	Detection methods	Effects (possible FMs)
Point defects	*Vacancies* – sites that are usually occupied by an atom but that are unoccupied. If a neighbouring atom moves to occupy the vacant site, the vacancy moves in the opposite direction. *Interstitials* – atoms which occupy a site in the crystal structure at which there is usually not an atom. *Frenkel defect* – a nearby pair of a vacancy and an interstitial. This is caused when an ion moves into an interstitial site and creates a vacancy. *Clusters (swirls)* – formed between different kinds of point defect (e.g. if a vacancy encounters an impurity, the two may bind together if the impurity exists as *extrinsic point defects* – more critical than intrinsic point defects, these involve foreign atoms, which usually come from dopants, oxygen or carbon – or as *dislocations* – linear defects around which some of the atoms of the crystal lattice are misaligned; these fall into two types: (i) edge dislocations (caused by the termination of a plane of atoms in the middle of a crystal) and (ii) screw dislocations (more difficult to visualise, these basically comprise a structure in which a helical path is traced around the linear defect (dislocation line) by the atomic planes of atoms in the crystal lattice and metals too large for the lattice)).	Photoluminescence [13], optical microscopy [14], preferential chemical etching and X-ray topography [15]	The presence of point defects is important in the kinetics of diffusion and oxidation. The rate at which diffusion of dopants occurs is dependent on the concentration of vacancies. This is also true for oxidation of silicon.
Line defects	*Dislocation loops* – these occur if a dislocation consists of an extra plane of atoms (or a missing plane of atoms) lying entirely within the crystal. The dislocation line of a dislocation loop forms a closed curve that is usually circular in shape, since this shape results in the lowest dislocation energy.	Transmission electron microscopy (TEM), field ion microscopy (FIM), atom-probe techniques, deep-level transient spectroscopy (DLTS)	Play a major role in the fatigue crack initiation phase. May also serve as sinks for metallic impurities and disrupt diffusion profiles (this may play a beneficial role in the removal of impurities from the wafer (gettering).

(*continued overleaf*)

Table 5.2 *(continued)*

Class of defect	Defect type	Detection methods	Effects (possible FMs)
Area defects	*Grain boundaries* – regions where the crystallographic direction of the lattice abruptly changes. This usually occurs when two crystals begin growing separately and then meet.	High-resolution X-ray diffraction, Nomarski microscopy, atomic-force microscopy (AFM), transmission electron microscopy (TEM), scanning electron microscopy (SEM), high-resolution electron microscopy (HREM), Raman and photoluminescence spectroscopy	Most common in EFG Si ribbons, used for manufacturing solar cells. General effects: (i) reduce the short-circuit current density as a result of a reduction in the minority carrier lifetime and (ii) reduce the voltage and the fill factor as a result of the introduction of a high density of space-charge recombination centres [16].
	Twins – electrically quiescent defects that do not introduce large stresses, and consequently do not accumulate impurities [16].		Oxidation-induced stacking faults (OSFs) may increase the reverse current in *pn* junctions, degrade the breakdown voltage and reduce the gain of bipolar devices.
	Stacking faults – typical defects induced by process phases (e.g. oxidation). The kinetics of growth is related to the local concentration of point defects. These defects are sinks for impurities [16].		
Bulk (volume) defects	*Voids* – small regions where there are no atoms; can be thought of as clusters of vacancies.		Typical defects induced by various process steps: voids by metallisation, precipitates by diffusion.
	Precipitates – where impurities cluster together to form small regions of a different phase.		

5.1.1.2 Passivation

Among the solutions for passivation, oxidation is the earliest. Today it is one of the basic steps in the fabrication of any active electronic component, for both Bi and unipolar metal-oxide semiconductor (MOS) technology. Because silicon is still the best choice among semiconductors for electron technology, silicon dioxide (SiO_2), with a relative dielectric constant of $k = 3.9$, is the material of choice for inter-layer dielectric (ILD) applications. Consequently, many of the silicon-based technologies that have evolved are based on the deposition, patterning and removal of SiO_2 during processing. This material has remarkable properties for device fabrication, including exceptional thermal, electrical and mechanical properties. In almost every important field (thermal stability, breakdown voltage, leakage current, mechanical properties, etc.), except for dielectric constant, the properties of any low-k materials (with k lower than SiO_2 1) may be expected to be inferior to SiO_2.

However, as device sizes decrease and device densities increase, lower dielectric materials become necessary for passivation, in order to prevent crosstalk between conductor lines and signal delays in the back-end-of-the-line (BEOL) interconnect wiring. Two important low-k candidates for the replacement of SiO_2 are organic polymers and silicates [17]. However, it seems semiconductor manufacturers have chosen inorganic-like materials, with which they had experience in oxide insulators. Consequently, Si-Me-containing organosilicate materials ($k = 2.7-3.0$) have been developed and are now manufactured for the current 90 nm technology [18]. Note that the methyl groups (necessary for the provision of low dielectric constant properties and hydrophobicity) reduce density, modulus, hardness and fracture energy [19].

Among the possible failure risks at passivation, one may note the formation of cracks or pinholes (which may lead to electrical breakdown and short circuit) and the non-uniformity of film thickness (leading to the lowering of breakdown voltage and the increasing of leakage current). The crystallographic defects also represent significant failure risks. The two most significant defects induced into silicon wafers during oxidation are dislocations and stacking faults [16].

A large variety of FMs can be induced by oxidation. Three typical ones are given in Table 5.3 and will be detailed below. Hot carrier injection (HCI) is one of the causes of interface state generation, but we have treated this FM separately, because it is one of the most important FMs in today's MOS devices.

Interface State Generation

The imperfections in the oxide (e.g. mobile ions Na^+ and/or K^+, fixed oxide charge or oxide traps), which are the result of poor fabrication processes, may represent a significant danger to the oxide as an efficient dielectric, because they can initiate interface states. The induced charge at the oxide–silicon interface and in oxide may alter the flat-band voltage. There are at least four distinct types of charge in the oxide–silicon system: (i) fixed-interface charge, (ii) oxide-trapped charge, (iii) interface-trapped charge and (iv) mobile charge resulting from alkali-metal ions, particularly sodium. But the interface states at the silicon–oxide interface can be generated even by the necessary electrical stress tests. There is still slight stress in the film, and dangling silicon bonds are usually passivated by hydrogen atoms and are not electrically active. The density of unpassivated silicon dangling bonds is negligibly low, below $10^{10}\,cm^{-2}$. However, under electrical stress, the weak Si-H bonds may break and interface states are created.

Negative-bias temperature (NBT) stress is another important cause of interface-state generation. The transistor degradation under this mode is called negative-bias temperature instability (NBTI) and occurs even when the circuit is in quiescent, if the *p*MOSFET happens to have its gate tied to high voltage. This is one of the most serious reliability concerns of all [20].

Hot-Carrier Effects

The term 'hot-carrier injection' describes the phenomenon by which the carriers (electrons or holes) under intense electric fields gain sufficient kinetic energy to overcome a potential barrier and be

Table 5.3 Main characteristics of oxide-related failure mechanisms

Failure mechanism (FM)	Short description	Possible failure modes (FMos)
Interface state generation	Charges induced at Si-SiO_2 interface and in the oxide may modify the flat-band voltage of MOS devices and increase the leakage in reverse-biased bipolar junctions.	In MOS devices, a drift of the threshold voltage occurs due to the modifications of the flat-band voltage. In bipolar devices, the reverse current increases and becomes instable.
Hot-carrier effects	High-energy electrons and holes are injected into the gate oxide, near the drain, as a localised oxide-charge-trapping and interface-trap generation. Hot-carrier-related degradation can occur in deep submicron devices at drain voltages as low as 1.8 V.	Significant reduction in drain voltage and transconductance, shifts in threshold voltage and decrease in drain-current capability.
Dielectric breakdown	Dielectric breakdown is the destruction of a dielectric layer, usually as a result of excessive potential difference or voltage across it, when the electric field strength surpasses the dielectric strength of an insulator. Two phenomena can be involved: (i) breakdown of gate oxide and (ii) time-dependent dielectric breakdown (TDDB).	Conductive or short-circuit paths through the dielectric or leakage at the point of breakdown.

injected into the gate oxide. The kinetic energy of microscopic particles is directly related to the temperature of the matter they constitute: the higher the temperature, the higher the (average) kinetic energy of the particles; hence the word ‘hot’.

Studying Degradation and Failure Phenomena by HCI

HCI occurs as carriers move along the channel of a metal-oxide semiconductor field effect transistor (MOSFET) and experience impact ionisation[2] near the drain end of the device. The damage can occur at the interface, within the oxide and/or within the sidewall space. Interface-state generation and charge-trapping induced by this mechanism result in transistor parameter degradation, typically as switching frequency degradation or breakdown [21]. A high gate voltage can also pull hot carriers into the gate oxide and trap them there before they even reach the drain region. Trapped carriers or charges in the gate oxide can shift the threshold voltage and transconductance of the device. The excess electron–hole (e-h) pairs created by impact ionisation can also increase substrate current, which in gross cases can upset the balance of carrier flow and facilitate latch-up.

There are four known mechanisms for HCI, which describe the conditions for carriers to enter the gate oxide [22]:

- **Substrate hot electrons (SHEs).** Electrons are thermally generated in the substrate and drifted by an electric field towards the interface. The substrate current produced by impact ionisation can induce Bi latch-up in complementary metal-oxide semiconductor (CMOS) structures and the hot carriers injected in gate oxide form interface states and create trapped oxide charge. In time, this

[2] Impact ionisation occurs when a high voltage is applied across the source and the drain of a MOS device and the channel carriers are accelerated into the drain's depletion region, causing them to collide with lattice atoms that result in e-h pairs. The displaced e-h pairs gain enough energy to propel some of them towards the gate oxide of the device and trap them there.

charge causes instabilities and parameter drift. These serious reliability problems increase with the decrease in device geometries. A correction method is to limit the source–drain voltage to values below the threshold for the generation of hot carriers.

- **Channel hot electrons (CHEs).** Carriers traverse the channel and undergo a low number of lattice collisions under the influence of the strong lateral electric field.
- **Drain avalanche hot carriers (DAHCs).** Carriers are created in avalanche plasma and undergo, due to the strong lateral electric field, a high number of impact ionisations; this is the most physically destructive HCI mechanism [23].
- **Source-side hot electrons (SSHEs).** This mechanism gives rise to a larger shift of the threshold voltage and a larger drift of the saturated drain current [24].

HCI produces noncatastrophic failures, which develop gradually over time and change the performances of the device. The effects of HCI are more prominent in nMOS devices compared to pMOS devices, because HCI requires 3.3 eV for electrons to overcome the surface energy barrier at the Si-SiO_2 interface and get injected into the oxide, compared to 4.6 eV for holes. In MOSFET with p channel, the degradation is caused by hot carriers injected into the drain side of the gate oxide; the type of trapped hot carrier depends on the bias conditions. Obviously, in MOSFET with n channel the degradation is caused by hot holes [20]. HCI generally occurs in logic circuits and not just in random access memory (RAM) cells. When MOS transistors are employed in digital logic, the logic steady states are regions of low stress, because either there is either a high field near the drain with a low gate and the channel off, or the electric field near the drain is low; in both cases there is no generation of hot carriers. Hot carriers are generated almost exclusively during switching transitions. The effects of the hot-carrier stressing can be determined by measuring a variety of device parameters, including assorted currents, voltages and capacitances for the device.

Degradation of device characteristics due to hot carriers also occurs in Bi transistors. This is a well-known phenomenon in which h_{FE} degradation occurs when a reverse bias is applied across the emitter and base. With the advanced shallow junction devices of recent years, there is a tendency towards increased reverse leakage current between the emitter and base, causing device-characteristic degradation to readily occur as a result of the hot-carrier effect. A significant increase in minimum noise figure NF_{min} and noise resistance R_n after hot-carrier stress (which cannot be explained by the change of the carrier density in the inversion layer alone) is observed [25]. It has been demonstrated that the presence of interface states at source side has a much greater impact on the degradation of NF_{min} and R_n. This provides strong experimental evidence that the local noise at source side plays a more important role in determining the channel noise.

HCI can be exploited to create a Trojan that will cause hardware failures [23]. The Trojan is produced not via additional logic circuitry but by controlled scenarios that maximise and accelerate the HCI effect in transistors. These scenarios range from manipulating the manufacturing process to varying the internal voltage distribution. This new type of Trojan is difficult to test due to its gradual hardware degradation mechanism. The danger lies in the fact that until the Trojan transistor has been in operation for some time, the circuit appears as if no malicious alterations have been inserted. The relationship between HCI effect and switching usage of a transistor makes the Trojan HCI transistors incredibly dangerous. There is a critical demand for detection techniques that can properly identify HCI Trojans.

Another application of HCI, possible after extensive knowledge of the physical phenomena has been acquired, is the NOR Flash memory, which is basically a floating-gate MOS transistor programmed by CHE and erased by Fowler–Nordheim tunnelling [26].

The scaling of process technology has increased the effect of HCI on CMOS degradation, because the benefits of higher electric fields saturate while the associated reliability problems worsen, since larger electric fields imply the presence of HCI. As a result, the operational lifetime of CMOS devices has been decreasing. One of the reasons behind degradation is the increased scattering, which leads to mobility degradation. In the following, some recent results from studies on HCI are gathered.

- The qualitative nature of the basic physical mechanisms responsible for hot-carrier-induced degradation in MOSFETs is discussed in [27], where mathematical models for each of the physical mechanisms (effects that are active during the normal operation of such devices and result in shifts in experimentally measurable device parameters such as threshold voltage and linear transconductance) are presented. These models can be used for numerical simulation of hot-carrier degradation in MOSFETs.
- Two degradation mechanisms for HCI behaviour of lateral integrated DMOS transistors have been identified: mobility degradation and source-side injection (SSI). Due to the first mechanism, the drift of linear drain current ($I_{d,lin}$) increases with increasing drain voltage and tends to saturate for longer times. The SSI mechanism gives rise to larger V_t shifts and large $I_{d,sat}$ drift, with a large statistical spread among different devices [24].
- HCI degradation in the presence of a substrate bias has been investigated and compared with the results obtained for conventional channel hot-carrier (CHC) regime [28]. Stress experiments and detailed characterisations (including spatial profiling of the damage) were carried out on state-of-the-art n^+-poly nMOSFETs and p^+-poly pMOSFETs. The results revealed that upon application of a substrate bias degradation becomes faster and more distributed towards the channel than in the channel hot-carrier regime.
- An anomalous hot-carrier effect was observed in vertically integrated trench-based (TB-MOS) power transistors [29]. The avalanche current reaches a maximum at intermediate drain voltage, and decreases for increasing drain voltage. The hot-carrier lifetime of the transistors yields a minimum at intermediate drain voltage, not at the maximum drain voltage. Charge-pumping experiments enable location of the degradation in the TB-MOS. A degradation model was proposed.

Methods for Avoiding Failure by HCI

In spite of the large effort made to explain the hot-carrier degradation effects and the remarkable progress over the last decade, these phenomena are still not well understood, particularly from the physical aspect. Nevertheless, as a result of many studies, methods for designing devices resistant to hot-carrier ageing under regular operating conditions have been elaborated (see Table 5.4).

Dielectric Breakdown (DB)

Methods Used to Study DB FMs

Typically, dielectric breakdown (DB) causes failure of the circuit element by the flow of excessive current. In identifying the location of the failure in a device (e.g. an integrated circuit, IC), the following techniques may be used: optical examination (including liquid crystal hot-spot detection) or electron beam-induced current (EBIC) examination in the SEM.

The current method for studying DB is to use test structures: MOS capacitors made by the studied dielectric. It is important to compare these estimations with the results of accelerated life testing for ICs (logic circuits). Emission microscopy, which detects very faint light emission from weak points in Si devices, has been routinely used to locate 'defects' in gate dielectrics, but it suffers from low spatial resolution and provides no chemical information about the failure defects. In Table 5.5, other methods for studying DB are gathered.

Studying Degradation and Failure Phenomena by DB

In electronic components, dielectrics are used for their insulating property; that is, very low current flows through the dielectric layer under operational conditions. The DB is the formation of a low resistive conduction path through the dielectric layer, which definitively compromises its insulating properties and generally causes a definitive failure of the component. DB is the main cause of irreversible device failure and therefore is a very important issue in modern microelectronics, especially in projecting device lifetime of operative electric fields.

Table 5.4 Methods for avoiding or eliminating the failure produced by HCI

Proposed method	Demonstrated result	References
Using drain engineering to diminish the peak of the lateral electric field located close to the drain edge, by modifying the drain doping profile through the introduction of source/drain extension implants using a lower dose.	Reduces the high electric fields that are generated by constant-voltage scaling.	Wittmann [30]
Design modifications: a larger channel length, double diffusion of source and drain or graded drain junctions.	Minimises HCI.	Wittmann [30]
Optimising the concentration of nitrate of the interface layer (the trap density is reduced) and of nitrogen (above the acceptable range leads to gate-oxide breakdown).	Combats the problem of gate-oxide degradation.	Wittmann [30]
Reducing the high drain field and separating the main current path away from maximum field.	Reduces hot-carrier generation.	Rubaldo [31]
Pushing impact ionisation region deep into silicon by positioning the injection inside the gate edge.	Reduces hot-carrier injection.	
Reducing the hot-carrier trapping by various techniques: gate-oxide thickness reduction, lightly doped MOSFET structure, deuterium post-metal annealing, double-diffused MOSFET structure and incorporating Si_3N_4 as the gate oxide.	Suppresses the hot-carrier effect.	Rubaldo [31]
Lowering channel implant dose (arsenic) and obtaining a smoother junction doping profile, and a wider and lower electrical field distribution.	By decreasing arsenic dose by a factor 1.5, the HCI reliability is improved by a factor 3.	Rubaldo [31]
A deep submicron CMOS process that takes advantage of phosphorus transient enhanced diffusion. Arsenic/phosphorus LDD *n*MOSFETs with and without transient enhanced diffusion are fabricated.	Improves HCI reliability of 3.3 V input/output transistors.	Wang [32]
Using deuterium instead of hydrogen for the optimised post-metal anneal process. The benefits of the deuterium anneal are still observed even if the post-metal anneal is followed by the final SiN cap wafer passivation process.	50–100 times improvement in transistor-channel HCI lifetime.	Kizilyalli [33]
SiO_2 passivation of AlGaN/GaN HEMT, which decreases the surface trapping and 2DEG depletion during hot-carrier stress, so that a passivated device exhibits less degradation than an unpassivated one.	Diminishes degradation of the electric characteristics (forward drain current).	Ha [34]
Technology computer-aided design (TCAD)-driven hot-carrier reduction methodology of 3.3 V I/O *p*MOSFETs design. The drain structures are successfully optimised in short time by applications of TCAD local models. Considering tradeoffs between HCI and I_{ON}, HALO/SDE of both core and I/O transistors can be totally optimised for photo-mask reduction.	Improves HCI reliability.	Miura [35]

A breakdown may be initiated in solid dielectrics with thickness of more than 1 μm placed in strong and super-strong electric fields. In addition to the breakdown, the so-called forming of dielectrics with thickness $d \leq 1\,\mu m$ can be observed. At the most general level, faulty design, improper use of a device, electrostatic discharge/electrical overstress (ESD/EOS) and weak sites in dielectric material (contaminants and structural imperfections) are possible sources of the high field strengths that cause the dielectric to break down.

Table 5.5 Methods for studying DB

Subject of the measurement/analysis	Methods and results	References
The structural and electrical evolution process of gate DB.	Conductive atomic-force microscopy (CAFM) with ultrathin SiO_2 films. The degradation mode is found to be quite different from that in the case of thick films. A DB transient is observed at higher electric field and current density than in 5 nm-thick SiO_2 films.	Zhang [36]
Two stressing methods are used for: (i) time-zero dielectric breakdown (TZDB) (the stress is increased in steps) and (ii) TDDB (constant stress).	(i) Voltage tests (a voltage is applied to the gate contact and the current is monitored by the dielectric layer).	Sirad [37]
	(ii) Current tests (a current is injected and the gate voltage needed to sustain such a current is measured.)	Sirad [37]
The chemical nature of the percolation path formed in ultrathin SiON layers.	Scanning transmission electron microscope (STEM) with high-resolution electron energy loss spectroscopy can be used.	Li [38]
Microstructural defects responsible for the breakdown in ultrathin SiO_xN_y and Si_3N_4, and HfO_2 gate dielectrics.	Transmission electron microscopy (TEM) reveals that the physical defects associated with the gate dielectric breakdown involve both the gate electrodes (poly-Si gate) and the Si substrate. High-resolution TEM and chemical/elemental analysis in TEM show the sources of gate dielectric breakdown failure defects.	Pey [39]
The post-breakdown degradation of ultrathin gate oxide Si MOSFET devices.	Electrical characterisation, cross-sectional TEM analysis and theoretical simulation. The physical location of the DB point is shown to be of critical importance in determining the type of DB and the post-DB electrical characteristics of the device.	Tung [40]

The DB can be intrinsic or extrinsic. Theoretically, the intrinsic phenomenon occurs when the electric field reaches the maximum field strength that the molecular bonding can withstand; for instance, 30 MV/cm in SiO_2. Experimentally, the DB always occurs at lower electric fields (6–12 MV/cm) in thermal oxide layers of good quality, due to the non-ideal SiO_2 structure (the higher values of the electric field at breakdown are obtained for the thinner oxides). The breakdown phenomenon at 8–12 MV/cm is called intrinsic breakdown. The extrinsic phenomenon occurs at electric fields of 5–6 MV/cm due to pinholes and weak spots, which can significantly decrease the dielectric quality [37].

Examples of DB include formations of short circuits between the plates of a capacitor and observed logic failures in CMOS devices caused by DB of the gate oxide. Such oxide breakdowns may be time-dependent; thus this phenomenon is known as time-dependent dielectric breakdown (TDDB).

MOSFET operates at voltages much lower than the DB voltage, but there is still a significant probability of breakdown. As shown experimentally, the lifetime before breakdown follows a Weibull distribution and is a major challenge to oxide scaling.

Some recently reported results in the study of two typical DB phenomena, oxide breakdown and TDDB, are presented below.

- The main physical mechanisms that have been proposed as causing defect generation under constant voltage stress conditions are [41]: (i) trap creation/hydrogen release, (ii) anode hole injection and (iii) thermo-chemical electric field. Some studies have attempted to delineate between the various physical models by analysing the dependence of breakdown time on the oxide electric field. The defect generation and breakdown of ultrathin oxides by substrate hot-holes stress is significantly different from that observed for constant voltage tunnelling stress. This suggests that defect generation by anode hole injection may not cause breakdown during constant voltage stress.

- The dielectric layer of MOS devices is put under stress until it fails. This failure occurs in two phases: (i) a wear-out phase in which the external stress slowly degrades the oxide and (ii) a breakdown phase where certain physical parameters are exceeded and the dielectric undergoes a runaway current flow. This high current flow causes the dielectric, shortly after the electrical breakdown, to breakdown thermally and to be permanently damaged.
- Gate oxide reliability assessment requires a thorough understanding of the involved statistics of failure and should preferably be based on a detailed knowledge of the physics of oxide degradation. Breakdown in SiO_2 occurs at approximately 8×10^6 V/cm^2. Thus, a 15 nm gate oxide will not tolerate voltages greater than 12 V without breaking down. Since ultrathin oxide devices sometimes still perform within specifications after the occurrence of oxide breakdown, and because reliability margins shrink with oxide thickness scaling, the study of the post-breakdown properties has been a subject of much interest in the last few years.
- The strong electric fields (10 MV/cm or above) induce DBs when applied to oxide films. However, DBs can occur when the relatively low electric field (approximately 3 MV/cm) found in the large-scale integration/very-large-scale integration (LSI/VLSI) operative condition is continuously applied to the oxide film. TDDB has been studied as a critical FM in LSI/VLSI reliability.

Some significant ideas proposed by recent studies of DB are presented below.

- Two competing models for TDDB were proposed, and it seems both could exist in parallel: (i) the time to breakdown is proportional to the electric oxide field E_{ox}; and (ii) the time to breakdown is proportional with the reverse of E_{ox} [42].
- For *n*FETs and *p*FETs, voltage, polarity and thickness dependence of time to breakdown were investigated under inversion and accumulation conditions. The voltage dependence in all cases is clearly described by the power law [43].
- Two methods of predicting device lifetime for fast reliability evaluation were proposed [44], based on the direct correlation using breakdown signature parameters, by: (i) the correlation between voltage data from a ramped voltage stress (RVS) test and lifetime data from a TDDB test; and (ii) the homogeneous correlation technique using near-time-zero leakage current data and the corresponding lifetime data, both from the same TDDB test.
- Conceptually, the TDDB process can be divided into two stages: (i) the build-up stage, possibly spanning years, in which the oxide is slowly damaged under electrical stress until localised high field/current density regions are formed as a result of charge-trapping or neutral defect generation; and (ii) the runaway stage, which is completed in a very short period of time, on the order perhaps of microseconds. The rapid acceleration of the damage leads to the formation of a permanent conductive path through the oxide. Hence, the time necessary to reach the runaway stage determines the lifetime of the oxide.

In the last few years, gates fabricated with high-k dielectric have proven to be effective in continuing the device gate-length scaling. The use of high-k as a new gate dielectric in combination with a metal electrode brings new reliability challenges for qualification of advanced technology nodes not previously encountered with conventional poly-Si/SiON gate stacks [45]. In addition to NBTI in *p*FET devices, positive bias temperature instability (PBTI) and stress-induced leakage currents (SILCs) in *n*FET devices, as well as DB, are the focus of metal gate/high-k reliability [53].

Silicon-based semiconductor devices have evolved from the micro- to the nano-regime over the last few years, the gate length now being about 30 nm, and the thickness of the dielectric separating it from the substrate about 1.5 nm. At these small dimensions, it is possible to identify FMs which differ significantly from those at micrometre scale.

Methods for Avoiding the Failures by DB

Some methods for avoiding the failure induced by DB are shown in Table 5.6, along with the results obtained.

5.1.1.3 Metallisation

In electronic components, the metal is used to contact and interconnect the various regions of the device. There are some basic requirements for the interconnect materials, such as: low/medium resistivity, good adhesion to adjacent layers, stability of mechanical and electrical properties, resistance to corrosion and EM, low film stress, controllable deposition and patterning methods. Possible materials for interconnection are: metal (low resistivity), polycrystalline Si (medium resistivity) and highly doped diffused regions in Si (medium resistivity). Basically, there are two approaches to lowering the contact resistance: to use highly doped Si as the contact semiconductor or to choose a metal with a lower Schottky barrier height. The main characteristics of the typical FMs at metallisation level are gathered in Table 5.7.

In the following, some FA methods are given for identifying the above FMs, together with possible corrective actions for avoiding (or eliminating) their results.

Electromigration (EM)

Methods for Studying EM FMs

In recent years, a huge number of research works have progressively improved our understanding of EM. This is a subject of great technological interest because the unwanted migration of atoms in electronic components (especially VLSI) leads to degradation of the metallic fine lines, ultimately resulting in device failure. Because the main subject of this book is FA, results reporting the use of various physico-chemical characterisation tools have been selected below (Table 5.8).

The standard wafer-level EM accelerated test (SWEAT) is one of a few highly accelerated stress tests of metal-line test structures that are used to monitor EM resistance. Various works have shown that when the JEDEC standard SWEAT method is used, unexpectedly large sigmas are observed which can sometimes be traced to the existence of bimodal failure–time distributions [66]. The cause of these anomalous behaviours is the algorithm in the JEDEC standard method that is used in the control loop to adjust the stress current, so that a constant target failure time is maintained.

Studying Degradation and Failure Phenomena in EM

The first theoretical works were performed by Fiks [67] and Huntington [68], who independently introduced the ballistic model to determine the driving force for EM. This model could be applied for an isolated impurity atom, but it is difficult to use it for a defect complex in a real metal, where electron band structure effects are important. For many years, aluminum (Al) seemed to be the best choice for metal lines. Extrapolation of EM lifetimes from accelerated testing to operational use conditions was made by using Black's law [69]:

$$\mathrm{t}_{50} = \mathrm{A}\mathrm{j}^{-n}\exp(E_a/\mathrm{kT}) \tag{5.1}$$

where t_{50} is the median lifetime of the population of metal lines subjected to EM, j the current density, T the temperature in degrees K and k the Boltzmann constant. The other three parameters of this equation have to be determined experimentally: (i) n is the current exponent (scaling factor with values from 1 to 7; usually, n = 2), (ii) E_a is the activation energy for EM (which is 0.5–0.7 eV for pure Al) and (iii) A is a constant based on metal line properties. Theoretically, this equation can be justified for specific values of n and E_a, but the general use of empirically generated values can be problematic.

Table 5.6 Methods for avoiding or eliminating failures produced by DB

Proposed method	Demonstrated result	References
Studying the impact of gate-DB-induced microstructural defects.	For poly-Si/SiON gate stacks, the effect of DB induced epitaxy (DBIE) dilation on the post breakdown degradation rate has been found: oxygen vacancy and Si-path formation in the breakdown path control the post-breakdown evolution. For metal-gate/high-k gate stacks, ultrafast degradation can occur with the formation of metal filament if the compliance current level exceeds a certain limit. However, the gate leakage current can recover from the breakdown damage through reverse electrical biasing.	Li [46]
Chosing the best high-permittivity insulator for possible gate dielectric applications.	Thin stacks comprising alternating layers of Ta_2O_5/HfO_2, Ta_2O_5/ZrO_2 and ZrO_2/HfO_2 are deposited on silicon substrates using atomic-layer deposition. The last ones show the highest breakdown field and the lowest leakage current.	Zhang [47]
Chosing the best silicide material as gate dielectric about the behaviour at breakdown transient.	Contact and gate silicide migration induced by gate dielectric breakdown transient has been observed in various silicide materials (Ni, Co and Ti silicides). The best choice seems to be the Ti silicide.	Tung [48]
Using an array-based test circuit to efficiently characterise gate dielectric breakdown.	The proposed circuit facilitates study of the statistical process and investigations of any spatial correlation of dielectric failures, and can monitor a progressive decrease in gate resistance.	Keane [49]
Improving device performance (subthreshold swing and threshold voltage shifts after stress) by using stacked ultrathin oxide/nitride (O/N) dielectrics with interface nitridation.	Experimental evidence shows more severe breakdown and device degradation in the threshold voltage, drain current and transconductance for shorter-channel *p*MOSFETs with O/N dielectrics. These degradations result from the enhancement of hole trapping in the gate–drain overlap region, as evidenced by a positive off-state leakage current, which leads to hard breakdown and the complete failure of device functionality.	Lee [50]
Studying failure of dielectrics by simulating the Cu ion concentration and internal electric-field profiles in a dielectric material and solving the transient continuity/Poisson equations.	It was assumed that failure in $Cu/SiO_2/Si$ devices occurs due to a pile-up of Cu ions at the cathode and a subsequent increase in electric field. A comparison with experimental data shows that polarisation of the dielectric due to the high field surrounding the copper ions contributes to acceleration of the breakdown and that breakdown in porous dielectrics is faster than in solid dielectrics due to field concentration around the pores of the dielectric.	Achanta [51]
Replacing SiO_2 as gate dielectric material with high-k dielectrics. The particular area of concern is the wet etching.	Wet etching of heat-treated atomic-layer chemical-vapour-deposited (ALCVD) zirconium oxide (high-k dielectric) in HF-based solutions has been studied. It is found that heat-treated material, while refractory to wet etching at room temperature, is more amenable to etching at higher temperatures when methane sulfonic acid is added to dilute HF solutions.	Balasubramanian [52]
Degradation of MOS transistors under RF stress by hot-carrier degradation, negative-bias temperature instability and gate dielectric breakdown.	The experimental results indicate that the existing models are strongly applicable into the gigahertz range for describing the degradation of MOS transistors in an RF circuit. The probability of gate dielectric breakdown appears to reduce rapidly at such high stress frequencies, increasing the design margin for RF power circuits.	Sasse [53]

Table 5.7 Main characteristics of metal-related failure mechanisms

Failure mechanism (FM)	Short description	Possible failure modes
Electromigration	Due to the momentum transfer from current-conducting electrons to metal atoms, the 'electron wind' produces a flux of metal atoms in the opposite direction to the conventional current flow, producing voids at the negative electrode and whiskers (if a passivation layer is present and the excess matter extrudes out) or hillocks (if no passivation layer is present and the metal acts towards minimising the total surface area). The effect is significant at high current densities ($>2 \times 10^5$ A/cm^2 in Al).	Open circuit (if voids produce the interruption of metal lines), leading to interconnect open or high resistance, resulting in malfunction or speed degradation. Short circuit (if hillocks and whiskers bridge between conductors).
Hillocks	Protrusions on the surface of metal films, formed during deposition or post-deposition heat treatments. Main possible causes: incorrect metallisation process, electromigration and thermal cycling.	Interlevel shorts (faults at wafer testing). Short circuit during usage (failures).
Voids	Over-alloying of Al/Si systems, coupled with improper cooling rates in subsequent processes. These produce tensile stresses in interconnections, which over time can create voids sufficient to lead to failure. The defects initiating this FM are difficult to detect during the manufacturing process; the only known method involves destructive physical analysis following the ageing of test structures.	Interruption of interconnections (open circuit).
Delamination	Separation of thin metal films from intended bonding surfaces if the substrate was contaminated prior to film deposition or if film materials were mismatched to the substrate.	Open circuit, high resistance, intermittent contact.
Step coverage	Thinner metal layer over a step than in a flat area.	Open circuit produced by metallisation interruption.
Corrosion	Wearing away due to chemical reactions, mainly oxidation when a gas or liquid chemically attacks an exposed surface, often a metal. It is accelerated by high temperature or by acids and salts.	Increased electrical resistance of the interconnections. The copper contacts may corrode together, causing shorts.
Other FMs	Various FMs have been gathered under this heading, such as improper thickness, non-ohmic contact, scratches on the metallisation layer produced by manipulation faults and so on.	Open circuit (increased resistance, as incipient form) or short circuit.

Table 5.8 Methods for studying EM

Subject of the measurement/analysis	Methods and results	References
Progress of EM damage in fully-embedded Cu interconnect structures.	In situ analysis with SEM equipped with a heating stage and electrical connections for the experiment. Focused ion beam (FIB) is used to prepare cross-sections of the samples, maintaining their electrical functionality.	Meyer [54]
Degradation studies to understand EM-induced transport processes in on-chip Cu interconnects.	Correlation of the real-time X-ray images with post-mortem SEM micrographs is used to discuss degradation mechanisms in Cu interconnect.	Schneider [55]
Cause of early EM failure in Cu vias. It seems pre-existing defects are essential to the formation of EM-induced void.	An infrared (IR) microscope technique has been developed to detect weak via before catastrophic failure.	Kim [56]
Investigation of EM lifetime and FMs for SnAg Pb-free solder bumps with two types of under-bump metallurgy (UBM).	SEM and energy-dispersive X-ray spectroscopy (EDX) analysis reveal a failure mechanism for solder bumps with Ni UBM caused by the dissolution of UBM as a result of Ni migration and subsequent solder-cracking or de-wetting.	Ding [57]
First direct experimental measurement of EM-induced stress in Al interconnects using a series of micro-rotating stress sensors.	Experimental verification using optical or X-ray techniques has proven to be difficult. The limited resolution of previous techniques restricts their ability to obtain a detailed characterisation.	Wilson [58]
The role of the sidewall barrier in the nucleation/growth of EM-induced voids in Cu interconnects.	SEM shows the steps of void growth in 120 nm line width and electron backscattering diffraction (EBSD) shows that voids are nucleated at the grain boundary.	Arnaud [59]
EM-induced degradation processes and failure in on-chip interconnects in microprocessors, based on experimental studies.	In situ microscopy (side-view) studies at embedded via/line dual inlaid Cu interconnect test structures show that void formation and evolution depend on both interface bonding and microstructure.	Zschech [60]
EM testing of passivated Cu damascene interconnects for the study of void formation.	In situ microscopy (top-down-view) studies	Thompson [61]
EM-induced stress evolution in dual-inlaid Cu interconnects.	EBSD is used to study stress evolution.	Sukharev [62]
Statistical analysis of EM lifetimes of inlaid copper interconnects.	In situ SEM and transmission X-ray microscopy studies of EM degradation processes.	Zschech [63]
Fabrication of reliable diffusion barriers from thin films made of refractory metals and their compounds, such as: Ta or TaOx as diffusion barriers in Si/Ta/Cu, Si/TaOx/Cu and Si/Ta-TaOx/Cu systems.	TEM, X-ray diffraction, X-ray photoelectron spectroscopy and secondary neutral mass spectrometry.	Lakatos [64]
The use of E-beam lithography and shadow evaporation through a stencil mask to fabricate nano-bridges made of gold and other metals. The bridges are then thinned by a controlled cyclic EM process in ultrahigh vacuum (UHV).	Scanning tunnelling microscope (STM) and SEM have discovered the phenomenon of tunnelling voltage-dependent deposition of additional material, possibly carbon, of up to 10 nm thickness.	Stoffler [65]

The current densities that can be achieved today in metallic nanojunctions approach 10^{15} A/m^2. At such high current densities, the forces on atoms caused by the passage of the current can be so large they may lead to failure of the wire in much less than a second, a phenomenon called 'current-induced embrittlement' [70].

A drastic solution for eliminating EM is the use of Copper (Cu) instead of Al. The resistivity of Cu is nearly 50% less than that of Al. An additional and important benefit of the use of Cu is an improved reliability due its higher melting point (1085 °C versus 660 °C for Al) and so improved EM resistance. However, the problem of Cu corrosion must be solved (to be discussed later in this section).

Another process improvement is the introduction of low-k dielectrics, which poses a greater integration challenge than the introduction of Cu metallisation. This is reflected by the fact that many device manufacturers have chosen to introduce Cu before low-k. Candidate low-k materials can be categorised by type (silicates, fluorosilicates and organo-silicates, organic polymeric, etc.) and by deposition technique (chemical vapour deposition (CVD), spin-on). Dielectric constant reduction is achieved by one or more of the following: (i) reducing polarisability, (ii) reducing density and (iii) introducing porosity, because porous low-k materials are available with dielectric constants ~50% lower than that of SiO_2.

Many EM studies indicate the wide variety of pathways available for EM damage formation. The candidates are: (i) an interface mechanism, (ii) the Blech effect, (iii) low-integration and (iv) Joule heating. The Blech effect, also known as stress-induced backflow, was first reported for Al lines [71]: as the metal ions move towards the anode end of the line, stress build-up occurs opposing the electron wind, thus constraining the void growth. A similar, even stronger, effect has also been observed in Cu interconnects. This is consistent with the interpretation that extended defects such as interfaces, grain boundaries and/or surfaces play a major role in mass transport.

An analysis of simulated time to failure (based on a numeric model that is able to simulate time to failure corresponding to EM tests on a wide range of current density) predicts that the error on Black's law parameters E_a and n may occur from both wafer level (high current density and Joule heating) and package level (low current densities). As a consequence, it is recommended that temperature gradients as well as the Blech effect be taken into consideration to correctly extrapolate lifetime [72].

The EM in Al and Cu was compared on the basis of the ratio of their melting points to the device operating temperature of +100 °C [73]. The dominant mechanism seems to be grain boundary diffusion for Al and interface diffusion for Cu [74]. In a multilevel interconnection, an EM force due to current crowding has been proposed to explain why many EM-induced damages occur away from the high-current-density region. In mean-time-to-FA, the time taken to nucleate a void is found to be much longer than the growth of the void in Al, but this is not the case for Cu interconnects. On accelerated tests of EM in Cu interconnects, the results gathered above +300 °C will be misleading since the mass transport will have a large contribution of grain boundary diffusion, which is irrelevant to EM failure in real devices induced by surface diffusion.

Reliability of Cu interconnects is a critical concern for the microelectronics industry, particularly with aggressive development towards 90 and 65 nm technology, where the feature size of Cu interconnects is also in the nanoscale range. Such ultrafine Cu interconnects are expected to be more prone to physical failures, not only because they are subjected to more strenuous use conditions, but because the critical size for fatal defects, either as processed or as developed, is smaller. All these conditions are especially problematic to EM reliability and are expected to pose significant technical challenges. From the investigation of the EM FM in various line widths of single-level, damascene[3] Cu interconnects, the study [75] concluded that the failure is controlled by two mechanisms. The first and primary mechanism determining failure kinetics is Cu–barrier interface EM. Although limited,

[3] 'Damascene' means that the underlying silicon oxide insulating layer is patterned with open trenches. A thick coating of copper that significantly overfills the trenches is deposited, and chemical-mechanical planarisation (or polishing) is used to remove the copper to the level of the top of the insulating layer. The remaining copper within the trenches of the insulating layer becomes the patterned conductor.

evidence suggests that the secondary factor is probably attributed to pre-existing defects or grain boundaries.

In recent years, Cu has become the main target for EM studies, partly because EM studies on Al have already been completed, but also due to the more and more frequent usage of Cu as interconnection material in modern electronic components.

Some recent results of the studies of EM are synthesised below:

- The early EM fails are either by voiding into the via (via depletion) or by voiding underneath the via (line depletion). While most of the early EM fails for via depletion are related to the liner quality in vias, for line-depletion EM the contact configuration between the via and the underlying line is critical to the failure characteristics [76].
- Apparent negative activation energy behaviour in EM lifetime is observed for Cu/low-k single damascene interconnects. This is explained by the fast FMo by thermo-mechanical damage at low temperatures around +250 °C. Mechanical strengthening of the via by thickening the sidewall barrier and surrounding it with strong ILD has successfully suppresses the thermo-mechanical failures and restores the activation energy behaviour to normal [77].
- In submicron Cu damascene interconnect, EM is mainly due to the diffusion at the interfaces of Cu with liner or dielectric capping layer. Two effective methods, enlarging via size and enhancing Cu/capping process, have been demonstrated to improve the EM distribution [78].
- A general EM model is proposed with special focus on the influence of grain boundaries and mechanical stress. The possible calibration and usage scenarios of EM tools are discussed. The physical soundness of the model is proved by 3D simulations of typical dual-damascene structures used in accelerated EM testing [79].
- The EM-induced voids in Cu interconnects are nucleated at grain boundary. This result was studied for an anisotropic physical vapour deposition (PVD) process and a conformal CVD process. Voids are nucleated at grain boundary [80].
- Recently, a perturbation method to the EM process has been employed for the first time in order to develop a general analysis of how morphological stability is correlated to microstructure-induced spatial variations in the effective diffusivity [81]. Of particular interest to this study is the relationship established between flux divergence, microstructure and void–hillock-pair EM damage.
- A quantitative analysis of the impact of Joule heating on the package-level EM test of copper interconnects is given in [82]. It is found that high test current density has a negligible impact on the estimation of activation energy and offers an opportunity to shorten the test time. However, the current density exponent is found to be a strong function of test current density, especially for interconnects with a larger line width.
- EM under bidirectional current has been studied on Cu interconnects for the 65 nm node [83]. Physical analyses confirm void location at both ends of the line and copper transport over long distances. Resistance evolution correlates to void healing/growth kinetics. Interest has been shown in bidirectional tests to study multimodal FMo.
- Multimodality EM behaviour of Cu interconnects is studied in [84]. Both superposition and weak-link models are used for statistical determination of lifetimes of each failure model (statistical method). Results are correlated to the lifetimes of respective failure models physically identified according to resistance time evolution behaviours (physical method). A possible mechanism for EM lifetime enhancement has been proposed, based on Cu-silicide formation before cap-layer deposition and adhesion of Cu–cap interface.
- The results of a study of EM under bidirectional direct-current (DC) stresses show that Cu is transported over long distances and that resistance evolution is in agreement with the healing/growth processes taking place under a reverse current [83]. Under a reverse current, and depending on the structure geometry, resistance can decrease down to its initial value; the void is then completely healed, and EM is fully reversible. The multistress technique and bidirectional tests have proved to be useful in studying multimodal population.

- Current-induced forces give rise to EM. A thermodynamic analysis of EM has been developed, which reveals the macroscopic free energy that drives the process and relates it directly to the microscopic force and its quantum-mechanical origin [85].
- As dimensions shrink, reliability considerations become more critical. In deep submicron devices, at certain stages of processing, even an atomic layer variation can be a defect. Studies on the physical FMs in submicron devices reveal that the major reliability concerns are the same as those before scaling. A comprehensive overview of the reliability issues in ultrathin gate dielectrics and copper interconnect material is given in [86], linking how the physical effects to devices can be a threat to long-term reliability.
- The effects of ESD, EM and their interdependence on Cu interconnects have been studied [87]. FA shows the gradual growth of failure sites appearing above the copper melting point (1084 °C) and leading the metal line to open failure when the critical temperature is reached. It has been shown for the first time that the latent damage caused by transmission line pulse (TLP) can degrade EM lifetime depending on the wafer temperature and the applied current during EM stress.
- The effects of EM, TDDB and hot-carrier degradation wear-out mechanisms on scaled technologies and product reliability have been investigated [88]. Accelerated stress testing across several technology nodes was performed, and FA was conducted to confirm the FMs.

Methods for Avoiding the Failures Produced by EM

Some practical ideas to diminish the effect of EM in Al interconnections were proposed many years ago: (i) to add Cu or titanium, allowing higher current densities before EM arises; (ii) to encapsulate conductors with dielectrics; (iii) to cover the Al conductor with a chemical–vapour-deposited SiO_2 layer or to grow an anodic layer on to the Al conductor; (iv) to use some surface treatments, which has proved to be very effective (e.g. oxygen plasma treatment in a barrel reactor, preceded or followed by an annealing at 450 °C for 30 minutes in forming gas led to significant improvements of the mean time to failure); (v) to restrict the diameters of the Al particles; (vi) to use a barrier metal under the Al layer; and (vii) to reduce the stress of the passivation film.

A more complex solution for avoiding EM in Al lines is given in [89]. A global interconnect structure is proposed, formed by Ti/TiN/Al(Si-Cu)/TiN, with the following results: (i) Cu in Al suppresses EM and hillock formation; (ii) Ti improves the grain texture of the Al for better EM protection and promotes adhesion; (iii) TiN is used as a barrier between the Al and the Ti in order to limit $TiAl_3$ formation; (iv) the Ti/TiN and TiN layers on the bottom and the top of each interconnect provide an electrical shunt in case of void formation in the Al; and (v) TiN on the top also acts as an antireflective layer for lithography. Many variations on this structure, involving the exact makeup and the order of materials in the interconnect stack, are used by different manufacturers, but the basic idea of using multilayer planar structures is very common. Very high test temperatures can be used for via chain tests because there is no EM in tungsten at temperatures less than approximately 1800 °C, the re-crystallisation temperature of tungsten.

Some methods for avoiding EM in Al and Cu lines are given in Table 5.9.

Hillocks and Voids

In any electronic component, the metal layers are used for the transport of electrons, analogous with a car driving on a highway. So, the metal surface must be as smooth as the highway's. Any protuberance or hole is detrimental to the traffic. There are many possible causes of such protuberances (called 'hillocks') or holes (called 'voids').

First, the deposition process, if not properly optimised, can induce hillocks and voids. If a thin metal layer, particularly a soft metal such as Al with a high coefficient of thermal expansion (CTE), is sputtered on to a substrate with a low CTE, such as Si or SiO_2, microscopic protuberances often appear in the surface of the metal layer, named 'growth hillocks'. The reverse problem of void formation occurs in the same soft metals when they are stressed in tension beyond the yield point. The material is selectively transported out of certain regions of the film to relax the stress elsewhere.

Table 5.9 Methods for avoiding or eliminating the failures produced by EM in Al and Cu lines

Proposed method	Demonstrated result	References
Doping with Al impurity in order to diminish the effect of EM on damascene Cu interconnects. The EM-induced Cu drift is suppressed as the Al concentration increases.	The EM lifetime is improved by the suppression of Cu diffusion due to the piled-up Al at the top surface of the Cu interconnects.	Yokogawa [90]
Electrolessly deposited cobalt tungsten phosphide (CoWP) film, often referred to as a self-aligned barrier (SAB), is employed to prevent EM along the top surface of the Cu line.	Although this SAB is effective at preventing EM, exposure of both Cu and CoWP to dilute hydrofluoric (dHF) permits a galvanic corrosion mechanism that ultimately consumes the CoWP layer.	Lauerhaas [91]
A diffusion-barrier metal is deposited after a conventional argon bombardment is used to clean Cu surface in order to determine the effects of a surface-clean process on stress voiding and EM of Cu metallisation.	Higher pre-clean bias power and shorter pre-clean time demonstrate remarkable low via resistance and excellent Cu-reliability performance. The pre-clean bias power of argon plasma should be kept high, but not too high, in order to avoid damaging the underlying metal, while the re-sputtering clean time should be kept short but still long enough to clean via bottoms.	Wang [92]
A modification of pre-clean before cap-layer deposition and Cu cap/dielectric materials in order to improve EM lifetime.	EM lifetime is enhanced. Cu-silicide formation before cap-layer deposition and the adhesion of the Cu/cap interface have been found to be critical factors in controlling Cu EM reliability.	Lin [93]
Failed via can be healed and provides an electrical path even after EM failure. This 'recovery' can be classified into two groups: short-term recovery and long-term recovery.	FA strongly implies that the short-term recovery occurs due to an accidental local melting and the long-term recovery obtains its driving force from compressive stress at the anode side. However, this is simply an electrical recovery, not a complete structural healing.	Kim [56]
Preventing Al-Si interdiffusion and resultant junction spiking in Al lines.	Introduction of 1–2 atomic-percent silicon into the aluminum. Copper is introduced in similar quantities to improve the electromigration resistance of the aluminum wiring. BPSG is used as a pre-metal dielectric and CVD SiO_2 as the inter-metal dielectric (IMD).	Kim [56]
A new via technology for improving EM reliability of Cu dual-damascene (DD) interconnection, called direct-contact via (DCV).	An early FMo of a conventional Cu DD structure is found as void formation at the via–bottom interface, where flux divergence of Cu ions is large due to diffusion barrier layer. The early failure mode is eliminated by the DCV technology and lower via resistance is obtained.	Kazuyoski [94]

The second source of hillocks and voids could be any post-deposition heating treatment, which might produce 'annealing hillocks/voids'. Some possible examples: (i) exposure of the electronic component to a high temperature during manufacturing steps and (ii) extended periods under temperature-cycling conditions during operation.

The third source of hillocks and voids has already been mentioned: EM (see section Electromigration EM).

In any case, the hillocks and voids are troublesome and can cause subsequent device failure. For example, hillocks can cause shorts between conductive layers in a device in areas where conductors cross over one another, or in elements of the device with two layers of conductors, such as integrated capacitors. More particularly, if an insulating layer is formed on an Al layer at a thickness of less than 1 μm, metal hillocks greater than 1 μm will protrude through the insulating layer and contact any subsequently deposited metal layers, causing a short circuit. The protrusion of the hillocks potentially causes short circuits between conductive layers in a device at positions where the conductors cross over each other, or in devices where there are two spaced layers, such as in capacitors, forming part of an IC. Hillocks are also found in Al layers formed in optical discs, where they can cause light-scattering in optical applications.

However, there are situations in which such protrusions have a beneficial effect. A method for creating nano-sized protrusions on insulating surfaces using slow highly-charged ions is presented in [95] and holds the promise of forming regular structures on surfaces without inducing defects in deeper-lying crystal layers.

Studies of Hillock and Void Formation

Obviously, much effort has been made towards explaining the formation of hillocks and voids. Hillocks often occur in the electron flow upstream from the area of voiding, but they grow faster at the downstream edge closest to the source of the migrating metallisation. Voiding and hillock formation sometimes occur on top of each other in temperatures in the neighbourhood of 140–200 °C. Hillock growth is more extensive in films deposited on room-temperature substrates than on heated substrates, indicating that the effect may vary with grain size. Hillocks form at random in Al films heated to temperatures around 400 °C during fabrication and can cause electrical shorts between adjacent lines and fracture of the overlying metal layers. Hillocks caused by EM can result in thin dielectric sites that are susceptible to subsequent breakdown of the intermetallic dielectric.

One mechanism proposed for hillock formation is stress relief. A high compressive stress is induced during thermal cycling, due to the thermal mismatch between Al (which has a high CTE) and the substrates (e.g. Si, SiO_2, which have significantly lower coefficients). The low yield strength of Al causes the film to yield. A study of hillock formation in Al films of thicknesses in the interval 0.25–2.2 μm and which have been deposited by electron-beam evaporation is reported in [96]. Hillock sizes, shapes, numbers and formation temperatures were determined, the latter on a heating stage in situ in an SEM. The internal structure of the hillocks was studied by cross-sectional transmission electron microscopy (X-TEM) technique. This study provides strong support for the idea that hillocks are formed by migration of material along grain boundaries, presumably at triple junctions, up to the surface, where it is deposited in a growing hillock. Initially, hillocks are separated from the original film surface by a grain boundary-like interface, but prolonged annealing will cause under-laying grains to grow into the hillocks, until they become integrated in the film.

A mechanism to explain hillock formation is proposed in [97]. This involves nucleation by silicon redeposited from the etch solution. The incidence of hillocks in this model is the result of a competition between the forward and reverse etch reactions.

A Cu hillock-induced interconnect FM is presented in [98]. The Cu hillock appears to damage the SiN capping layer and results in copper corrosion during via etch. The corrosion generates Cu particles inside via holes and the defects are found to make subsequent metal depositions incomplete during via formation. Based on the observations, a Cu hillock-induced defect model is proposed and a new copper process is suggested to reduce Cu hillocks.

Voids have been observed to form within several microns of a growing hillock literally emptying aluminum by a river-like mechanism into the hillock structure not constrained by grain boundaries or other defects [99]. Changes in growth patterns (hillocks and voids) were noticed when the circuits were exposed to ambient air for short periods of time. It was observed that the presence of a surface oxide on the walls of a void greatly retarded void growth, indicating that a primary mechanism for aluminum transport occurs along the oxide free wall structure.

Another source of voids is stress migration (also called stress-induced voiding, SIV): vacancy migration driven by the hydrostatic stress gradient. Large voids may lead to open-circuit or unacceptable resistance increase that can affect the IC performance. For instance, high-temperature processing of copper dual damascene structures leaves the copper with a large tensile stress due to a mismatch in CTE of the materials involved. The stress can relax with time through the diffusion of vacancies, leading to the formation of voids and ultimately open-circuit failures [100].

The metal interconnects are dielectric passivated. When the dielectric cools from its deposition temperature, the metal will contract more that the silicon substrate and passivation layer, placing the metal under tensile stress orders of magnitude greater than its yield stress. Some dislocation motion will occur to relieve this stress, but the encapsulation prevents further relaxation. Stresses in an Al line can be as high as 400 MPa. Some relaxation also occurs by vacancy flow and this can result in stress voids that reduce the cross-section of the line. Study of this phenomenon requires accurate strain measurements in individual grains along a line with and without current flow in the line. The only tool that can accomplish both these objectives on passivated interconnects is X-ray diffraction using suitably prepared X-ray beams. With the availability of modern third-generation high-brilliance synchrotron sources, micron and submicron X-ray beams have been produced with sufficient intensity to carry out meaningful orientation and strain measurements on individual interconnect lines [101].

A new FM in AlCu and AlCuSi metallisation is described in [102]: inter-level metal short-circuiting occurs between two or more levels of metal, caused by theta-phase (Al_2Cu) hillocks which nucleate and grow during high-temperature vacuum heat treatment and processing. Hillock growth occurs at high-energy sites, such as Si precipitates and grain boundary nodal points. The growth of Al_2Cu hillocks depends on the heat-treatment/processing temperature and Al film purity. The growth kinetics indicate that grain boundary diffusion is the dominant mass-transport mechanism. A layer of pure Al deposited beneath the Al-Cu layer acts as a sink for Cu and delays hillock formation. Also, increasing film Cu content reduces hillock formation: theta-phase hillocks, up to 1.3 μm in height, are observed in films with 1 wt% Cu, whereas negligible (<0.2 μm in height) hillock formation is observed in 11 wt% Cu films.

Recently, the effect of quenching in different media on hillock formation and electrical resistivity has been studied in Au-Pd layers [103]. Oxygen was released from substrate by a substrate relaxation process. It was suggested that hillocks appeared on the triple junction grain boundaries. However, lower electrical resistivity was observed in the sample which quenched in the air. It was concluded that grain boundary scattering decreases the conductivity of the quenched films due to a higher density of dislocations.

A comprehensive explanation of hillock and void formation was given by Burton [104], starting with the definition of 'creep'. This is a deformation mechanism defined as a tendency of a solid material to slowly move or deform permanently under the influence of stresses. It occurs as a result of long-term exposure to levels of stress that are below the yield strength of the material. Creep is more severe in materials that are subjected to near-melting-point heat for long periods. It always increases with temperature. As a rule of thumb, the effects of creep deformation generally become noticeable at approximately 30% of the melting point for metals, and 40–50% of melting point for ceramics.

Hillock growth is produced when stress relaxes itself by transporting film material to the surface. The transport apparently occurs by diffusion of the film material along grain boundaries. Cobble creep is a form of diffusion creep that occurs through the diffusion of atoms in a material along the grain boundaries, producing a net flow of material and a sliding of the grain boundaries. Diffusion is known to be faster there than the bulk of the grains, and in the case of Pb, for instance, the known value of the grain boundary diffusion coefficient has been found to agree with the rate of hillock growth. Another clue is that hillocks tend to occur over grain boundaries.

During creep, damage accumulates in the form of internal voids, which appear to grain boundaries normal to the tensile stress. Atoms diffuse from the void surfaces into grain boundaries, and the voids grow, faster and faster, until they link, leading to creep fracture.

It is interesting to note that hillocks and voids can also arise in non-metal layers. A GaN film with a thickness of 250 μm was grown on a GaN/sapphire template in a vertical hydride vapour phase epitaxy (HVPE) reactor [105]. The full-width at half-maximum (FWHM) values of the film were 141 and 498 arcsec for the (0 0 2) and (1 0 2) reflections, respectively. A sharp band-edge emission with a FWHM of 20 meV at 50 K was observed, which corresponded to good crystalline quality of the film. Some almost circular-shaped hillocks located in the spiral growth centre were found on the film surface, with dimensions of 100 μm, whose origin was related to screw dislocations and micro-pipes. Meanwhile, large hexagonal pits also appeared on the film surface, which had six triangular facets. The strong emission in the pits was dominated by an impurity-related emission at 377 nm, which could have been a high-concentration oxygen impurity.

Methods for Avoiding Failures by Hillocks and Voids

First, because EM is one of the causes of hillock and void formation, all the curing methods described for EM are valid in this case too. Obviously, there are many other methods for avoiding hillock and void growth. Hillock formation is remedied by using alloyed metal, such as Al-Cu, and by improving the technologies of passivation and packaging. It may also be necessary to increase the thickness of a dielectric overcoat to ensure adequate coverage to prevent these failures. CVD dielectrics conform to the hillocks but sputtered insulators probably afford less protection. Resist may be thinned over hillocks and thus erode prematurely during reactive ion etching (RIE). Hillock formation during deposition depends on the deposition rate and the substrate temperature. When the temperature is increased, at a constant deposition rate, the density of the hillocks decreases, but their average size increases. As the deposition rate is increased, the hillock density and average size decrease. In general, hillocks form when films are deposited at low temperatures and low rates.

Other methods are shown in Table 5.10.

Delamination

The reliability of metallisation is linked to the ability of the thin metal films to remain adhered to substrates during the component lifetime. The mechanical integrity of these films, controlled by the adhesion energy of the interface, is shown to be influenced by strain relaxation and diffusion. Understanding the time-dependent behaviour can provide insight into adhesion energy trends for the long-term reliability of film systems. Thin metal films will separate from intended bonding surfaces if the substrate was contaminated prior to film deposition, or if film materials were mismatched to the substrate. The failure types associated with delamination of thin films are: open circuit, high resistance and intermittent contacts. Generally, the tool used to verify delamination is optical microscopy (from 10–100×).

The effects of elevated temperature and humidity on the interfacial fracture of the Au-SiO system have been investigated [117]. This film system was chosen since it is currently used in MEMS mirrors and nanoscale interconnects – applications that require long-term reliability. As thinner gold films are fabricated, transitioning from micrometre to nanometre thicknesses, changes in composition that were previously limited by slow diffusion rates and long diffusion paths may become more important to device lifetime.

The thin films may be delaminated from a substrate by contacting with dusts and other mechanical devices. In [118] an intermolecular force (van der Waals force) is introduced into the fundamental mechanism of adhesive force. A boundary element method (BEM) is used to analyse the delamination of thin film. The thin film is pushed up using an indenter from the substrate. The behaviour of thin

Table 5.10 Methods for avoiding or eliminating failures produced by hillocks and voids

Proposed method	Demonstrated result	References
Adding impurities such as Si, Cu, Ag and Au to the Al to immobilise the grain boundaries in the metal layer.	Avoids hillock formation in thin Al layers.	Chaudhari [106]
Treating the layer to form a boehmite (AIO-OH) layer on its surface.	Reduces hillocks, specifically in Al layers.	McMillan [107]
Adding an alloying element, tin (with a small diffusion coefficient, a large binding energy with a lattice vacancy and a large atomic diameter) and manganese (with a relatively small solid solubility) to Al. It has been proved experimentally that Al-0.06 wt pct Sn alloy and Al-0.1 wt pct Mn alloy, which were selected for the above-mentioned reason, have a marked effect in suppressing the growth of hillocks.	Suppresses hillocks, which grow on the surface of deposited Al conductors after they have been subjected to thermal cycling (200 °C to room temperature) or high-temperature heat treatment (400 °C).	Sato [108]
The Al film, deposited on an oxidised silicon wafer and sintered, is stripped off in orthophosphoric acid. The wafer is thoroughly cleaned in DI water and no HF is used. This treatment forms an $A1_2O_3$ + Si dipole layer on the oxide surface and makes it aluminumphilic. Further aluminisation and sintering generate the adherent composite layers of Al, Al_2O_3 + Si on the oxide surface.	Eliminates hillocks in IC metallisation.	Singh [109]
A sandwich layer is grown between the Al film and the underlying SiO_2. This layer must have a thermal-expansion coefficient between those of SiO_2 and Al. The technique has been demonstrated with WSi_2 as the sandwich layer.	Eliminates hillocks in IC metallisations.	Cadien [110]
Pretreatment and deposition by a high-power ASTeX microwave plasma chemical vapour deposition (MWP-CVD) apparatus leads to growth of homoepitaxial diamond films with high crystalline quality, yielding only band-edge emissions as the main peak in CL spectra at low temperatures (80 K).	Suppresses hillock growth on homoepitaxial diamond surfaces.	Wang [111]
The fine-grained film structure caused by the addition of Ta.	Suppresses hillocks in Al-Ta alloy films for interconnections of thin-film transistor-LCDs.	Iwamura [112]
The use of PECVD-PSG film on the metal line before heat treatment of 400 °C or of any type of PSG film before 350 °C.	Suppresses hillocks in metal lines.	Eungsoo [113]
Al wiring (formed into multi-layer structure with each layer containing an element that is not solidly solubilised with aluminum) is used to reduce line resistance in a display unit. The elements are preferably rare earth metal such as Nd, high-melting-point transition metals such as Ta and noble metals such as Pd. Intermetallic compounds of Al and the element are reduced at an interface of the multi-layer wiring and grains of Al are prevented from enlarging to form hillocks.	Prevents hillock formation.	Ueno [114]

(*continued overleaf*)

Table 5.10 (*continued*)

Proposed method	Demonstrated result	References
Al-Y alloys are proposed as candidates for the interconnection materials in ICs; in Al-Y alloy films, as the solid solubility of Y in Al is very small, almost all Y precipitates at grain boundaries and suppresses the grain growth on the film.	Hillock formation is significantly suppressed in small-dimension interconnections.	Lee [115]
The suppression of hillocks in diamond (001) substrates with misorientation angles of around 2° is achieved by the enhanced lateral growth on high-θ substrates.	A hillock-free heavily boron-doped diamond (001) film is grown with a concentration of 6×10^{20} atoms/cm^3, making an ideal bottom layer for diamond electronic devices.	Tokuda [116]

film and substrate in the delamination process of thin film is simulated and the force required for delamination is examined.

Thin films are usually bonded on thicker elastic substrates or assembled to form multilayer materials. In such structures, the mismatch of the material properties and loading effects of different physical origins (thermal, electric, drying, etc.) induces important internal stresses and leads to the formation of complex crack patterns. Cracks may propagate at the interior of each layer or at the interface between different layers (delamination). In applications, the comprehension and prediction of the underlying fracture mechanism is of key importance not only to avoid material failure, but also to drive the self-assembly of small structures through the controlled formation of ordered crack patterns. Despite the effort devoted to the field by leading research groups from the fracture community, reliable analysis and simulation tools are still missing. One of reason is the difficulty the classical theories have in predicting crack initiation, multicracking and complex crack patterns.

Some recent results on delamination in thin films are gathered below.

A finite element (FE) method is used to study the effects of the interface adhesive properties and the thickness of the thin film on the onset and growth of the interface delamination and buckling [119]. The interface adjoining the thin film and substrate is assumed to be the only site at which cracking may occur. Both the thin film and the substrate are taken to be ductile materials with finite deformation. A traction-separation law, with two major parameters – interface strength and interface energy – is introduced to simulate the adhesive and failure behaviour of the interface between the film and the substrate.

Laser-induced dynamic delamination of a patterned thin film on a substrate has been studied [120]. Controlled delamination results from the insertion of a weak adhesion region beneath the film. The inertial forces acting on the weakly bonded portion of the film lead to stable propagation of a crack along the film–substrate interface. Through a simple energy balance, the critical energy for interfacial failure is extracted, a quantity that is difficult and sometimes impossible to characterise by more conventional methods for many thin film–substrate combinations.

A novel experimental method has been proposed to evaluate the fracture properties of thin brittle films on compliant substrates for flexible optoelectronic devices [121]. Based on our understanding of the FMs, a mechanical calculation has been provided to estimate the thin film fracture toughness and film delamination toughness. These values serve as design parameters for the device flexibility and reliability.

Under tension, an elastic thin film may fracture by growing channel cracks. Moreover, channel cracking may be accompanied by interfacial delamination. The conclusions of a recent study made on many substrates [122] are as follows: if the substrate is relatively compliant, critical interface toughness is predicted, which separates stable and unstable delamination, but when the substrate is relatively stiff, a channel crack grows with no delamination, where the interface toughness is greater than a critical value. An effective energy release rate for the steady-state growth of a channel crack is defined to account for the influence of interfacial delamination on both the fracture driving force and the resistance, which can be significantly higher than the energy release rate assuming no delamination. When the film is under compression, it tends to buckle. So it seems that the interfacial properties are critical in controlling the morphology and failure of elastic thin films on compliant substrates.

A procedure to avoid delamination in thin film transistors TFTs (at a non-adhesive interface between a metal pattern and an organic layer) is to make some apertures in the metal pattern. Through these apertures, adhesion between the organic layer and organic material at the surface of the substrate can be brought about [123].

In modern devices, chemical mechanical polishing (CMP) is frequently used. However, the delamination of copper and low-dielectric interlayer has been noticed during CMP. Voids are formed in the copper wiring layer and the migration resistance is lost if the present high stress of the copper layer remains and the stress relaxation by annealing lowers the adhesion strength. This problem can be solved by depositing a low-stress copper-wiring layer. With respect to the adhesion strength, TaSiN or WSiN film is necessary as a barrier layer instead of the present Ta and TaN. In addition, in order to improve the adhesion strength of the film, it is necessary to raise its critical pressure [124].

Step Coverage

Most metallisation processes result in a thinner metal layer over a step than in a flat area. Generally, the target is $>50\%$ step coverage. To achieve this, many experimental techniques have been developed. An example is the use of an oblique angle physical vapour deposition (OAPVD) technique, to obtain conformal thin films with enhanced step coverage on patterned surfaces such as trenches and vias [125]. OAPVD combines a conventional PVD system with a tilted and rotating substrate. The substrate tilt angle (deposition angle) is chosen such that the particles obliquely incident to the patterned substrate can uniformly reach the sidewalls and bottom corners of the trenches/vias. The substrate rotation allows the uniform coating of all the trenches/vias on the substrate. OAPVD allows us to obtain: (i) films that are deposited in a conformal way with enhanced step coverage, avoiding overhangs and voids; and (ii) a conformal coating that provides the complete void-free filling of the trench/via by a single deposition step. The tilt angles required to coat trenches and vias by OAPVD are typically small, making the invention applicable to many PVD systems. Additional technical features such as the in situ control of the deposition angle, the application of an electrical bias to the substrate and ionisation of the incident flux can further increase the step-coverage efficiency of the OAPVD technique.

Corrosion

When Cu began to be used as an alternative to Al for connection lines, the problem of corrosion became critical, leading to chip failure. Corrosion can increase the resistance of the interconnections, nullifying the advantage of the lower resistance of Cu. Furthermore, corrosion can cause a semiconductor device to malfunction. The copper contacts may corrode together, causing shorts. Corrosion can also cause rapid thickness loss of the copper being corroded. Copper corrosion particularly occurs during CMP. This is especially problematic, because CMP is used to expose metal contacts for subsequent connecting. Where the metal contacts are copper, this means that corrosion and its deleterious effects may occur.

Many solutions have been proposed to diminish Cu corrosion. For instance, the patent [126] proposes the following method: a semiconductor device includes an insulating layer, a metal line, one or more corrosive metal components and one or more sacrificial corrosive metal components. The metal line is situated within the insulating layer. The one or more corrosive metal components are situated within the insulating layer and connected to the metal line. The one or more sacrificial corrosive metal components are situated within the insulating layer and connected to the metal line. The presence of the sacrificial components substantially reduces corrosion of the nonsacrificial components. The metal line acts as a cathode, whereas the corrosive metal components and the sacrificial metal components act as anodes. Corrosive current therefore flows from the cathode to the anodes. However, the sacrificial metal components substantially absorb the corrosive current emanating from the metal line. This substantially reduces, if not eliminates, corrosion of the nonsacrificial metal components.

But corrosion can also be found in Al lines. In this situation, one solution is to use an organic moisture barrier (epoxies, silicones and parylene), but the moisture still penetrates through this layer. Another solution is a multilayer metal process, which is more effective: a thin barrier such as titanium or tungsten is laid over the exposed Al, then coated with a noble metal such as gold. But noble metals are susceptible to dendritic formation in moist environments, which might be as destructive to the microcircuit as Al corrosion. A possible solution is an improved process which renders such exposed Al much less susceptible to such corrosion, by using zinc chromate to passivate the surface of the Al that is exposed to the corrosive environment, especially at the contact pads of a semiconductor element. This zinc chromate treatment is applied to Al without using organic binders to hold the chromate to the Al surface [127].

Other Failure Mechanisms for Metallisation

Under this heading, various FMs have been gathered, such as faults of the metallisation process and manipulation faults. Only a few things can be added concerning faults: they are induced by human errors made at the design phase (improper thickness, wrong process parameters, weak adhesion strength, etc.) or during fabrication (scratches on the metallisation layer produced by manipulation faults, etc.). Elaborating the corrective actions is important in delimitating the source of the errors between design and manufacture, by carefully checking the work instructions.

5.1.1.4 Other Wafer-Level Processes

Contamination

Each manufacturing step induces a failure risk, which may be increased by the synergy of subsequent steps. These failure risks depend on a large variety of factors, such as: quality of materials, contamination, quality of chemicals and of the packaging elements, and so on [128].

Particle contamination is a good example of technological synergy, with two effects inducing failure risks for the future device:

- **The physical effect.** The particles mask an area of the chip, hindering the deliberate impurity doping process or producing the breakdown of the processed layer.
- **The chemical effect.** The particle contaminant diffuses into the crystal, producing electrical effects such as degradation of junction characteristics (soft I–U characteristics or premature breakdown); electrical effects may appear later, after the contaminant has migrated into the active area, during device function.

For the physical effect, a failure risk synergy is obvious at subsequent manufacturing steps:

- At photolithography, the dust particles reaching the transparent area of the masks transfer their images on to the wafer.
- At etching, metallisation and ionic implantation, the particles may produce short circuits, open circuits, needle holes or localised areas with different electrical properties.

For the chemical effect, a failure risk synergy comes out because the contaminants containing alkali ions become active at the thermal processes (oxidation, diffusion). Localised regions with ionic contamination arise and the ions migrating to the active areas of the device produce an increase in the leakage currents or a drift in the threshold voltage (for MOS devices). Some corrective actions may be used to remove these effects:

- Contamination prevention, by identifying and avoiding the contamination sources.
- Wafer cleaning, with sophisticated methods for removing the particles reaching the wafer.

The main FA method used to identify the failures produced by contamination is visual inspection performed before sealing, or after various reliability tests such as temperature-cycling, high-temperature-storage and high-temperature-reverse-bias (HTRB).

Diffusion-Induced Defects

Diffusion aims to control the type and concentration of impurities in a specific area of the semiconductor crystal. The impurity atoms, intentionally introduced by solid-state, create *n* and *p* areas in the semiconductor (which is the intended effect), but they may perturb the equilibrium-point defect concentration (the detrimental effect). Two other possible effects are the localised stress generated in the crystal and the formation of precipitates. Moreover, the thermal effect is involved, because the diffusion process occurs at high temperature. The result of all these processes is a diffusion-induced generation of defects, as detailed below.

Basically, the defects of interest are dislocations (line defects) and precipitates (bulk defects), as described in Table 5.2.

In Si, the most used dopants are phosphorus, boron, arsenic and antimony. The dislocations may be induced by three main processes [16]: (i) thermo-mechanical stress during diffusion; (ii) stacking faults induced by previous oxidation processes; and (iii) prismatic punching of dislocations by SiO_2 precipitates in the crystal. Precipitation is possible when the solid solubility limit at the diffusion temperature is exceeded, the excess diffusant precipitating as an element or a compound with silicon. Generally, this phenomenon occurs near the surface, where the highest concentration of diffusant can be found.

The effect of diffusion defects is a degradation or instability of the device characteristics, leading to failure.

Obviously, several methods for controling the diffusion-induced defects have been developed, focused on [16]:

- Control of quality of wafer surface – a highly defect-free surface suppresses the heterogeneous nucleation of dislocations upon diffusion.
- Minimisation of the precipitation effects – the idea is to minimise the surface concentration by growing a barrier formed by a thin oxide layer on the silicon, prior to diffusion.
- Reduction of the diffusion-induced stress induced in silicon lattice, by stress compensation – that is, the sequential or simultaneous diffusion of more than one dopant. For instance, the simultaneous diffusion of phosphorus and arsenic has proven to be effective in reducing the density of misfit dislocations.

Last but not least, the procedure called 'gettering', which uses a diffusion process, must be mentioned. This is used to create a sink for undesired impurities (e.g. heavy metals) at the surface of the wafer: a heavily doped glassy layer (containing a high concentration of phosphorus, arsenic or boron) is deposited on the wafer surface, followed by a high-temperature anneal. The result is an increase of the minority carrier lifetime and improvement of *pn* junction characteristics due to the marked reduction in the metallic impurity content of the wafer.

Implantation-Induced Defects

Ion implantation has been promoted as an alternative to diffusion for doping a semiconductor area, by injecting the ionised dopant atoms into the semiconductor surface, accelerating the atoms to the desired energy and impinging them upon the wafer, the atoms being slowed down and stopped at a certain depth. During implantation, defects are generated by the primary and secondary collision of the ions with the atoms of the host crystal. The ion implantation is followed by an annealing treatment, aimed to electrically activate the dopant atoms and to reduce the degree of radiation damage in the crystal. However, some dislocations may remain in the semiconductor [16]. The main problem in ion implantation is to carefully optimise the implant dose. Moreover, the implantation process is followed by oxidation, which may lead to the generation of a high density of stacking faults.

During the fabrication process of MEMS devices, the single-crystal (and polycrystalline) silicon wafers are implanted with boron ions to improve their mechanical and tribological properties, which are essential for MEMS with moving parts. The effects of ion implantation on the crystallinity, microstructure, nano-hardness, friction and wear properties have been studied and it was found that silicon remained crystalline after ion bombardment at doses up to 2×10^{17} ions cm^{-2}, but with a large amount of defects. It was also found that there was a small increase in the nano-hardness due to boron-ion implantation and the ion-bombarded single-crystal silicon exhibited very low friction and low wear factor. The coefficient of friction of bombarded silicon in dry air and dry nitrogen was found to be even lower. The coefficient of friction and the wear factor of silicon bombarded at high doses were found to be lower at low humidity and dry nitrogen than at high humidity. The fracture of silicon surface and its oxidation during sliding was believed to play a significant role in the friction and wear process. Generally, ion implantations improve the friction and wear properties of surfaces, but the benefit depends on the operational environment [129].

Masking and Etching Defects

The electronic components are delimitated in the semiconductor crystal by the processes of masking and etching. During masking, various human errors may occur, producing scratches, cracks and scars in photo mask or misalignment. These defects may lead to abnormal photoresist patterns (abnormal line widths or intervals), pinholes and so on, the ultimate effect being the degradation of characteristics due to parameter drift, open circuits or short circuits.

The etching process may produce various defects, such as improper etching of oxide film, undercut of the metallised layer, spotting (stains) and non-uniform etching. The effect of these defects is device failure by open or short circuit, or by parameter drift.

For all these defects, the best control method is visual inspection before sealing. This could be preceded by a sequence of tests (e.g. temperature-cycling, high-temperature-storage and operational-life tests), which may activate the possible failures initiated by these defects, allowing the weak items to be removed.

5.1.2 Packaging

The package has an essential role in the operation of electronic components, based on three main functions:

- To interface the die with the external circuit.
- To remove the heat generated by device operation.
- To protect the die from the external environment (mechanical integrity, protection from temperature, radiation, moisture, ions and so on, chemical isolation from harsh environment).

The package must find the best compromise between electrical, thermal and mechanical performances and physical dimensions to meet product-specific applications, reliability and cost objectives. There are two main categories of package: *single-chip packages* (SCPs) and *multichip packages* (MCPs). As the name suggests, there is only one silicon die in an SCP. The earliest ICs were packaged in ceramic flat packs, which continued to be used in military applications for their reliability and small size for many years. When multiple dies are put in one package, this is called *system in package* (SiP). When multiple dies are combined on a small substrate, often ceramic, this is called *multichip module* (MCM).

In recent years, a new category of packaging technology has begun to be used: *wafer-level package* (WLP). An example is *chip-scale package* (CSP), which entered the industry's lexicon in 1994 [130], and is defined as a package with a perimeter that is no more than 1.2 times the perimeter of the die it contains. Such packages combine the best features of bare die assembly and traditional semiconductor packaging, and reduce overall system size, something that is to be desired in portable electronic products. WLP is used for the technologies of packaging an IC at wafer level, instead of the traditional process of assembling the package of each individual unit after wafer.

Basically, almost all previously mentioned packaging technologies comprise the same processes:

- **Dicing.** The wafer is cut into rectangular blocks, called dies, each containing one device.
- **Die attach.** Each die is firmly attached on a substrate or lead frame. Die attachment performs several critical functions: (i) it creates a good thermal path between the die and the package base, which is itself usually attached to a heat sink to remove the heat generated in the die by device operation; (ii) it creates a good electrical contact between the backside of the die and the package; and (iii) it maintains these two critical roles over the lifetime of the device and through the environmental conditions required for the mission.
- **Wire connection.** There are three possible variants: wire bonding (interconnections are made, by using metallic wires, from the pads usually found around the edge of the die to the leads of the package); tape-automated bonding (TAB); and flip chip (FC) (or controlled collapse chip connection).
- **Sealing.** The whole ensemble is sealed hermetically (by metal or ceramic cap) or nonhermetically (by plastic material).

The main FMs induced by each of the above processes are detailed below.

5.1.2.1 Dicing

An improper dicing process may produce cracked or chipped dies, which may lead to open circuits. To avoid this, a visual inspection (with optical microscope) before sealing, which may be preceded by a series of tests (temperature-cycling, vibration, mechanical-shock, thermal-shock), eliminates weak items before assembly.

5.1.2.2 Die Attach

Die attach may be carried out by epoxy, eutectic or glass frit attach. The common epoxy attach method bonds the chip to the lead frame using epoxy. Eutectic attach uses a thin film of gold on the backside of the wafer, alloyed with the metallised surface of the lead frame. Glass frit attach uses a mixture of silver and glass in an organic medium to attach chips in a ceramic package [131].

An important FM induced by die attach is **die lifting**, which is the detachment of the die from its pad or cavity. A die that undergoes die lifting is commonly referred to as a 'lifted die'. There are two possible causes for die lifting [132]:

- Fracture within the die attach material itself (cohesion failure).
- Delamination between die backside and die attach material or between die attach material and the die pad or cavity (adhesion failure).

The first action in FA is to identify the type of die lifting that is responsible for the failure.

The *cohesion failure* could be produced by material mismatch, excessive voids, insufficient fillet formation or inadequate bond line thickness (BLT). The fracture strength of the die attach material is diminished, which can lead to failure once the unit is subjected to thermo-mechanical stresses (temperature-cycling, high-temperature-storage, constant-acceleration tests). When this happens, cracks or peeling appear, then the die attach material fractures in the middle, resulting in die lifting, leaving die attach material still sticking on both the die backside and the die pad. The voids can lead to degradation by overheating.

The degradation of the mechanical strength of the die attach material can also be due to chemical degradation with time, or can be produced by external factors: moisture, temperature and so on.

The above-mentioned causes are also responsible for *adhesion failures*, but there is a specific cause for this FM: the presence of contaminants on the die backside, which can lead to die-attach-to-die delamination, while contaminants on the die pad can lead to die-attach-to-die-pad delamination. Either way, the resulting delamination can lead to die lifting. Eutectic die attach delaminations may also be due to inadequate scrubbing, incorrect perform size and improper equipment settings.

Die lifting may be accelerated by the solder heat-resistance test (SHRT), temperature cycles and thermal shock and can be identified by visual inspection.

Die cracking is another FM induced by die attach. Basically, this is caused by poor bonding of die to header. If eutectic is used for die attach, inadequate die attach fillet formation and excessive die attach voids can act as stress concentrators, which can also result in contiguous cracks at the backside of the die. These cracks can propagate to a point where the upper part of the die is separated from the bottom part [132]. This FM can be identified by shock and vibration tests before sealing, followed by visual inspection.

Other FMs are initiated by **deficiencies in the material** used for attaching the die. This may include silver-filed epoxies, gold silicon eutectics and solder types. In this case, the FMos are short circuit, loose die and loose conductive particles. Short circuits occur through the excess application of die attach material where it comes in contact with bond pads or wires. On the other hand, the weakening of a die mount comes from a lack of sufficient die attach material. For epoxies, insufficient curing times and temperatures can lead to deleterious effects associated with decomposition, where the release of moisture in a sealed package can induce corrosive byproducts affecting Al wires and causing thermal resistance changes that affect the heat-transfer characteristics of the epoxy meant to regulate thermal conditions of the die. Moreover, if a sealed package containing epoxies is overheated (typically at temperatures exceeding ~150 °C) during assembly, testing or use, the epoxies will begin to decompose, with water as one of the decomposition products.

In order to identify the failures, various methods, such as optical microscopy, SEM, X-ray and scanning laser acoustic microscopy, can be used. Decomposition of epoxies can be determined by observing the moisture content found in residual gas analysis (RGA) testing of a sealed package in accordance with MIL-STD 883E Method 1018.2. Thermal decomposition products of epoxies are typically H_2O, CO_2 and hydrocarbons such as methane and methanol.

Some methods for avoiding the FMs produced by die attach are presented below:

- Wafer backside coating (WBC) is an adhesive for die attach, supplied as a specifically designed paste that is applied to the back of a wafer and dried. This method has a number of extremely desirable practical advantages: (i) paste is 20–30% cheaper than film; (ii) BLT can be controlled to customer specs; (iii) fillet control is very similar to that of a film product; (iv) dispense operation is eliminated, which affords higher units per hour (UPH); and (v) coated wafer can be stored until required [133].
- Gold-based alloys were successfully used as die attach materials (eutectics), due to superior thermal performance. The die attach temperatures are much higher relatively: ~320 °C for AuSn and ~420 °C for AuSi; but the thermal conductivity is significantly better, with AuSn performing at ~70 W/mK and AuSi at >150 W/mK. The disadvantage of this approach is that it is a very tricky and sensitive process with a narrow processing window that needs to be monitored carefully. Another possible variant is the use of gold-based catalysts, which have significant performance

benefits in many reactions (selectivity, activity), but applications have been limited due to processing concerns [134].

- For wafer thicknesses of 50 and 25 μm, a more robust and less uncertain process than current die attach paste technologies is needed. The solution seems to be the use of dual structured films, taking the form of 'die attach films' (DAFs) [135].
- Another report about DAFs underlines their good control of paste bleed, creeping effect to die edge and consistent BLT at desired thickness. A new generation of DAF tape incorporates wafer dicing tape and adhesive in one, termed the 'dicing die attach film' (DDAF), which is mounted on to the back of the wafer. DDAF has a lower reaction rate and lower dynamics modulus and is better at ensuring its fluidity and gap-filling characteristic for better adhesion and reliability performance in the package, by eliminating delamination during the die attach process [136].
- A recent paper reported the effect of repeated heat cure during assembly processes on DAF material properties. The affected area experienced weak die bonding and delamination. Die attached condition and DAF material selection were evaluated to find the required reliability performance in the manufacturing of the 3D quad flat no-lead (QFN) stacked die package. Stress can be relieved by having a higher die attached temperature with an adequate bonding force and time; numerical simulation indicates that this is important [137].

5.1.2.3 Wire Connection

There are three main technologies for creating interconnections in semiconductor packages: wire bonding, TAB and FC, as detailed below.

Wire Bonding

Historically, wire bonding was the first method of interconnecting the pads of the die with the leads of the package and it is still the preferred assembly technique. The wire material could be Al, Cu or Au, with diameters ranging from 15 μm to hundreds of micrometres (for power devices). In Table 5.11, a comparison between the characteristics of the three metals used for wiring is shown.

The main FMs induced by wire bonding are detailed below:

- **Wrong bonding process.** Under this name, a collection of FMs induced by design or manufacturing faults may be mentioned: excessive or poor bonding strength, improper bonding method or control, insufficient bonding area or intervals, improper bonding arrangement, unremoved tail wire, cracked or nicked die, cuts, notches or scratches in lead, excessively looped lead, excessive or insufficient drooping length of lead, material mismatch, contaminated bonding pad. The typical failures are: broken or disconnected wires, intermittent or total open circuits or short circuits. Generally, these FMs are identified by applying a sequence of stress tests to the whole batch of devices: mechanical stresses (constant acceleration, shock, vibration) and thermal stresses (high-temperature storage, temperature cycling), followed by electrical tests and visual inspection (if the device is not sealed) or radiography (for sealed devices).
- **Au-Al intermetallic growth.** The two possible compounds are: Au_5Al_2 ('white plague', a white compound with low electric conductivity, leading to an increase of electrical resistance, and eventually to total failure) and $AuAl_2$ ('purple plague', a brittle, bright-purple compound, leading to creation of voids in the metal lattice). A recent X-ray photoelectron spectroscopy (XPS) study of early stages of Al-Au interface formation can be explained on the basis of comparison of the thermodynamic stability of the Al_xAu_y intermetallic compounds [138].
- **Kirkendall voiding.** An interdiffusion process arises at lower temperatures (400–450 °C) and forms compounds with different compositions, from Au-rich to Al-rich, with different growth rates. The denser, faster-growing layers consume the slower-growing ones and form cavities, leading to both increased electrical resistance and mechanical weakening of the wire bond.

Table 5.11 Comparison of the characteristics of metals used for wire connections: Al, Cu and Au

Metal	Wire characteristics	Typical use	Reliability issues
Gold (Au)	Diameter larger than 12.5 μm	Ball-bond to the chip, then stitch-bond to the substrate; wedge bonding.	The main failure risk is 'purple plague' (a brittle gold-aluminum intermetallic compound). The mechanical properties and thermal stability are improved by doping Au with controlled amounts of beryllium or copper.
Aluminum (Al)	Diameter larger than 100 μm	Ultrasonic bonding.	Al has to be doped with controlled amounts of magnesium to provide greater drawing ease to fine sizes and higher pull-test strength. By using Al instead of Au, 'purple plague' is eliminated.
Copper (Cu)	Large diameter (up to 250 μm)	Ball bonding. Replacing Al wire with high current-carrying capacity.	Cu is harder that Al and Au, so bonding parameters have to be optimised carefully. Oxide formation may lead to reliability risks (special wire protection is needed).
	Small diameter (up to 75 μm)	Ball bonding. Replacing Au wire, giving comparable performances and lower material cost.	

In **TAB**, the die is attached directly to the printed circuit board (PCB) by a polyamide tape, the bonding sites of the die being connected to fine conductors on the tape, which already contains the actual application circuit. Then the bare chip is encapsulated in epoxy or plastic material. TAB is used mainly for liquid crystal display (LCD) driver circuits. The polyimide tape takes the form of a roll, with widths of 35–70 mm and thicknesses of 50–100 μm. The main failure risks in TAB are linked to irregular bond height or shape, insufficient bonding pressure or energy, or tape contact contamination. All of these result in weak bonds [139].

FC (also named controlled collapse chip connection or C4) is a procedure of die and wire bonding developed for ICs and MEMS.[4] The connections to the external circuit are made by solder bumps, which are deposited on to the top side of the chip pads. In the mounting process, the die is flipped over, so that its top side faces down, its pads being aligned with the matching pads of the external circuit. To attach the FC to the circuit, the solder bumps are placed on to connectors on the underlying circuit board and re-melted by ultrasonic or reflow solder processes to achieve electrical connection. FC assembly has the advantage of being much more compact than traditional systems (no peripheral space for wire bonds is needed) and allowing higher-speed signals and efficient heat removal (an external heat sink can be added directly above the chip to remove heat).

In October 2008, a high-speed mounting methodology for FC packages, named 'MicroTape', was launched through a cooperation between Reel Service Ltd and Siemens AG, and proved to be easier to handle than wafers or other adhesive-based tape systems, providing for less component attrition due to handling problems [140].

FC could be included in WLP, which has expanded in recent years to a wide range of semiconductor devices applied in a cross-section of industries from automotive to mobile phone, from sensors to medical technology. WLP uses standard surface-mounted technology (SMT) platforms.

[4] The first commercial flip chip was used by IBM, in the 1960s.

Chip interconnect (solder to silicon) reliability is one of the critical elements in the qualification of FC bumping technologies. Since the interconnect materials, structures and processes vary between different bumping technologies, the strength and reliability must be evaluated for each design. As lead-free solders are used in the systems, the intermetallics associated with the lead-free solders and the under-bump metallurgy (UBM) also have an influence on the interconnect reliability. In addition, the stress that an interconnect experiences during thermal cycling depends on the properties of the solder alloy used in the interconnects. Different solder alloys require different interconnect strengths to achieve good reliability in thermal cycling. The interconnect reliability has been studied by comparing the interconnect strength and the working stress in the interconnect during qualification and application in several lead-free solder systems, including Sn/Ag, Sn/Ag/Cu and Sn/Cu solders and Ni-Au and TiW-Cu UBMs [141]. A simple stress model has been developed to determine the interconnect stress during thermal cycling and a testing methodology has been established to determine the interconnect strength.

Another significant failure risk of FC is related to the mismatch of CTE between the silicon die and the organic substrate, which leads to premature failures of the package. A method of improving the package reliability has been proposed: a novel underfill material, obtained by a nano-filler technology, which provides a previously unobtainable balance of low CTE and good solder joint formation [142].

Cracks in interconnections can lead to failure, so methods to fight against this FM have been created. A test chip module for testing the integrity of the FC solder ball interconnections between chip and substrate has been proposed [143]. The interconnections are thermally stressed through an array of individual heaters formed in a layer of chip metallurgy to provide a uniform and ubiquitous source of heat. Current is passed through the interconnection to be tested by a current-supply circuit using a signal I/O interconnection and the voltage drop is measured by a voltage-measuring circuit connected through another signal I/O interconnection. Stress-initiating cracking and degradation at the interconnection creates a measurable change in voltage drop across the interconnection.

An efficient procedure for reliability improvement is the introduction of underfill in FC packaging, which gives solder interconnect technology an unforeseen mechanical robustness and increased FC solder fatigue resistance [144].

5.1.2.4 Wafer-Level Packaging

In the last few years, a subset of FC technology called wafer-level packaging has been developed. This is also called chip-scale packaging, because the resulting package has practically the same size as the die. However, in recent years, the trend has been to increase the package size, by including fan-out connections [145]. With WLP, device interconnection and protection (attaching the top and bottom layers and the solder bumps) are executed during wafer processing, before wafer dicing.

There are two major factors limiting the reliability of WLP, especially for die sizes larger than 5×5 mm: (i) interconnect fatigue due to stresses generated by the CTE mismatch between the die and the PCB and (ii) packaging cost. For WLPs that require a redistribution layer (RDL) for I/O redistribution and a compliant layer for reliability, the electroplating process and the dielectric layer make up a large portion of the overall packaging cost. The costs of RDL for a WLP with two metal layers are even greater. Recently, Tessera has developed a new compliant WLP that dramatically lowers cost versus previous compliant WLPs [146]. The compliant layer is more cost-effective than spin-on polyimide, benzocyclobutene (BCB)[5] or silicone dielectrics, which are used in conventional WLPs. The copper conductor is etched (which is cheaper than electroplating) to form traces and protected with solder mask. Wirebonds are used to connect the individual die pads to the copper traces, and eventually encapsulation and solder ball attach complete the packaging process.

Underfilling (UF) is used to solve reliability problems, because UF reduces the effect of CTE mismatch between the silicon chip and the substrate. Also, UF protects the chip against impurities

[5] Employed as a filling material for encapsulation.

and makes the structure mechanically stronger by minimising the stress levels or fatigue in the solder joints. However, the adhesion of the die side passivation (polyimide) to UF is critical to the reliability of FC assemblies [147]. Weak interfaces between the die polyimide layer and the UF resin can result in yield loss during thermal cycling or when exposed to a highly accelerated stress test environment of heat and humidity. UF materials absorb moisture, and accumulation of moisture at the interfaces can lead to:

- Crack propagation caused by swelling stress.
- Weakening mechanical support.
- Die-level interconnect failures.
- Corrosion due to ionic contaminants resulting in metal migration failures.

Hence, with decreasing trends in the form factor, the adhesion of the UF polyimide is critical to the reliability of FC assemblies and is becoming more of a focus in the assembly world [148]. New reliability challenges are brought by the move to finer pitch. Reliability is affected because the cross-sectional area of the joint between the package and the board tends to be smaller, so can handle less stress and often fails during reliability testing. As pitch decreases and the distance between adjacent contacts is reduced, it becomes difficult to eliminate bridging or electrical shorting during the assembly processes. A new technology is needed to solve these issues.

In [149], a study of the possible packages used by WLP is presented. Thin film redistribution and bumping, encapsulated package, compliant interconnect and wafer level underfill are discussed, with an emphasis on the challenges and processes of the wafer-level underfill. The WLP integrated with wafer burn-in, test and module assembly shows great attraction due to the dramatic cost reduction, with cost-effective methods of building wafer-level test and burn-in being detailed.

Two new trends in WLP are towards the third dimension (3D package) and small dimensions nano-wafer-level packaging (nano-WLP).

A **3D package** contains two or more chips stacked vertically so that they occupy less space. The concept was developed for IC, but could be applied to other electronic components. The names used are SiP or chip stack MCM. The wiring of the stacked chips is done along their edges, increasing the length and width of the package and usually requiring a supplementary layer between the chips. The solution to avoiding such an increase of package dimensions is to use through-silicon vias (TSVs) to wire the chips. TSVs are vertical connections through the body of the chips. Consequently, a TSV 3D package (also called TSS – through-silicon stacking) can also be flatter than an edge-wired 3D package, but is smaller in length and width. TSV technology has a wide range of applications: cell phones, MP3 players, notebooks and digital still cameras. Being a new technology, the reliability issues of TSVs are not yet completely understood.

3D interconnects offer an attractive option to reduce the energy dissipation and propagation delay of long on-chip wires (51 and 54% reduction in latency and energy dissipation respectively at 45 nm node). Also, optical interconnects offer reduced latency compared to scaled Cu/low-k technologies, but do not offer significant improvement compared to other technologies like WLP interconnects. A good solution might be the carbon nanotube (CNT) interconnects, which are compared favourably with scaled Cu/low-k interconnects in terms of latency, with a 42% reduction in delay [150].

Recently, a new concept, called **nano-wafer-level packaging**, was proposed by a joint group from the Institute of Microelectronics, Singapore and the Georgia Institute of Technology, United States, led by Prof. Rao Tummala [151]. This packaging system aims to incorporate a number of convergent technologies, both at IC and at systems level, so as to be much smaller and more affordable. The new nano-WLP has the potential to bring time and cost savings in manufacturing, and to enhance electrical performance (due to the short nano-interconnections), but much work is still needed before this concept can become a commercial one. Obviously, a careful analysis of the reliability risk is required. But most existing interconnect technologies today, even with dramatic improvement, cannot address the reliability of ultra-fine-pitch interconnections. Hence, the thermo-mechanical design requirements, FMs and reliability of such interconnects must be studied and materials will be characterised for tensile strength, hardness, fracture toughness, fatigue and electrical properties. Finally, an integrated wafer-level testbed prototype to demonstrate the reliability of the 100 μm-pitch WLP system will have to be built [151].

5.1.2.5 Encapsulation

The final operation of device construction is the encapsulation (or sealing) of the die in ceramic, plastic or epoxy to prevent physical damage or corrosion. Many FMs are related to this operation, being produced by:

Contamination Issues

The current assembly technology relies heavily on the control of contamination at the package level to increase product reliability. In the assembly world, surface analytical techniques are primarily used as problem-solving tools as part of the framework of FA. Fourier transform infrared spectroscopy (FTIR) provides unique chemical information and some surface-specific information by attenuated total reflectance, which is a special case and not generally applicable to most types of sample in assembly. The primary surface-analysis tools are XPS and time-of-flight secondary ion mass spectroscopy (TOFSIMS) [147].

The improper atmosphere in packaging may lead to the degradation of characteristics attributed to the inversion layer or channelling. To identify weak items, stress tests (operation-life, HTRB, high-temperature-storage and temperature-cycling tests) are used.

Process Errors

The main possible process errors are:

- Faults in the seal glass (cracks, voids or migration), leading to leakage – intermittent or open circuit – to be identified by stress tests (seal, electrical, high-temperature-storage, temperature-cycling and high-voltage tests).
- Incomplete hermetic seal (for metallic or ceramic packages), producing characteristic degradation or short circuit due to chemical corrosion or humidity. A seal test is needed to identify the failure risks.
- Dielectric particles floating in the package that may produce intermittent or short circuit. The recommended stress sequence for eliminating these failures is: constant acceleration, vibration (monitored), radiography, shock (monitored) test.
- Broken or bent external lead, which leads to open circuit and can be identified by visual inspection followed by lead-fatigue test.

5.1.2.6 Tin Whiskers

Pure tin-plated finishes provide a protective coating that is resistant to oxidation and corrosion and possesses good solderability characteristics. The typical FM is *tin whisker* growth, which causes system failures in both earth- and space-based applications as well as missile systems. Tin whiskers are very thin, single-crystal fibres, with large length-to-diameter ratios and constant cross-sectional area. These whiskers can grow only from the surface of pure tin. When internal stress exists, annealing alone may not be sufficient to totally relieve it. Such failure experiences and root-cause determinations have led to the prohibition of pure tin-plated components by the military and NASA [152].

The main FMo produced by tin whiskers is the short circuit, which is dependent upon the spacing away from the adjacent components/conductors and the whisker growth potential [153].

A study on tin whiskers has clarified some of their characteristics [154]. Growth rates are increased if the samples are held in an oven at 50 °C. Bulk samples have been prepared in a similar way by depositing thicker tin layers on to larger brass substrates. Judicious choice of the pressure in the sputtering chamber allows samples to be prepared in which the macroscopic stress in the film is tensile, compressive or neutral. The whisker is electron-transparent and it is possible to examine its structure. If the nucleation site is close to the edge of the hole then this region can also be examined. The whisker has been found to be monocrystalline, single-phase and to contain no extended defects. The absence of dislocations implies that dislocation-mediated growth models are not applicable. Diffraction patterns obtained from the whisker base indicate intermetallic formation that causes a biaxial stress in the tin film. It is suggested that compressive micro-stresses coupled with diffusion lead to formation of whiskers. The use of appropriate buffer layers may prevent their nucleation.

The drive to eliminate lead from electronics is pushing component manufacturers to consider pure-tin coatings as an economical lead-free (Pb-free) plating option. This change in plating material is further motivated by the reported industrial problems caused by contamination of Pb-free solders by the Pb contained in protective solder coatings. However, it has been shown that tin whiskers increase in the absence of lead in solder. Alloying of the tin will eliminate the potential growth of tin whiskers [155].

It is not easy to study tin whiskers, because they can lead to field failures that are difficult to duplicate or are intermittent (sometimes referred to as 'could-not-duplicate' or 'no-fault-found' failures): at high enough electrical potentials, the conductive particle can vaporise, thus removing the failure condition. Alternatively, disassembly or handling may dislodge a failure-producing whisker [156]. An even more insidious factor is the large unpredictable variation in the incubation or dormancy period for tin-whisker formation. For example, while studies report whisker formation within a period of days to months, in other cases, field failures are described that occurred more than 20 years after the components were manufactured [157].

With the reduction in spacing in electronics, the probability of a conductive whisker bridging the gap between interconnects and producing a short increases. In addition to miniaturisation, the voltage used in many electronics has been reduced. At lower voltages, a conductive whisker is unlikely to be destroyed if it does successfully create a short. As a result, persistent shorting failure may occur. Further, vibration screens and handling of an electronic assembly may cause surfaces with tin-whisker growth to shed. The shed whiskers can then produce shorts within the electronic system. Unfortunately, existing screens may not find whiskers. Whisker growth in fielded product represents a potential failure time bomb. At present, there is no known method to guarantee whisker-free surfaces on pure-tin finishes. The growth mechanisms are unknown; diffusion processes within finish or on surface are likely involved, but what drives diffusion into specific grains and launches them out from the surface? The fundamental research is incomplete; whiskers are not dendrites [156].

The factors that may influence metal whisker growth (increasing stress or promoting diffusion within the deposit) are: plating chemistry, plating process, deposit characteristics, substrate and environmental parameters; however, many experiments show contradictory results for these factors. A solder joint intermittent failure case study is shown in [158]. The no-fault-found observations are discussed and guidelines for assessment of intermittent failures are provided.

As mentioned before, the best method for avoiding tin whiskers is to use tin-allowing instead of pure-tin plating. But if this option is not available, the following approaches may also be considered to reduce risk [159]:

- Solder-dip the plated surfaces by using a tin-lead solder to completely reflow and alloy the tin plating.
- Replate the whisker-prone areas.
- Apply conformal coat or foam encapsulation over the whisker-prone surface.
- Evaluate application-specific risks.

5.1.2.7 Plastic Package

The *plastic package* (a non-hermetic package) was proposed by General Electric, in 1962, for Bi transistors. Used initially only for mass consumption, without taking into account the reliability issues (which were solved only by hermetic packages, whether metal or ceramic), the market for plastic packages quickly increased. The main advantages of plastic package are high resistance at mechanical stress and at aggressive liquids and gases, good surface isolation of the incorporated die, good precision of the mechanical dimensions and reduced costs. There were some problems linked to the free ions, especially at high temperatures, but plastic materials with a much reduced number of free ions, and with dilatation coefficients close to those of the metal or silicon, were soon obtained.

Due to extensive reliability studies, the failure risks were removed one by one, and a significant improvement in the performance of plastic packages was obtained, especially in the early 1990s. In a study from 1996, performed by the Reliability Analysis Centre (RAC), field failure rates from one-year-warranty data were analysed [160]. It seemed that for both hermetic and non-hermetic devices a decrease of more than 10 times in the failure rate was found between 1978 and 1990. In another study, reported in 1993, a 50-times decrease in the failure rate of plastic-encapsulated microcircuits (PEMs) over the period 1979–1992 was found. These results are confirmed by many other industry studies and can be explained in two main ways [1]:

- Covering 97% of worldwide market sales, the plastic-encapsulated semiconductor devices were the most studied devices.
- The absence of the severe controls of military standards allowed a continuous process improvement, leading to the given results.[6]

Eventually, a major cultural change arose in the procurement politics for military systems. Known as the Acquisition Reform, this new approach encouraged the use of plastic-encapsulated devices in US military equipment and, consequently, in the military systems of all countries.

The material used for plastic encapsulation is thermo-reactive resin: a combination of phenol and epoxy resins or silicone resins. The moulding material contains a basic resin, a drying agent, a catalyst, an inert material, an agent for firing delay and a material facilitating the detachment of the package after the moulding operation. If chosen properly, the moulding material may diminish the action of various FMs produced at wafer level, such as intermetallic Au bond–Al pad interface, which may lead to corrosion and failure at operation above 180 °C. It is necessary to choose the right moulding compound alternatives, such as environmentally friendly halogen-free compounds, to mitigate high-temperature corrosion. A study focused on this subject offered a characterisation of bromine-related wirebond weakening processes, establishing the high-temperature reliability of halogen-free moulding compounds [161].

An ideal moulding compound has: low permeability to moisture, high strength at elevated temperatures, a high glass transition temperature and excellent adhesion. Much research on moulding compounds has been directed towards reducing their moisture permeability and raising their temperature gradient, in order to increase their high temperature strength. Newer biphenyl resins have been developed with filler content approaching 90%, considerably reducing the moisture permeability of the moulding compound. Though the strength of the moulding compound also decreases, the compensation derived from the reduction of moisture absorption counterbalances this at reflow temperatures.

It is important to determine the appropriate size of the spherical silica filler particles included in the package body to ensure thermal cycling-related reliability in plastic-encapsulated packages [162]. This is because the thermal shrinkage of the plastic package body can cause serious damage to the active pattern of the device due to the compressive stress resulting from the fillers pinned by the lead

[6] This is a good example of what could be done by continuous process improvement, without standards (e.g. ISO 9000 series) aimed to restrict such activity.

frame. In particular, the model suggested in this work indicates that the combined action of a large filler and smaller fillers can become a failure-causing factor in the plastic package. Thus the adoption of an appropriate filler size in the plastic encapsulation might allow a greater reliability margin for the modification of the lead-on-chip package structure.

The content of ionic impurities in the resin material is important for the moisture resistance. Ionic impurities in the resin are evaluated using the hot-water-extraction method. Of the various ionic impurities, Cl^- ion is thought to have an especially pronounced effect on moisture resistance. Formation of a parasitic MOS due to the accumulation of ions is one of the typical mechanisms of device degradation caused by ionic impurities in the resin [163]. Using this phenomenon, it is possible to evaluate the ionic substance in the resin through the ion accumulation rates on the gate oxide film.

Generally, **ion contamination** may diminish the current gain of a transistor, increase the leakage current and produce the corrosion of the Al metallisation, accelerated by the ionic impurities from the moulding material, especially in a humid environment. The moisture penetrates into the package, reaching the die, especially along the contact area between the moulded material and the metallic frame. There are two possible sources of ion contamination: (i) the encapsulant (even if the moulding compounds are considered ion-free, there is a typical ionic residue level of less than 10 ppm) and (ii) the external sources (e.g. salt mist, industrial atmosphere and corrosive solder flux).

A series of methods aiming to diminish the effects of ionic contamination have been proposed [1]:

- Recovery of the wires after bonding with a high-purity protective resin.
- Die passivation with a silica glass, which mechanically protects the die. The phosphorus concentration must not exceed 2% in weight, in order to avoid a catastrophic increase in Al corrosion.
- Impregnation of the package, after moulding, with resins liable to fill the holes or the micro-cracks which may exist at the frame–moulding material interface.
- Replacement of the Al layer by a multilayer (titanium/platinum/gold) passivated with silicon nitride. In this system, silicon nitride assures the junction hermeticity, the titanium layer improves the adherence of the dielectric, platinum is a barrier layer for diffusion and gold constitutes the conducting layer.
- Use of silicone gels as plastic encapsulants for high-reliability ICs.

Corrosion may be chemical, galvanic or – with an external bias – electrolytic. The time till the appearance of a short circuit depends on temperature, relative humidity (RH), presence of ionic contaminants, geometry of Al interconnections and the type, plastic purity and mechanical design of the package. Galvanic corrosion may occur in plastic packages if the following elements coexist: a bimetallic couple (most often Au-Al, present in the gold bond wire to the aluminum metallisation pad), free (mobile) ionic contamination (usually chlorine, potassium, bromine and/or sodium) and moisture (diffused from the atmosphere), to form an electrolyte.

PEMs with Al triple-track structures were exposed to mixed flowing gas conditions to simulate and accelerate possible environments during long-term storage [164]. No increase in resistance was measured and no corrosion products were observed after 800 hours of accelerated exposure. Further experimentation indicated that chloride gas reacts with surface moisture in microscale and macroscale voids within the encapsulant, creating chloride ions. These ions become strongly bound to ion-getters present in the epoxy moulding compound (EMC), trapping the chloride ions within the bulk encapsulant and effectively retarding the diffusion process, which can lead to corrosion at the surface of the die.

In a plastic-encapsulated leaded surface-mount package, the FM known as *creep corrosion* will only be a reliability concern if the corrosion product is electrically conductive and bridges across two electrical paths, such as leads. The FMo is generally current leakage [165].

The main parameters characterising the resistance to humidity of a plastic package are: relative hermeticity, dilatation coefficient of the moulding material, quantity of hydrolysable contaminants in the moulding material and die resistance to corrosion. The most significant (but also most controversial) accelerated test for the evaluation of resistance to humidity is to study ageing in function at high temperature (+85 °C) and in a humid environment (RH 85%, deionised water). The bias must lead to a minimum dissipation on a die, but with a maximum voltage gradient between the neighbouring Al conductors. The penetration of the moisture depends on the partial vapour pressure. However, it should be emphasised that for this kind of test the ions essentially arise from the plastic package itself, while in an operational environment they are brought from the outside, by the moisture.

Moisture sorption (absorption and desorption processes of moisture) experiments were conducted on a set of PEM samples with a common type of encapsulant material in order to: (i) characterise sorption behaviour; (ii) compare weight-gain measurement to the measurement of moisture concentration using a moisture-sensor device at the die surface; and (iii) assess the moisture-sensor measurement method. In PEM samples tested in this study, simple Fickian diffusion was shown to agree closely with the experimental results [166]. The calibration constants determined for the sensors were found to be significantly different from those collected by the manufacturer prior to the encapsulation of the devices. The problem is believed to be degradation of the moisture sensor's sensitivity due to exposure to high temperatures and storage conditions.

Moisture sorption was also studied using experimental measurement and FE simulation in [167], which found that the water molecules inside the plastic material were chemically bonded with polymers by hydrogen bonds in the micro-holes formed by the polymer molecule chains. The delamination of FC packaging inspected by C-mode scanning acoustic microscopy (C-SAM) during a high-temperature and high-humidity accelerating test was explained by the state change of the water in plastic material space. The delamination recovery resulted from the increase in content of liquid water. The bonding of water molecules and polymers reduced the adhesive strength at the interface between epoxy material and die, and the delamination on the interface was initiated.

When PEMs underwent a few hundred hours in a steam pressure pot (SPP) test, a harsh moist environment, high leakage currents were noticed. Two possible causes were identified: (i) mould compound and (ii) the polyimide tape used for coplanarity of lead-frame fingers [168]. It seems however that the leakage current is independent of the frame and is not caused by the mould compound, but rather by the ionic content and acrylic-based adhesive layer of the polyimide tape. Ionic leakage current may occur as a result of moisture penetrating the package and accumulating at the chip. Moisture can reach the chip via penetration along the plastic–lead frame interface, through pores and cracks, as well as via vapour diffusion through the EMC. Polyimides, like epoxies, can absorb several weight per cent of moisture, affecting both their mechanical and their electrical properties. The solution proposed for eliminating the high leakage current is to use polyimide tape with low ionic content and non-acrylic-based adhesive.

Failures in plastic packages caused by **thermo-mechanical stress** may occur at die or plastic level. The lead frame can initiate failure in the die or plastic, leading to an increase in the thermal resistance of the package. Die-related failures include: metal shift, die cracking, electrical failure, filler particle point failure and passivation damage. Plastic-related failures are concerned with the formation of cracks in the body of the package. Plastic cracks are usually derived from the delamination of the plastic from the plastic–silicon interface or the plastic–die paddle interface, which can give rise to the popcorn cracking [169].

The combination of moisture absorbed in the plastic and thermal stresses caused by the different expansions of the metal lead frame and the plastic may produce cracks in larger plastic surface-mounting packages, initiated by internal stresses during soldering. Some results show that a critical amount of moisture absorption may lead to cracking, which can be diminished by baking procedures [170].

The mismatch between the CTEs of the plastic material and those of the other constituent parts (frame, gold wires and die) may lead to open or intermittent contacts by **interfacial delamination** between the epoxy and the lead frame, which is a loss of adhesion at an interface that provides an entrance for moisture into the package. This moisture can transport ionic contamination to form an electrolyte, and therefore corrosion can result. Parylene and urethane conformal coats have been widely used to provide additional protection to assembled circuit boards through potential incidental contamination and moisture ingress due to handling.

Delamination occurs in packages because of: (i) surface contamination that degrades the interfacial strength; (ii) excessive shear stresses induced by the thermal treatment of the post-assembly processing; (iii) partial degradation of the interface due to moisture absorption – analysis by scanning acoustic microscope, conducted on packages that were preconditioned to various levels of moisture, revealed a change in signal intensity with higher moisture content; and (iv) the familiar 'popcorn' effect encountered during assembly (see below).

To minimise the delamination, a multifaceted approach is required: selecting the proper epoxy resin chemistry, improving the resin wettability to the package components and reducing the hygroscopic nature of the resin. The proper resin chemistry incorporates the optimal concentration of coupling agent (to enhance adhesion of the matrix to the fillers), adhesion promoter (to enhance adhesion of the matrix to the die and the lead frame) and release agent (to facilitate removal of the packages from the mould cavities). On the other hand, resin wettability can be improved by lower resin and hardener viscosities. Finally, moisture uptake is reduced by using hardeners with higher functionality or less hygroscopic groups in the polymer chains.

The interface of an EMC–copper lead frame was studied [171] by conducting a series of button shear tests to evaluate the interfacial adhesion between EMC and copper. In each test, the failure load acting on the EMC of the button shear sample was measured at different shear angles and an FE model was used to evaluate the stresses at the EMC–copper interface. An energy-based failure criterion was proposed, which could be applied by deriving the interfacial strain energy density to the tensile and shear FMos across the chosen interface.

Interfacial delamination may lead to **cracks** at the interface, but in real electronic packages the size and location of the cracks and/or delamination cannot be predicted. Cracking is more prevalent in PEMs with larger CTE mismatches, since the magnitude of the thermal stresses developed during reflow is higher. The interface with the largest CTE mismatch is usually the first to delaminate. The chance of delamination or cracking is lowest when the CTE of the moulding compound is between those of the die and the lead frame, which minimises the total thermal stresses developed during reflow. Thinner packages show a greater tendency to crack as the capacity of the moulding compound to withstand reflow stresses decreases. Kitano *et al.* [172] found that package cracking depends not only on the level of moisture saturation but also on the hysteresis of moisture absorption. The moisture content at the interface of the die paddle and the moulding compound has the greatest influence on the generation of vapour and package cracking, and does not correspond to the average level of moisture saturation.

Cracks may vary with: die size, distance from the paddle edge and thickness of moulding compound under the paddle. The effect of structural weaknesses caused by poor bonding, voids, micro-cracks or delaminations may not be evident immediately in the electrical performance characteristics, but may cause premature failure. The C-SAM is an excellent tool for nondestructive FA of IC packages [173]. To gain insights into the nature of images obtained from acoustic microscopy, the acoustic wave parameters must be understood and used to interpret the data (images and waveforms) obtained via acoustic microscopy.

Sometimes, **intermittent contact** in plastic packages may cause serious troubles. As the solder joints and the connection wires are completely included in plastic material, the moulded components are extremely resistant to vibrations and mechanical shocks, even if a fracture (or a discontinuity) arises in the connection wire. The two discontinuous elements remain hold together as long as the moulding environment continues to exercise the same compression force on these two parts. This force does however have the tendency to weaken the contact, and eventually the electrical connection is broken. But as soon as the temperature changes the contact is restored. This is an intermittent contact. If the ambient temperature does not restore the electrical contact, an open circuit arises. Such failures are generated by the thermo-mechanical stress produced by the changes in the package dimensions.

This kind of failure arises mainly at the user, during the infant mortality period, and is hard to detect. FA allows accumulation of important knowledge on the physical and chemical mechanisms of the failures, leading manufacturers to improve the reliability of plastic packages.

Popcorning is another potential reliability problem in PEMs, as a moisture-related FM arising during the reflow process: a small amount of moisture is heated during the wave solder/reflow process and turns to steam. Resultant internal cracking is initiated by a combination of thermal-expansion mismatch of the package and the moisture-vaporised pressure acting on the crack surface inside the package [174]. The damage originates in the die attach and is propagated along the weakest interfaces in the package. At the lower ramp rate, a much smaller amount of delamination is observed. This suggests that there is a critical ramp rate below which popcorn cracking can be inhibited. An optimum temperature ramp rate must be attained to eliminate this reliability risk in board assembly [174].

Popcorning failures may arise during power-up of the circuit board, making them more difficult to trace as such failures are intermittent [175]. The component damage is internal and may not be recognised by visual-inspection procedures. Only X-ray inspection allows a nondestructive inspection of components and presents a clear image of the component's internal construction. However, cracks in the die are difficult to see because silicon is transparent to X-rays and provides few density or contrasting differences. Conditions of popcorning may occur during repair and re-work of circuit-board assemblies as any device will be at risk of popcorning if it has not been properly handled and stored [176]. Popcorning can be prevented by: dry nitrogen storage, increasing 'popcorn' resistance of the moulding compound, modifying the PEM manufacturing process or using an optimal reflow temperature ramp rate [177].

5.1.3 Operation

5.1.3.1 Component Mounting in Systems

In systems, the electronic components are mounted on PCB, which ensures mechanical support and maintains electrical connections on conductive pathways or tracks etched from copper sheets laminated on a nonconductive substrate. There are basically two ways of mounting the components: (i) through-hole type (the component leads are inserted in holes) and (ii) surface-mount type (the components are placed on pads or lands on the outer surfaces of the PCB). In both types, component leads are electrically and mechanically fixed to the board with a molten metal solder. Many FMs are associated with component-mounting on PCB. In recent years, research has focused on new packages, like the ball grid array (BGA), and on lead-free assemblies, because lead (Pb) induces major risks of environment pollution.

The defects at Pb-free BGA solder joints (SAC387, 95.5 wt%Sn/3.8 wt%Ag/0.7 wt%Cu) in PCB placed under different thermal-cycle conditions have been investigated [178]. By using a digital radiography X-ray technique, the soldering defects in the packages, including bridges, voids, ball missing, poor wetting and tilted balls, can be identified. Subsequently, the thermal mechanical behaviour of the selected defective BGA packages are investigated by high-temperature Moiré method. Based on the shear strain values obtained from Moiré tests, the fatigue life of defective and nondefective BGA packages subjected to thermal-cycling tests are predicted and compared. Different defects will have different influences on the thermal-mechanical behaviour and reliability of a BGA package. The decrease in lifetime of defective packages depends on the temperature and defect classification. The experimental results achieved will provide valuable data for further optimisation of Pb-free BGA design.

Pb-free PCB assemblies were studied using different footprint designs on PCBs, solder-paste deposition-volume and reflow profiles [179]. Lead-free SnAgCu plastic-ball-grid-array (PBGA) components were assembled on to PCBs using SnAgCu solder paste. The assembled boards were subjected to the thermal-cycling test ($-40 \pm 125\,^{\circ}$C) and crack initiation and crack propagation during the test were studied. Failures were not found before 5700 thermal cycles and the characteristic lives of all solder joints produced using different process and design parameters were more than 7200 thermal cycles, indicating robust solder joints produced with a wide process window. In addition, the intermetallic interfaces were found to have Sn-Ni-Cu. The solder joints consisted of two Ag-Sn compounds exhibiting unique structures of Sn-rich and Ag-rich compounds. A crystalline star-shaped structure of Sn-Ni-Cu-P was also observed in a solder joint. The intermetallic thicknesses were less than 3 μm and the intermetallics growth was about 10% after 3000 thermal cycles. However, these compounds did not affect the reliability of the solder joints. The findings in this study were compared with those in previous studies, proving its validity.

A model for solder under impact-loading was proposed in [180]. The model was to be used in the simulation of electronic packages under drop impacts. As such, it was necessary to incorporate the effects of strain-rate sensitivity in the model. Dynamic material properties of Sn63/Pb37 solder were obtained by testing it using split Hopkinson pressure bars (SHPBs). The accuracy of the material models was evaluated by subjecting single solder balls to impulsive loads on the SHPBs. Examination of the fractured surfaces showed that there is a transition from ductile to brittle fracture as strain rate is increased. SHPB tests on single solder balls also showed that the stiffness of the solder balls is strongly dependent on rates of deformation.

5.1.3.2 Radiation Field

In a radiation field, the electronic components fabricated by Bi technology are affected by various FMs. The rapid neutrons produce current-gain degradation and increase of saturation voltage for Bi transistors, by creating defects in the crystalline structure. The ionisation radiation generates photocurrents in all reversely biased *pn* junctions, producing modifications of the logic states [1].

However, MOS technology seems to be more robust in this respect. A study of DC and radio frequency (RF) performances of devices obtained by a 0.12 μm CMOS technology undergoing a 63 MeV proton irradiation (an equivalent total gamma dose of 1 Mrad Si) has shown that this technology is appropriate for the manufacture of devices aimed to be used in a radiation field, without any additional radiation hardening procedures [181]. Only slight degradations were noticed for both DC parameters (DC current-voltage, low-frequency (l/f) noise, S-parameters and broadband noise) and RF ones (S-parameters, cut-off frequency, broadband noise).

For diodes, the parameters measured to check the radiation effect are forward voltage drop, reverse current and breakdown voltage. The forward voltage drop is the most sensitive to radiation and the most commonly measured parameter. For Zener diodes, the breakdown or reference voltage is the most important for applications.

Temperature is an additional factor in radiation, the combined effect being more important than the sum of the individual effects. The available data about diodes in radiation fields indicates degradation similar to that experienced by switching diodes: the minimum fluency at which the reverse current increases by a factor of 10 or more is approximately 2×10^{14} n/cm^2 (E > 10 keV). Gamma irradiation of reference diodes using cobalt-60 as a radiation source has caused essentially no change in reference voltage with total exposures of 8.8×10^5 rad (C) [182].

5.2 Failure Modes and Mechanisms of Passive Electronic Parts

FA of passive components is a very broad topic because there are numerous device types and constructions. Passive components, such as resistors, capacitors, inductors, transducers, switches, relays, connectors, fuses and so on, outnumber semiconductor devices in most electronic equipment. Considering the preponderance of passive components in the field and the relatively high failure rates of passive components (compared with semiconductor devices), one might conclude that passive component failures contribute to most field failures of electronic equipment. However, failed passive components are often discarded after replacement, and no feedback is given to the component manufacturers, who, lacking information to the contrary, assume that their products are infallible.

In the following section, the reliability issues of the main types of passive component will be discussed. In each case, component construction and general applications will be presented, followed by the typical FMs, together with methods for diminishing the failure risks.

5.2.1 Resistors

A resistor is a two-terminal element which exhibits a voltage drop that is directly proportional to the current passing through it. Resistors are widely used in electronic systems. Their most important constructive characteristic is the position of the leads. We distinguish four possibilities: (i) leads leaving the body axially (a 'through-hole' component); (ii) leads coming off the body radially; (iii) SMT types; and (iv) one of the leads being placed into a heat sink (power resistors). The numerous material types and technologies for resistor manufacture have led to a large range of resistor families, including: carbon-composition/carbon-film resistors, metal-film resistors, wirewound resistors, thick-film resistors, thin-film resistors and variable resistors.

Being a dissipative element, the general FMo for most types of resistor is the open circuit. This is not always the case for power wirewound resistors, where an overheating condition can cause the material inside the resistor to fuse across adjacent turns of the resistor. Resistor failures can be explained by one or more of the following:

- Material degradation (e.g. fatigue; interruptions due to the degradation of inorganic materials, produced by impurity migration in the substrate layer and by oxidation of constitutive elements of the resistors after uninterrupted utilisation across several years).
- Design errors (rarely encountered in operating products).
- Manufacturing errors (which may appear if the producer employs a new material without a sufficient testing).
- Inadequate utilisation (can only be reproached to the user; hence derating is the best method to enhance resistor reliability).

In general use, various causes of resistor failure may be identified:

- Non-homogeneities of the film composition, stress relief and ion migration may induce risk failures. Resistor failure depends not only on the load power, but also on the rate of load increase. This is why during the design of thin-film resistors under high load, the possibility of catastrophic failure should be considered. These effects are less dangerous for heat-conducting substrates.
- Excessive current flow may cause catastrophic failure by damaging, melting or fusing open the resistive element or by damaging the contact between the resistive element and the terminals.
- Resistors can exhibit significant levels of parasitic inductance or capacitance, especially at high frequencies.
- Temperature is an important factor of failure risk, because excess heat accelerates the drift mechanism in resistors and may ultimately result in circuit failure. The fundamental material properties such as conductivity, permittivity or permeability – which determine the parameters of passive elements – exhibit a dependence on temperature to a greater or lesser degree. The normal FM of a resistor can create circuit difficulties if not carefully considered in advance. Resistor materials show little oxidation effects at temperatures lower than a threshold temperature. At temperatures higher than threshold, the resistor material may be oxidised, which leads to a change in resistance over time. All resistors tend to change slightly in value with age. The temperature coefficient of resistance (TCR) is the relative change of a physical property when the temperature is changed.
- High humidity may be an important environmental factor in some applications. Because the resistor value changes appreciably in a humid environment, specific methods are required to diminish this failure risk.

- The packaging paint plays a critical role in the production of resistors. There are special property requirements, with curing time less than two minutes at 170 °C and the varying ratio of resistance ($\Delta R/R$) less than 0.1% after experiment on high-humidity resistance [183].
- The noise produced by resistor operation is detrimental to circuit operation. There are two types of noise to consider: (i) thermal noise, due to the random motion of electrons within the resistive conductor – the voltage developed by thermal agitation sets a limit on the smallest voltage that can be amplified without being lost in the background; and (ii) the current noise, which is the bunching and releasing of electrons associated with current flow, and is present in resistors to varying degrees depending upon the technology employed.

The typical FMs for various categories of resistor are detailed below, and corrective measures aimed at reducing the failure risks are presented.

5.2.1.1 Carbon-Composition Resistors

Initially, these resistors were made from solid cylindrical pieces of carbon, but later the solid carbon was replaced by composite materials with variable and controlled resistivity. They have low parasitic inductance due to the lack of any spiral wound element, but, unfortunately, a high TCR and poor stability. Their resistance tends to increase with age.

Usually, FMos are open circuit or value change, and occur when power levels exceed the resistor's rated specifications. Resistors can also fail due to manufacturing defects (e.g. poor lead-to-body contact), a defective resistive element or corrosion. Excessive current flow may cause catastrophic failure by damaging, melting or fusing open the resistive element or by damaging the contact between the resistive element and the terminals.

5.2.1.2 Carbon-Film Resistors

These are a new and improved version of the carbon-composition resistors, with an improved TCR, fabricated by depositing a thin carbon film on to a nonconductive former, which is then fitted with end caps, to which the leads are welded. An epoxy coating is usually applied to seal the device. TCRs of 500 ppm or more are typical for these resistors, with power ratings from 1/8 to 2 W.

At normal temperature, carbon-film resistors do not fail, if the nominal charging capability is not overreached by short circuit or interruption. The same is true for small-value resistors, whose drift failure rate is smaller than that of high-value resistors [1]. Typically, carbon-film resistors fail as short circuits. The main causes of failure are:

- Impurifying the ceramic support, the carbon film or the encapsulation with ionic substances, which, in a humid environment, may lead to ion migration, and to indirect destruction of the film resistors, as a result of electrolytic phenomena.
- Irregular feldspat local concentration on the surface of the porcelain lowering the thickness of the carbon film, leading to strong local heating and eventually to the destruction of the film by early failures. The best preventive measure is to under-heat the resistor.

5.2.1.3 Metal-Film Resistors

Metal-film resistors contain a resistive metal film deposited on to a nonconducting ceramic core. The resistor value is trimmed by laser during manufacture, cutting a spiral groove through the resistive film. Metal-film resistors are usually coated with nickel chromium (NiCr) or with any cermet[7] materials. The predominant reliability risk factor is oxidation, which is in accordance with Arrhenius law. Other

[7] Cermet is a composite material made of ceramic (cer) and metallic (met) parts.

parameters that influence the stability characteristics include: surface roughness, chemical reaction between the materials used, proportion of alkaline ions and so on.

These resistors exhibit good stability but have a small inductive component due to the spiral form of the resistive element. TCs are typically 50–200 ppm/°C. Precision-film resistors have greatly improved accuracy with TCs up to 2 ppm/°C and accuracies as good as 0.01%. The metal film is thinner than the layers utilised for the manufacture of carbon film resistors, therefore the probability for fissures and 'hot spots' is higher and can lead to failures at interruption (the most frequent phenomenon for early failures). Non-homogeneities may arise in the resistive layer due to the defective disposition of the helixes and of the bad cut grooves, and can lead to intermittent contacts between two neighbouring helixes and – as a consequence – to instability and high noise level. Some metal-film types have significant inter-lead capacitance, which may be detrimental at high frequencies.

Experiments were carried out to identify the principal physical mechanisms in metal thin-film resistive networks contributing to the degradation of these devices, the main candidates being homogenisation of the film composition, stress relief and ion migration [184]. Large increases in resistance of some metal-film resistors (due to contamination) led to many failures. Corrosion of the resistive film by residual chlorine for a particular resistor vendor's cleaning process was also responsible. A statistical designed experiment indicated that the resistors were not all degraded in the same manner. As a result, the failures were attributed to poor process controls.

5.2.1.4 Wirewound Resistors

Wirewound resistors are commonly made by winding a metal wire around a ceramic, plastic or fibreglass core. Due to the metal coil, they have a higher inductance than other types of resistor, which may be minimised by winding the wire in sections with alternately reversed directions. The typical FMos and FMs are synthesised in Table 5.12.

Variable resistors become electrically noisy as they wear. For the fixed types, the most frequent failure causes are short circuits between two neighbouring helixes and bad contact between wire and terminal, especially for thin wires. For variable types, the failures are induced by contact interruptions produced when corrosion occurs between cursor and resistance wire, because at high temperatures the wire can be strongly oxidised [1].

The derating factors indicate the maximum recommended stress values and do not preclude further derating; when derating, the designer must determine the difference between the *environment specifications* and *operating conditions* of the application, derate – if possible – and then apply the recommended derating factor(s).

5.2.1.5 Thin-Film Resistors

Thin-film resistors contain an extremely thin layer (hundreds of Angstroms) of resistor material that is deposited, using a sputtering process, over the entire surface of a substrate surface made of Si,

Table 5.12 Possible causes of the main FMos of wirewound resistors

Failure mode	Possible failure mechanisms
Open circuit	(i) Excessive wire tension; (ii) wire–end-cap separation; (iii) break in the wire; (iv) poor end-cap–wire weld; (v) ionic contamination/corrosion (between cursor and resistance wire, because at high temperatures the wire can be strongly oxidised; (vi) lead–end-cap separation; (vii) excessive force in lead-forming; (viii) bubbles or voids; or (ix) mechanical deformation of contact wiper.
Short circuit	(i) Corrosion or (ii) contamination.
Parametric shifts	(i) Mechanical deformation; (ii) intermittent contact; (iii) contamination and current leakage; (iv) partial corrosion; or (v) poor end-cap crimp.

GaAs or alumina. Usually a conductor layer is deposited on top of the resistor layer. Using a photolithographic process the substrate is patterned and the two layers are etched away independently, so you have a part with both a conductor and a resistor pattern. Possible resistor materials include: tantalum nitride (TaN), nickel chromium (nichrome), lead oxide (PbO) and ruthenium dioxide (RuO_2), but many combinations of metal-alloys are possible. The sputtering process allows accurate control of the thickness of the film. A limited range of sheet resistance is possible, from perhaps 5 to 250 Ω^{-2}. After fabrication, the resistance is usually trimmed to an accurate value by abrasive or laser trimming.

Thin-film resistors are high-reliability components, with sub-ppb levels of failures. There are two main FMos:

- Open circuit (total failure), caused by over-current pulses or by corrosion of the resistive layer if the humidity breaks through the protective coating.
- Resistance drift, caused by ageing of the thin film at high current or high temperature levels.

The thin-film resistors limit the magnitude of the parasitic currents generated by irradiation and thereby increase the radiation tolerance of the electronic systems.

The effects of corrosion, oxidation and inter-diffusion on the reliability of thin-film Ni-Cr resistors have been studied since the 1970s [185], the main results being:

- The Ni-Cr thin films are corrosion-resistant to salt solutions and to most acids, with the exception of HF solutions and Ni-Cr etches.
- From 'water drop' tests it was found that unpassivated resistors were subject to anodic dissolution at potentials above 2.5 V. The Al metallisations were dissolved at potentials above 0.5 V. Passivation with 10 kÅ of SiO_2 protects the nichrome from anodic dissolution.
- The EBIC mode of the SEM revealed two phases in the films. It was concluded that oxidation is a non-uniform process.
- Al diffuses into Ni-Cr resistors with an activation energy of 96 kcal/mol.
- No resistance discontinuities were observed up to 24 hours, at +500 °C.
- The highly accelerated test conducted on nichrome test patterns proved that a good passivating glass is capable of protecting the film resistors.

A recent study of the electrical behaviour at low temperatures (4.2–300 K) of a thin-film resistor made of a metal-alloy (Ni-Cr-Cu-Al-Ge) coated with an alumina (Al_2O_3) layer obtained by atomic layer deposition (ALD) is reported in [186]. It was experimentally demonstrated that the protective dielectric alumina coating improves the long-term stability and repeatability of high-value, thin-film resistors, in the range 100–500 kΩ. The drift rate of the resistance due to the native oxidation at room temperature was reduced from 2.45×10^{-6} hour^{-1} for an uncoated resistor to 3×10^{-8} hour^{-1} for an alumina-coated resistor. It was shown that the additional 15 nm-thick alumina coating does not significantly change the thermo-electrical properties of the metal-alloy, thin-film resistors.

Thin-film resistors (also called embedded resistors) manufactured by organic printed wire boards (PWBs) and inorganic low-temperature co-fired ceramic (LTCC) were tested [187]. Samples of each group were exposed to four environmental tests and several characterisation tests to evaluate their performance and reliability. Both technologies performed favourably. The resistors were not embedded deep into the substrate structures but were placed on the surface and coated. This served two purposes: (i) the resistors could later be trimmed if they resided on the surface and (ii) it represented the worst case for the protection of the resistive elements in reliability testing, mainly moisture exposure. Note that the environmental protection could be compromised if a pin-hole or damaged area were to arise. During the moisture environmental testing, a resistor in the PWB technology failed due to corrosion. The level of concern for this FM is elevated only for laser-trimmed resistors, where the coating would be opened and an additional coating applied following the adjustment. The resistor failed at the 1000-hour read point of 85%RH/85 °C and the failure was not an open circuit, but an increase in resistance.

Thin-film resistors may be used for IC fabrication, especially when the TCR must be lower than 25 ppm/°C or the noise parameter is important. For common application, the thick-film chip resistors are cheaper. ICs are frequently exposed to high-voltage pulses as a result of static discharge during handling and from electrical transients associated with system power supplies or interconnecting networks. Such electrical transients can cause device failures by burning out the metallisation or the thin-film resistors and producing open circuits or large changes of resistance value; the failure could not be a short circuit.

In CMOS technology, TaN resistors are commonly used in RF IC applications. Deposition and integration of the films are controlled to produce a high-precision resistor, and the TCR characteristics of the film make it ideally suited for application across a large temperature range. While the time-zero characteristics of the device are well understood, of equal importance are its reliability properties. A study based on stress data and an Avrami phase-change degradation model is presented in [188]. TaN thin film was determined to be reliable over the normal IC use conditions. It was concluded that the temperature rise of the resistor due to the resistive joule heating is the limiting factor for maximum allowed current.

An investigation of the 'wear-out' characteristics of thin-film resistors was performed [189] with the following main goals: (i) to find out what happens to the resistors under accelerated stress and (ii) to generate better reliability-based design rules as a function of current density, temperature and resistor geometry. The changes in resistance and TCR were explained as a result of accelerated to highly accelerated test conditions, using thermal and constant current stresses (CCSs), and activation energies for some of these changes were calculated. It seems that resistance increases and resistance decreases can be produced, depending on resistor process and stress levels, with an activation energy of about 1 eV for resistance increase for one of the resistor process types at one fabrication site and of about 3 eV for the resistance decrease for the other three process type and fabrication-site combinations.

In nanotechnologies, resistors fabricated in nanoscale crossbars are observed to be nonlinear in their current versus voltage (I–U) characteristics, showing an exponential dependence of current on voltage. These devices are called *tunnelling resistors* [190]. The introduction of nonlinearity can either improve or degrade the voltage margin of a demultiplexer circuit, depending on the particular code used. Therefore, the criteria for choosing codes must be redefined for demultiplexer circuits built from this type of nonlinear resistor. It seems that for well-chosen codes, the nonlinearity of the resistors can be advantageous, producing a better voltage margin than can be achieved with linear resistors.

5.2.1.6 Thick-Film Resistors

Launched in the 1970s, thick-film resistors became popular for use in surface-mount device (SMD) technology. Like with thin-film resistors, the resistance of thick-film resistors is trimmed to an accurate value after manufacture, via abrasive or laser trimming. Thick-film resistors may use the same conductive ceramics as thin-film resistors, but they are mixed with sintered (powdered) glass and some kind of liquid so that the composite can be screen-printed. This composite of glass and conductive ceramic (cermet) material is then fused (baked) in an oven at about 850 °C. Thick-film resistors are widely used in consumer and industrial products such as timers, motor controls and a broad range of high-performance electronic equipment, as well as in hybrid circuits, current sensing, power resistor and power conversion [191].

The main FMos for surface-mount thick-film resistors are: open circuit, short circuit and parametric changes. An open circuit may be caused by: (i) delamination; (ii) thermal cycling; (iii) micro-cracks; (iv) CTE mismatch; (v) placement force fracture; (vi) silver leaching by Sn/Pb due to poor nickel layer; (vii) EOS/excess current; or (viii) stress in conformal coating. The cause of a short circuit may be: (i) silver migration; (ii) ionic contamination; or (iii) dendrite growth. Parametric shift can be produced by: (i) ESD damage; (ii) external contamination; or (iii) thin generated micro-cracks. Other possible FMs are:

- Electrically overstressing; this is usually accompanied by visible signs of overheating.
- Delamination between layers, allowing intrusion of solder flux, cleaning solvents or water; these foreign materials may support corrosion or EM, or may cause parametric changes.
- EM and tarnish, due to silver-based terminations; these are prevented by cleaning.
- Copper dendrite growth, silver migration, sulfur atmosphere corrosion and crack due to moulding compound mechanisms; these are all enhanced by temperature and humidity.

Deviation of the resistor parameters from the designed values can occur during any stage of processing or any stage of the subsequent environmental, mechanical and electrical tests required for quality conformance approval of a circuit. The FMs identified in a bismuth ruthenate-based resistor system used in thick-film technology are reviewed in [192], including: firing, substrate, termination conductor, laser trimming, component attachment and packaging, mechanical stresses, environmental tests, power loading or refiring, and high voltages.

In [193] the results of FA carried out on RuO_2 thick-film chip resistors failed in field are presented. Microscopic investigation performed on virgin, degraded and open-circuit devices showed the FM involved: evidence for mechanical cracks may be related to the observed degradation and failures. On the failed and degraded devices some bad adhesion of the resistive film on the metallised alumina was observed, which may also be consistent with a thermo-mechanical hypothesis for the FM.

5.2.2 Capacitors

A capacitor is a device designed to store an electric charge. It is formed by two conducting surfaces, separated by free space or a dielectric material. Charge is stored when a voltage difference is applied between the two surfaces. In recent decades, the dielectric material has changed from paper to all-poly-propylene film, the electrode from plain foil to metallised film, metallised paper or embossed foil, and the fluids from polychlorinated biphenyls to those based on hydrocarbon or phthalate ester chemistries. As a rule of thumb, the life of a capacitor decreases by a factor of 2 for each 10 °C rise in temperature. The effects of ageing are perhaps more relevant to electrolytic tantalum and ceramic-type capacitors.

The important factors in a capacitor are: (i) the area of the conducting surfaces; (ii) the type of dielectric used; and (iii) the separation between the surfaces. Failures in capacitors occur with environments of excessive voltage, temperature, moisture, vibration and external pressure.

According to the nature of dielectric, capacitors are classified into two broad categories:

- **Polarised capacitors.** The electrical characteristics are dependent on bias polarity (values larger than 1 μF). Examples include electrolytic capacitors and tantalum bead capacitors.
- **Unpolarised capacitors.** Electrically symmetrical devices (values smaller than 1 μF) employing the same metal for both electrodes, with a dielectric material (air, paper, ceramic, glass, mica or polymer film) sandwiched between them (e.g. ceramic capacitor).

The FMs of capacitors can be divided into four categories:

- Material interaction-induced FMs.
- Stress-induced FMs (directly attributed to either poor device design or poor and careless device application).
- Mechanically-induced FMs.
- Environmentally-induced FMs (covering a wide spectrum of possible environmental conditions, such as humidity and hydrogen effects).

For electrostatic capacitors, the breakdown voltage is inherently related to the properties of the dielectric, with the important parameters being the dielectric field strength, which is related to the dielectric constant, and the dielectric thickness. These are not necessarily related to the capacitance value and the rated voltage, but generally the larger values of capacitance have lower breakdown voltages. Foil and wet-slug electrolytic capacitors can withstand conduction current pulses without apparent damage (in either direction for foil types). For solid tantalum capacitors, damage occurs whenever the capacitor charges to the forming voltage.

The most appropriate FA methods for capacitors are:

- **Nondestructive inspection.** FA begins with a review of the failure report, followed by a visual inspection (using a stereomicroscope with a 10–70× magnification range) of the component. Any external signs (e.g. small cracks in the ceramic) are identified and explained. Ultrasonic examination or scanning acoustic microscopy (SAM) can be used to better characterise these types of flaw.
- **Electrical characterisation.** Leakage current, dissipation and capacitance are the specified electrical parameters. Modern circuits are resistant to these parameters, so degradation is often invisible to the user. Hence, the most commonly detected problems are short circuits and open circuits. The components are characterised by a curve tracer, selecting a low voltage range (<10 V DC) with a limiting resistor to maximise the measurement. A good capacitor is characterised by a loop centred on the horizontal or vertical axis; a shorted part is indicated by a degraded or vertical trace. Investigating the leakage current with voltage and temperature variation is a good technique for modeling the mechanism. Sometimes it may be necessary to measure capacitance versus frequency, leakage current over time, dielectric absorption, dielectric ageing and other characteristics.
- **Destructive analysis.** This involves cross-sectioning of the capacitor, after any encapsulation or surface coating has been removed. Localisation of the failure is performed using another three-dimensional technique (monitoring the leakage current during cross-sectioning). The electron beam of an SEM can be used as a probe and conductive paths can be highlighted using a conventional voltage-contrast technique.

Typical FMo and FMs of the main types of capacitor are described below.

5.2.2.1 Electrolytic Capacitors

Electrolytic capacitors contain a thin film of oxide on an Al foil. An electrolyte is used to make contact with the other plate. The two plates are wound around one another and then placed into a can, often made of Al. They are polarised, and care should be taken to ensure they are placed in circuit the correct way round. If they are connected incorrectly they can be damaged, and in some extreme instances they can explode. They are widely used in audio applications as coupling capacitors, and in smoothing applications for power supplies. Electrolytic capacitors are available in both leaded (tubular Al can, each end being marked to show its polarity) and surface-mount formats (rectangular packages).

The most common electrolytic capacitors are half-dry electrolytic wound capacitors, formed from an oxidised Al foil (anode and dielectric) and a conducting electrolyte (cathode). A second Al foil is utilised as a covering cathode layer. They are available with two formed nonpolarised foils and they have large loss factors (frequency- and temperature-dependent), a limited useful life and a relatively low reliability: the failure rate is of 10–50 FIT, typical FMos being drift of parameters, short circuits and open circuits. It was shown that the reliability increases with the size of the case: the smaller the capacitor, the shorter the useful life and the higher its failure rate. The typical FMs are synthesised in Table 5.13.

The main factors that influence reliability are oxide layer (various hydrate modifications are possible), impregnation layer (the conductivity of the impregnating electrolyte works directly on the loss

Table 5.13 Possible causes of the main FMos of electrolytic capacitors

Failure mode	Possible failure mechanisms
High loss factor and capacity-diminishing	As a consequence of diffusion, ageing or decomposition, the active electrolyte quantity of a capacitor system diminishes and leads to a growth of the loss factor (tan δ) or to a diminution of the capacity. These modifications are important in the case of fluid electrolytes; for solid semiconductor electrolytes, the changes are insignificant and generally do not lead to capacitor failure.
Open circuit	(i) Corrosion of foils; (ii) presence of halide ions; or (iii) chemical intrusion through seals. At high temperatures, the degradation and failure process is accelerated.
Short circuit	(i) Evaporation of electrolyte; (ii) high voltage (leading to high leakage current and, eventually, to breakdown); or (iii) reverse-polarity exposure. At high temperatures, the degradation and failure process is accelerated.
The time variation of the residual current	Especially determined by the metal impurities of the Al foil and by the dielectric porosity. The leakage current dependence on voltage and time variation is determined by the applied voltage during the oxidation of Al foil, by impurities and by the exchange effect between dielectric and impregnation electrolyte.

factor of impedance, on the chemical combinations and on the stability of electrical values) and foil porosity. The quality of the electrolyte may induce additional failure risks. In the mid 2000s, the tendency of some system manufacturers to use ultra-low-cost electrolytic capacitors in the construction of their products in order to keep costs down has led to the so-called 'capacitor plague': a lot of malfunctions after one year of operation. Because of the low-quality electrolyte used, hydrogen gas bubbles built up between the capacitors' plates during usage. This increased the internal pressure within the sealed cans, leading to the escape of an electrolyte/gas bubble mix [194].

Temperature has a higher influence on reliability than operating voltage. Note that the total temperature is given by the environmental temperature plus self-heating due to loading in alternating current. A characteristic of electrolytic capacitors is the difficulty of identifying the components with failure risks at the final electrical control. Extensive accelerated life testing involving abnormally high ripple currents and high operating temperatures is needed to identify the weak components. This contrasts strongly with most electronic components, which are much less subject to spontaneous failure after assembly.

All electrolytic capacitors have a finite life, measured in thousands of hours. There are usually no external signs that a capacitor is nearing the end of its life. However, it is possible to determine whether a capacitor is serviceable or not by measuring the equivalent series resistance (ESR), which is the sum of in-phase AC resistance, including the resistance of the dielectric, plates, electrolytic material and leads, at a particular frequency. As the name implies, ESR acts just like a resistor in series with the capacitance. Towards end of life, a capacitor's ESR begins to increase as its dielectric losses increase. To test a capacitor's ESR, you need an ESR meter, because an ordinary capacitance meter usually won't indicate a problem [195].

In recent years, manufacturers have replaced liquid electrolyte with an organic semiconductor. The FMos and FMs of surface-mount Al film capacitors are similar to those of the leaded version; the dominant FM is open circuit.

5.2.2.2 Tantalum Capacitors

The tantalum (Ta) capacitor is a metal-oxide rectifier used in its blocking direction (this explains its polarisation). It is characterised by reduced dimensions (much smaller than Al electrolytic capacitors), good stability of electric parameters, very good high-frequency properties, long lifetime and a large temperature domain. Like electrolytic capacitors, Ta capacitors are polarised and are very intolerant of being reverse-biased, often exploding when placed under stress. They are available in both leaded and surface-mount formats.

Experience shows that the failure rate is less than 10^{-8} hour^{-1}, with a confidence level of 90%. The high intrinsic reliability of solid Ta capacitors is explained by the lack of wear-out mechanisms, such as loss of electrolyte. However, there are two restrictions:

- Small value of the working voltage (smaller than 35 V).
- Reduced reliability of solid electrolyte Ta capacitor in pulse operation, for example for circuits with small impedance, where the over-voltage can lead to blackout failures.

The key element of a Ta capacitor is the Ta_2O_5 dielectric, which is inherently thermodynamically unstable; this instability becomes more pronounced with an increase in operating temperature, working voltage and electrical field in the dielectric. Stabilisation of the dielectric and the Ta-Ta_2O_5 interface can be accomplished by kinetic means. Extension of high stability and reliability to higher temperatures and voltages (a real challenge!) requires a deep understanding of these thermodynamic and kinetic factors. With a manufacturing technology that is focused on stabilising the Ta_2O_5 dielectric and its interface with anode and cathode, this goal is achievable [196].

The basic FM of solid Ta capacitors is field crystallisation of the essentially amorphous dielectric oxide. The growth of higher-conductivity crystalline oxide during operation of the capacitors causes an increase in leakage current and may result in catastrophic failure. The effect of field crystallisation can be minimised by using high-purity Ta to reduce the number of crystallisation nucleation sites. Since crystalline growth is primarily dependant on applied voltage, high-voltage capacitors are much more susceptible to failure than low-voltage units.

In Table 5.14, the possible causes of the the main FMos of Ta capacitors are shown.

The solution for avoiding electrical breakdown is to use a conductive polymer (CP) as dielectric. Such CP capacitors have shown a slightly different current conductivity mechanism than standard Ta capacitors. The breakdown of CP dielectrics is similar to avalanche and field-emission breaks. It is an electromechanical collapse due to the attractive forces between electrodes, electrochemical deterioration, dendrite formation and so on. However, some self-healing of the cathode film has been reported. This can be attributed to film evaporation, carbonisation or re-oxidation. Not all breakdowns of CP capacitors lead to self-healing or an open-circuit state. Short circuits can also occur [197].

Table 5.14 Possible causes of the main FMos of tantalum capacitors

Failure mode	Possible failure mechanisms
Open circuit	(i) Fatigue of cathode layers or (ii) anode wire weld break.
Short circuit	(i) An impurity creates a thin spot in the dielectric; when a surge occurs, more current passes through the thin spot than through its neighbouring sections; if the local heat generated is high enough, the capacitor will scintillate or momentarily short. (ii) Surge failure depends on the existence of thin spots in the MnO_2 layer; the thin area in the MnO_2 layer results in a lower local resistance and a higher local current density, creating a hot spot at the thinned area that will crystallise the dielectric, causing it to eventually crack and fail. (iii) MnO_2 crystals damage the dielectric as they expand and contact due to the effect of the transient. (iv) Electrical breakdown produced by an increase of the electrical conductance in channel, due to an electrical pulse or voltage level, leads to capacitor destruction followed by thermal breakdown; in the reverse mode, thermal breakdown is initiated by an increase of the electrical conductance by Joule heating at a relatively low voltage level.
Degraded parameters	Degradation can occurs if there is a negative voltage component on the part due to the combined effect of DC and AC bias components.

Recently, a new series of Ta capacitors, called NeoCapacitors, based on the use of CP in the cathode layer, was proposed [198]. CP has higher electrical conductivity than the previously used manganese dioxide material and it allows the ESR to be reduced significantly. The development of capacitors with higher withstanding voltages of 20–25 V or more has become necessary in order to meet demands from the server and LCD module markets that necessitate even higher functions and reliability for design optimisation and the elimination of turning-back as an approach to the development of high-voltage products.

The fundamental problem, which has been well documented over the years, is irregularities or weaknesses in the delicate constituent layers, resulting in high-leakage current paths introduced during the manufacturing process, mounting or application. Irregularities can be formed by impurity sites that are not totally eradicated during reform, dielectric cracks, delaminations, thinning of the dielectric or manganese dioxide layers, voltage breakdown or popcorning of the manganese dioxide (caused by moisture ingress during storage, boiling off during solder reflow). This list is by no means exhaustive. High-leakage current paths result in localised heating, which will cause a chemical reaction to occur in the manganese dioxide layer if the temperature rise exceeds approximately 450 °C. MnO_2 has relatively poor conduction characteristics (2–6 Ω/cm), but this is generally overlooked due to a more valuable property that allows conversion from the semi-conductive state (MnO_2) to the highly resistive state (Mn_2O_3) when the appropriate is energy available. The Mn_2O_3 formed literally plugs the high-leakage path, a process referred to as 'self-healing'. Without this ability to self-heal, the solid Ta capacitor would not be a viable proposition, due to severely degraded reliability and yield [199].

Solid Ta capacitors have not been so popular in space applications, leading to some restrictions in power-supply filtering circuits where high-current handling capability and low impedance are prerequisite. The over-voltage spikes severely degrade the reliability of Ta capacitors [200].

Fast charging and discharging of solid Ta capacitors cause localised heating due to energy dissipation in surface areas of MnO_2 having lower resistance or insufficient thermal exchange with the case. In some cases, overheating can cause combustion of the Ta. The failure generally occurs only during charging. Until now, wear-out processes for Ta capacitors have not been observed. In most cases, a time decrease of the failure rate has been noticed. In the case of Ta chip capacitors, using standard capacitors at 150 °C will lead to increased failure rates and decreased lifetime in terms of leakage current and impedance. The reason for this thermal breakdown is leakage-current overload and an override of degradation temperatures of the contact layers [201].

An interesting result was reported in [202]: the working temperature of high-reliability Ta capacitors was extended from +125 to +200 °C, and the rated voltage to 125 V, by modification of the chemistry of the Ta-polymer capacitors. Another attempt to improve the reliability was made by adding a moisture-barrier layer, which proved to be extremely effective [199].

5.2.2.3 Film Capacitors

Film capacitors consist of alternating dielectric and electrode plates. Generally, film capacitors are available in both surface-mount and leaded styles, with foil or evaporated-film electrodes. Various plastic films are available, the most popular being polyester, polystyrene and polypropylene. The possible causes of the main FMos of film/foil capacitors are given in Table 5.15.

Polyester-film/foil capacitors contain a non-inductive wound section of Al foil with a polyethylene terephthalate (PET) film, protected by a hard, water-repellent, self-extinguishing lacquer. The leads are of solder-coated copper wire, crimped and cropped. Polyester-film capacitors are used where cost is a consideration as they do not offer a high tolerance (5 or 10%, which is adequate for many applications). They are generally only available as leaded electronics components.

A variant of polyester-film capacitor is the *metallised-film capacitor*, where the polyester films themselves are metallised. The advantage of this process is that because their electrodes are thin, the capacitor itself can be contained within a relatively small package.

Table 5.15 Possible causes of the main FMos of film/foil capacitors

Failure mode	Possible failure mechanisms
Open circuit	(i) Broken epoxy seal and corrosion; (ii) end-termination damage; (iii) lead-attach damage; (iv) fusing of the interface between the electrode and termination, when the current flow in the capacitor is too large for the electrode thickness; or (v) high AC ripple current, when the bond between the electrodes and the device termination is poor enough that its resistance is abnormally high.
Short circuit	(i) Embrittlement of film dielectric with time and temperature exposure, the dielectrics being sensitive to defects in the film (cracks, pinholes, wrinkles, contamination); (ii) entrapped moisture, causing dielectric decomposition; (iii) over-voltage punch-through; (iv) ionic contamination on the body; or (v) electrode-spacing change.
Winding resistance and quality factor Q outside the specification range	Corrosion of the wire can lead to failure by parameter drift: it is possible for turns to have been severed by corrosion while continuity still exists, because of an electrically conductive path through the corrosion products.

Polystyrene capacitors are a relatively cheap form of capacitor. They contain an Al foil serving as a coating and polystyrene as dielectric. The terminal connections are joined with the coating (thickness: 10–40 μm), so that the capacitors can be used at high frequencies. Unlike the classical paper types with various impregnations (with wax, oil, chloride naphthalene and epoxy resin), no degradations by the ageing of parameters are noticed for plastic-foil capacitors. Other remarkable characteristics are the small losses at high frequencies, high capacity constancy, insensitivity at overcharge and mechanical overstresses, well-defined temperature coefficient, and relative insensitivity to humidity and temperature.

The *polypropylene capacitors* (with a polypropylene film for the dielectric) are used when a higher tolerance is needed. One advantage is that there is very little change of capacitance with time and voltage applied. They are also used for low frequencies, with 100 kHz or so being the upper limit. They are generally only available as leaded electronics components.

A typical flow is a pinhole or a wrinkle in the dielectric film; if enough current is available, the metallisation surrounding the pinhole can evaporate, thereby isolating the flow and eliminating the failure indication (the so-called 'self-healing' process). Therefore, if a reported short cannot be verified, it is necessary to inspect all the dielectric surface area in the capacitor to locate possible clear flaw sites. The drawback to this process is that there may not be any way to determine whether a cleared area was pre-existent during testing at the manufacturing site or was related to a specific event.

The most commonly observed capacitor failure is simple breakdown of the dielectric caused by the application of excessive voltage. Failure is also observed due to degradation of dielectrics with pre-existing manufacturing defects, usually voids or thin regions.

5.2.2.4 Ceramic Capacitors

The ceramic capacitor contains alternating layers of metal and ceramic. The dielectric is the ceramic and typically a titanate compound (barium titanate, neodymium titanate, magnesium titanate) is used. All these dielectric materials have a high dielectric constant and stable electrical performance across a range of temperatures and voltages. External contact is made through a series of interfaces to a solderable surface of a lead-frame system. There are various shapes and styles, such as: disc shape (the classical ceramic capacitor, launched in the 1930s), multilayer ceramic capacitor (MLCC, surface mount), bare leadless disc (used for UHF applications), tube shape and so on. Generally, ceramic capacitors are less reliable than Al electrolytic capacitors, because Al electrolytic may self-heal. However, the ceramic capacitors are preferred for applications operating at temperatures higher that

Table 5.16 Possible causes of the main FMos of ceramic capacitors

Failure mode	Possible failure mechanisms
Open circuit	(i) Lead electrode separation; (ii) excessive handling force; (iii) CTE mismatch; (iv) poor alignment; (v) silver leaching by solder; (vi) end-termination damage; or (vii) loss of contact between layers (heat).
Short circuit	(i) Dielectric damage and breakdown; (ii) electrode material migration under high voltage; (iii) lead material migration under high voltage; or (iv) ceramic fracture due to thermal shock.

+65 °C. Ceramic capacitors, ranging from a few picofarads to around 0.1 μF, are normally used for RF and some audio applications.

As for others types of capacitor, the possible causes of the main FMos are gathered in Table 5.16.

The most common FM of ceramic capacitors is **dielectric failure**, which can result from the following causes:

- **Over-voltage.** When a sufficient over-voltage is applied, current injection at a weak point in the dielectric can cause the dielectric to break down, and the resulting energy release can cause thermal runaway. This often causes localised melting and cracking of the dielectric and adjacent electrode areas, and it may cause any encapsulation to char. Surges with low energy and high voltage can increase current leakage. Thermal stress can crack the dielectric and may also result in increased leakage or shorts. A high-energy surge may crack the ceramic and let in moisture, providing a conductive path [203].
- **Mechanical damage.** Mishandling can cause the dielectric to crack, creating a path between electrodes that can arc or facilitate metal migration.
- **Dielectric flaws.** Internal flaws (voids, crack, etc.) may provide a path between opposing electrodes, resulting in failure due to either arc-over or metal migration. In addition, chemical defects in dielectric are important failure risks.
- **Other external factors.** During assembly, components pass through a series of processes, all of which can generate stresses – direct mechanical stresses and those derived from the difference in the CTE between the capacitor and the board. Lead-free assembly will further increase the stress on vulnerable MLCCs. Extra care will be required to prevent a surge in these types of failure. Some potential sources of damage caused by the mounting process were identified in [204]: (i) board support pins used for second side printing (or placement) making direct metallic contact with parts on the underside; (ii) insufficient board support during second side placement, leading to excessive board flexure; (iii) incorrect height of board support pins, or over-clamping, leading to the board being bowed upwards during second side placement or printing; (iv) damaged placement nozzles; (v) over-pressure and/or nozzle over-travel during placement cycle; (vi) board sag during wave or reflow soldering, caused by poor support, followed by a subsequent straightening process; (vii) boards becoming caught in conveyors; (viii) separation of individual circuits from master panels ('break-out' or 'de-panelling'); and (ix) careless operator board handling, especially when pressing a board into retaining clips, connectors or boxes.

The technological variant of ceramic capacitors for SMT is MLCC, which contains: the dielectric (a mixture of differently formulated and processed ceramic compounds), the electrodes (their compositions vary from lead to silver–palladium to palladium–gold–platinum) and the terminations. The capacitor is placed in an encapsulant undercoat and potted or moulded with encapsulant. MLCCs are very stable in their electrical properties. Electrical lifetime testing is usually performed with a much higher voltage than rated. But they are brittle, and cracking due to mechanical stresses is a very common FMo, leading to capacitance loss and increased leakage currents. Stresses developed adjacent to the solder pad of SMDs are generally relieved by creep of the solder. Repeated thermal cycling may result in fatigue failure of the solder joint and an electrical open of the device.

In recent years, improvements in ceramic technology have reduced the incidence of cracks, at least as far as well-made components are concerned. However, several other factors may also lower the insulation resistance, in particular a poor choice of cleaning solvents, or the use of solvents containing large amounts of dissolved flux residue. More sensitive tests for micro-cracks and delamination apply a mobile ionic material such as methanol to the part, measuring infrared (IR) changes; but the effects can be similarly obscured by contamination. An alternative (and nondestructive) method is SAM, which uses an ultrasonic transducer, typically operating at 10–100 MHz, with the component immersed in fluid (usually water) to couple it to the transducer. The ultrasound travels through the material until it reaches interfaces or discontinuities, which reflect some of the energy.

MLCCs have a failure risk that is substantially higher than other commonly used active or passive components [205]. Additionally, the relatively small ceramic bodies are prone to mechanical damage. Their proportionately high numbers, sensitivity to mechanical stress and difficulty in isolating to a specific failing device on PCB (since many of these parts exist in parallel with many other identical capacitors), all combine to make the successful isolation and analysis of the root cause of failure particularly difficult for the failure analyst. Often, the cause of failure is misdiagnosed, or the evidence is compromised by the methods used to perform the analysis. A novel nondestructive analytical technique to detect cracks in the ceramic and the metallic layers within an encapsulated MLCC is described in [206]. A complex FA performed on MLCCs failed during operation in an automobile engine control unit led to the conclusion that migration and avalanche breakdown were the dominant FMs [207].

Avoidance of failures in MLCCs is strongly driven by the ability of the designer, both electrical and mechanical, to follow guidelines based on an understanding of how surface-mount ceramic capacitors fail. The transition to Pb-free has required a change in materials and processes, potentially requiring changes in these guidelines. To understand how and when these guidelines must be modified, a diligent listing of potential FMs is provided in [208]:

- **Thermal shock** may initiate cracks, because the ceramic capacitor is unable to temporarily relieve stresses during transient conditions. The most common signature of thermal shock is a 45° micro-crack emanating from the termination of the end cap. If the capacitor is exposed to varying levels of voltage or temperature, the crack will eventually grow, cutting off the electrodes from the termination, the dielectric material, the solder pad geometry and the parameters of the soldering process. Under a relatively high humidity–temperature bias, silver migration along the cracks is most responsible for the subsequent electrical failure of capacitors. The presence of micro-cracks can be detected by subjecting the component to a high-voltage insulation resistance test at 85 °C and 85% RH. The heat-induced local melting of internal electrodes can lead to blow-out or charring of the capacitor. By employing mechanical cross-section and emission-microscopic analysis, precise localisation of failure sites is possible even without any observable physical signs [209].
- **Internal defects** introduced during the manufacturing or assembly processes can cause internal shorts that lead to explosions due to the large amounts of energy stored in capacitors. This not only destroys the MLCC, and any evidence of root cause, but can also cause damage to surrounding components, the printed board, adjacent circuit card assemblies, and in the worst case lead to catastrophic fires.
- **Flex cracking** occurs when there is excessive flexure of the PCB. Experimental studies have demonstrated that ceramic capacitors assembled with Pb-free solders consistently show similar or improved robustness to flex cracking compared to capacitors assembled with SnPb solder. Cracks due to excessive flex tend to propagate at 45° angles from the termination of the end cap. Both types of crack can range in size, but flex cracks tend to be larger, propagating through the ceramic until the crack reaches the end cap [210]. To determine the event that initiated flex cracking, one

must determine the strain in the printed wiring board necessary to cause failure. Large capacitors are more sensitive to printed wiring-board flexure. Thick printed wiring boards can be more resistant to flexure but also transfer greater stresses into the capacitor, negating this benefit.

- **Dendritic growth**, also known as *electrochemical migration*, is the migration of metallic filaments under bias through an aqueous solution. It typically requires the presence condensed moisture or contaminants. The presence of condensed moisture can be eliminated through case design or the use of conformal coating and is independent of SnPb and SnAgCu. The presence of contaminants, however, can be very dependent upon the solder material and flux composition being used. Pb-free solders and board plattings are much less solderable than SnPb and therefore require fluxes with higher activity to ensure sufficient wettability.

In recent years, high-k dielectric thin films of barium titanate ($BaTiO_3$, also known as BTO) have been used in MLCCs. As always, a new material means new reliability issues. BTO is piezoelectric in nature, which results in elastic expansion and contraction with changes in the applied electric field. Typically this displacement is on a microscale and has minimal influence on capacitor behaviour. However, at certain resonant frequencies, the capacitor can begin to vibrate along its length. This vibration, if at sufficient magnitude, can induce scattered internal micro-cracking, resulting in a decrease in capacitance and an increase in leakage current. The frequencies of concern are typically in the hundreds of kilohertz to tens of megahertz. The frequency of concern decreases with increasing case size and increasing capacitance. It is currently unknown whether exposure to Pb-free reflow will change this behaviour as most characterisation of resonance behaviour is performed on loose parts.

In 2003, another new FM of MLCC was identified, produced by the replacement of silver palladium electrodes by nickel, for economic reasons. Diligent qualification procedures predicted very long life. However, usage uncovered an Achilles heel: the capacitors fail when exposed to high humidity ($>85\%$ RH) and unpowered for extended periods [211].

5.2.2.5 Metal-Insulator-Metal Capacitors

A metal-insulator-metal (MIM) capacitor has two metal plates sandwiched around a capacitor dielectric that is parallel to a semiconductor wafer surface. The top capacitor metal plate is formed by a planar deposition of a conductive material, which can be lithographically patterned and etched using an RIE process, for example. The patterning of the top metal plate requires the use of a mask, and there can be alignment problems with underlying features and vias in connecting to interconnect layers. MIM dielectric material has to be carefully selected, due to its potential interaction with or diffusion of the metals (such as copper) used for the metal plates. MIM dielectric material restriction may result in limited area capacitance. The use of copper (which has a lower resistivity than aluminum, titanium nitride and tungsten, for example) for the top and bottom metal capacitor plates improves the high-frequency capability and produces a MIM capacitor with higher quality factors (Q-values).

TaN and TiN were investigated as bottom electrode materials for MIM capacitor applications [212]. Atomic vapour-deposited HfO_2 films were used as high-k dielectric. The influence of the interfacial layer between HfO_2 and the bottom electrode on the electrical performance of MIM capacitors was evaluated. It seems that the capacitance density and the capacitance voltage linearity of high-k MIM capacitors are affected by the electrode material. TaN and TiN also have an impact on leakage current density and the breakdown strength of the devices.

MIM capacitors are rather large in size, being several hundred micrometres wide, depending on the capacitance, which is much larger than that of a transistor or memory cell, for example. Typically, they are used to integrate many different functions on a single chip, such as decoupling capacitors for microprocessor units, RF capacitors in high-frequency circuits and filter and analogue capacitors in mixed-signal products. The drive towards high-speed and high-density silicon-based ICs has necessitated significant advances in processing technology. The requirement to reduce passive chip space has led to active research into MIMs with high dielectric-constant (κ) film. An overview of MIM capacitor integration issues with the transition from AlCu BEOL to Cu BEOL is given in [213]. The key to MIM capacitor electrical properties is optimised dielectrics.

For MIM capacitors, surface roughness and particles have been proposed as major sources of defects, but there have been no reported methods for visually elucidating how such defects interact with their surroundings or correlate with the FMo. Because electrical analysis can easily destroy a defect through interlayer fusion and voiding, defect visualisation can turn out to be crucial in finding the origin of defects and improving the process. A nondestructive analysis technique is proposed in [214], using TEM/EDX analysis. A comprehensive FA has successfully revealed for the first time the TiN nano-dendrite growing from an Al whisker (10 nm in diameter), thus elucidating its detrimental mechanisms in causing the high failure risks in MIM capacitors.

A study of MIM capacitors has shown the importance of carbon contamination on component performances [215]. The AC barrier heights (calculated using an electrode polarisation model from capacitance–voltage characteristics) are 0.58 eV for the HfO_2 film with high carbon contamination and 0.95 eV for the HfO_2 film with negligible carbon contamination. The DC barrier heights extracted from current–voltage characteristics are 0.26 eV for the HfO_2 film with high carbon contamination and 1.1 eV for the HfO_2 film with negligible carbon contamination. All of these experimental results show that the increase in defect density in HfO_2 films generated from carbon impurities results in the degradation of barrier heights and poor performance of the MIM capacitor.

Because temperature is an important factor in increased failure risk, many studies have focused on this subject. Original structures have been designed to estimate the self-heating limit of MIM capacitors [216]. It seems this value is very high, showing that self-heating cannot take place in MIM capacitors, because of the much too small nominal voltages. However, for high temperatures, current instability has been observed and has been attributed to ionic diffusion in dielectric. It follows the space-charge-limited theory. This phenomenon dramatically increases the TDDB, which is a key aspect of some applications [217].

MIM capacitor degradation behaviour under a wide range of CCS conditions is another reliability issue. It has been found that capacitance degrades with stress, but the behaviour of the degradation strongly depends on the stress-current density [218]. At high stress levels, the capacitance increases logarithmically with the injection charge, until DB occurs. At lower stress conditions, the degradation rate is proportional to the stress current, and reverses after a certain period of time. A metal-insulator interlayer was observed to explain this reversal phenomenon.

Conduction mechanism in Ta_2O_5 MIM capacitors was found to follow the space-charge-limited theory, and the asymmetry between positive and negative polarisation was attributed to an inhomogeneous spatial distribution of traps. Based on a complete characterisation of the leakage current at time zero, the instability of the leakage current in Ta_2O_5 during electrical stresses was proved [219]. The cause seems to be the strong influence of the oxygen vacancies profile on the current instability as well as on the leakage current asymmetry between positive and negative polarisation.

The major challenge of recent years in MIM capacitors has been to shrink their dimensions, as required by the trend in IC development. A novel device, the nanocapacitor, is presented in [220]; it has nanoscale dimensions in longitudinal and transverse directions. The effects of dielectric constant, dielectric strength and quantum electrical phenomena on achieving relatively high capacitances and capacitance densities in nanocapacitors are discussed. Nanocapacitors may find applications in low-power, high-bandwidth and real-time applications such as biochemically powered telemetry nanocircuits and quantum-charge pumps powering biomedical implants. They may also find uses in high-fidelity, high-sensitivity proximity sensors, motion detectors and actuators for nanoelectronic circuits and nanoelectromechanical systems (NEMSs). These can range from capacitive and pressure-wave detectors to gravitational-wave detectors used to detect elusive but predicted gravitational radiation.

5.2.2.6 Ferroelectric Capacitors

Semiconductor digital memories require capacitors that can be charged to retain the necessary memory intact. Dynamic random access memories (DRAMs) must be refreshed periodically to prevent data from being lost. The period between required refreshes depends upon the capacitance of certain capacitors within the DRAM IC, which is directly proportional to the surface area and thickness of the capacitor, and also to the dielectric constant of the material used to separate the plates of the capacitor. Since there are practical limits to the maximum area and minimum thickness of the capacitor (moreover, the tendency is towards smaller physical dimensions), the method for increasing capacitance is to utilise a material with a high dielectric constant. The ferroelectric materials are good candidates, having very high dielectric constants. For example, lead zirconate titanate (PZT) and lead lanthanum zirconate titanate (PLZT) are particularly attractive in this regard, as thin films of these materials may be deposited on ICs with dielectric constants higher than 100.

A typical ferroelectric capacitor consists of a bottom electrode, a PLZT dielectric layer and a top electrode. The electrodes are typically constructed from platinum. The top surface of this structure is normally coated with SiO_2, which provides protection from scratching and acts as an ILD in isolating metal interconnects from the top and bottom electrodes. Metal interconnects to the top and bottom electrodes are typically provided by etching via holes in the SiO_2 layer.

The main reliability issues of ferroelectric capacitors are produced by the initiation of cracks in the SiO_2 layer when placed in contact with platinum electrodes. This layer induces important failure risks for the deposited metal interconnections. A complex work based on both electrical and microstructural studies of elementary capacitors, test vehicles and chips has shown the degradation of ferroelectric properties of elementary capacitors under electrical stresses [221]. X-ray irradiation was revealed as an accelerating factor of degradation. The advanced technological devices, synchrotron techniques and electron microscopy associated with reliability testing enabled the FMs to be correlated with the capacitor microstructure, its geometry (planar or 3D) and the integration steps.

A solution for overcoming the abovementioned failure risks is to use chemical-solution deposition to fabricate dielectric multilayers of continuous ultrathin (20 nm) PLZT films [222]. A multilayer capacitor structure with as many as 10 dielectric layers was fabricated from these ultrathin PLZT films by alternating spin-coated dielectric layers with sputtered platinum electrodes. Integrating a photolithographically defined wet-etch step to the fabrication process enabled the production of functional multilayer stacks with capacitance values exceeding 600 nF. Such ultrathin multilayer capacitors offer tremendous advantages for further miniaturisation of integrated passive components.

5.2.3 Varistors

The term 'varistor' is derived from its function as a variable resistor. The metal-oxide varistor (MOV) is a nonlinear device that operates as a nonlinear resistor when voltage exceeds the maximum continuous operating voltage (MCOV). The resistance of an MOV decreases as voltage increases. The MOV acts as an open circuit during normal operating voltages and conducts current during voltage transients or an elevation in voltage above the rated MCOV.

Basically, an MOV is a multijunction device with millions of grains which acts as a series–parallel combination between the electrical terminals. The voltage drop across a single grain is nearly constant and is independent of grain size. Generally, an MOV contains a ceramic mass of zinc oxide grains, in a matrix of other metal-oxides (such as small amounts of bismuth, cobalt and manganese) sandwiched between two metal plates (the electrodes). Because of the finite resistance of the grains, they act as current-limiting resistors and consequently current flow is distributed throughout the bulk of the material in a manner which reduces the current concentration at each junction [223].

An MOV is designed to limit the voltage across the terminals in case of surges and high-voltage transients. A typical application is as a lightning-protective device. MOV is the most common, economical and reliable device for low-voltage and telecommunication-system lightning protection [224].

Varistor manufacture must be carefully controlled: if the body of the device is not uniform, heat is not uniformly distributed across the volume and some parts heat up more than others. Excessive heat can lead to one of the following FMs: electrical puncture, thermal cracking or thermal runaway [225]. Initially, varistors fail in the short-circuit mode when subjected beyond their peak current/energy and voltage ratings. If system over-current protection or varistor thermal protection does not exist, the varistor will continue conducting, increasing its temperature to the limit where separation of the wire and disk at the solder junction is achieved [226].

Typical MOV ratings extend from 2.5 to 3000 V and currents up to 70 000 A, but the energy capacity extends beyond 10 000 J for larger units. It is possible to connect multiple MOVs in parallel to increase current ratings or to connect them in series to provide higher or special voltage ratings. Typical response times are in the order of 500 ps.

MOVs are affected by ageing, essentially due to the number and amplitude of stresses, and other factors such as overheating, pollution and humidity. A method, based on probabilistic arguments, for evaluating the ageing process of MOVs has been proposed [227]. The expected life can be used to decide when a MOV must be changed before its failure occurs, since the main standards do not give definitive indications about such features. Investigations at different temperatures and applied voltage ratios (AVRs) show that the logarithm of lifetime is in a linear relation to reciprocal ambient temperature. The slope of this curve is virtually constant for zinc oxide and it can be attributed to activation energy. When subjected to severe electrical, environmental and/or mechanical stress beyond the specifications, varistors may fail in either a short-circuit or an open-circuit mode. This results in a burn-out, smoke or flaming. It has been observed that the leakage current of ZnO varistors increases under the voltage stress at elevated temperatures with ambient humidity content. The change in the leakage current corresponding to a fixed electric field with respect to the initial current is taken as the dimensionless degradation index. This index is monitored in [228] at various experimental conditions in conjunction with the curing condition of the epoxy resin powder. It has been found that the diffusion process of the moisture into the ZnO varistors plays a key role in the degradation process, provided that these varistors had excellent properties to begin with. The ionisation of the moisture at the interface between the ZnO block and the epoxy resin coating leads to the increase in the leakage current. Furthermore, the role of the ambient pressure corresponding to the elevated temperatures is considered the variable to the degradation process.

Five FMs were identified where surges or momentary over-voltages arise during MOV operation [229]:

- **Puncture** occurs in the centre of the ZnO varistors, resulting from the non-uniform distribution of the temperature and the current density caused by the higher temperature and the larger current grown alternately at the centre. Large fault current can create plasma inside the ceramic, with temperatures high enough to melt it. This phenomenon may be caused by long-duration over-voltage, for example by switching from a reactive load or thermal runaway of the MOV connected to the AC mains. Open-circuit failures are possible if an MOV is operated at steady-state conditions above its voltage rating. The exponential increase in current causes overheating and eventual separation of the wire lead and disk at the solder junction [225].
- **Cracking** is produced by the very large stroke current during lightning surges, due to the higher thermal tensile stresses.
- **Fuse blowing** can cause immediate destruction of the varistor at the first occurrence, if the clamping level is set too low.
- **Decreasing thresholds of thermal runaway** are directly related to the selected clamping level, with ageing accelerated by a low clamping level selection. Thermal runaway is the result of a continuous operating voltage following lightning surge, because the surge arrester does not cool sufficiently and the temperature of the arrester is increased by the larger leakage current caused by the degradation of the ZnO varistors.

- **Repeated conduction of currents** are associated with momentary system over-voltages ('swells'), which lead to an increase in leakage current. Repeated conduction is limited to a highly thermally activated low-current region where the performance of the varistor is determined by the parameters of the potential barriers. The degradation is a result of processes leading to the movement of ions and the deformation of the potential barriers. Therefore, dielectric spectroscopy (testing of the dielectric response in a wide frequency band) may be useful in investigating degradation changes [230].

The above failure risks can be avoided by designing equipment with a reasonably high surge withstand capability, so that retrofit using protective devices with very low clamping voltage will not be necessary. For those situations where a close protection is required, a very careful consideration of all factors becomes imperative, rather than cookbook application of protective devices.

Other possible FMs of MOV are:

- **Mechanical** degradation, such as partially failed bonds or broken dice, which may lead to corresponding increases in the forward voltage drop, resulting in electrical degradation. Such local degradations often lead to local thermal runaway and total failure.
- **Misalignment**, resulting in a very small insulation path, where moisture or ion concentration may lead to high leakage. High and often unstable leakage currents may occur as a result of the oxide passivation being bridged by effects such as *purple plague*.
- **Degradation**, which is caused by large and short pulses that can affect the structural properties of the ZnO varistors. A decrease in gain was observed in [231] due to a non-uniform distribution of the energy in the varistor. An increase in the leakage current was noticed for the samples presenting high porosity. The degraded sample shows oxygen deficiency causing a decrease in the height of Schottky barriers. The electric discharges cause a bismuth oxide phase transformation and a slight displacement of the main peak. The varistors fabricated with uniform grain size and electrodes avoid the concentration of heat which can lead to failure (puncture and cracking types). In general, the varistor presents two classical failures: puncture (for long pulses) and cracking (for short pulses); some samples present both failures. Degraded MOVs have been found to have smaller average grain sizes and to change diffraction peak position compared to a new sample. The non-uniform temperature distribution in the material is due to the development of localised hot-spotting during current impulse and dissolution in some other phases.

5.2.4 Connectors

A connector provides a separable connection between two elements of an electronic system, without inducing signal distortion or power loss due to the resistance it introduces. The connector must create an electrical and mechanical connection that is easy to detach. Most connectors are made from copper or copper alloys, with beryllium copper and phosphor bronze being the most common base materials due to their high electrical conductivity, low stress relaxation and competitive cost. The connector assembly consists of several functional components: connector housing, contact pins, contact plating, spring, retainer and seals.

The reliability of a connector depends on the specification requirements and the degradation mechanisms. FA is the basic tool for continuous reliability improvement. Some FA issues are presented below: the methods are given first, followed by details of the typical FMs for various types of connector.

5.2.4.1 Methods Used for FA of Connectors

Some methods used for FA of connectors are given in Table 5.17.

Table 5.17 Methods for studying FMs of connectors

Subject of the measurement/analysis	Methods and results	References
Visual inspection of connectors	High-quality stereomicroscope with a 10–70× magnification range.	
Deeper investigation of the connector surface	X-ray fluorescence method, neutron radiography and SAM, to analyse small cracks and other flaws in connectors.	
Thermal behaviour of small power connectors	Thermal analysis and thermal diagnosis, to identify an increase in contact resistance due to Joule heating, and that increased contact resistance has produced more Joule heating; this mutual action causes the connector to lose efficiency.	Wang [232]
Analysis of dust particles	X-ray energy spectroscopy (XES), to examine the composition of the particles. The adhesion of dust on contact surfaces is due to the static-electric charges carried by the particles.	Gao [233]
	Millikan testing method, to inspect failed contact surfaces, which shows that gypsum, organic materials, mica, titanium oxide and so on, which are present in small amounts in the dust, may contribute to the contact failure.	Zhang [234]
Analysis of contact interfaces	SEM and X-ray energy dispersive spectroscopy (XEDS), to analyse contact interfaces.	Zhou [235]

5.2.4.2 Typical FMs of Connectors

The common FMs in contact systems are [236]: corrosion, edge creep and pore corrosion, fretting corrosion (resulting from thermal and/or vibratory environment), diffusion and migration, and wet and dry oxidation mechanisms.

Corrosion is an important degradation mechanism in connectors. The two basic classes of contact finish, precious metal and tin, differ in the kinetics of corrosion. Gold finishes degrade through ingress of corrosion products from various sources on the contact spring to the contact area. Palladium and palladium alloys are more sensitive to general corrosion than is gold. Related primarily to the contact interface and the contact finish, corrosion increases contact resistance by two mechanisms: (i) a series contribution due to films at the interface and (ii) a reduction in contact area due to penetration of corrosion products into the interface. Three general types of corrosion must be considered:

- Surface corrosion (formation of corrosion films over the entire surface of the contact, such as tin oxide and oxides and chlorides on palladium and palladium alloys). Can cause contact resistance increases.
- Corrosion migration (movement of corrosion products from sites away from the contact interface into the contact area). Very sensitive to the operating environment, and predominately of concern in environments in which sulfur and chlorine are present.
- Pore corrosion (a small discontinuity in the contact finish). The pores themselves do not affect contact resistance; only if the pores become corrosion sites is contact resistance degraded.

The effects of operating environments on contact resistance depend on the contact finish. For tin contact finishes, surface corrosion is the dominant mechanism, but with a specific kinetics comes the so-called *fretting corrosion* [236]. Fretting is considered a low-amplitude (<10 μm amplitude) motion at the interface of a connector. In time, the fretting process wears the surfaces of the connector, leading to a build-up of wear debris, which can result in eventual connector failure. With contact plating, there are essentially three types of fretting that can occur in a contact: mixed regime, mixed stick-slip (partial-slip) regime and gross-slip regime. The primary mechanisms that determine which regime is present are the fretting amplitude and the applied normal force [237]. The working lifetime of the connectors (>25 years) is limited by the effects of fretting. Poor fabrication and assembly at installation are the most common causes of failure.

A forced motion test was used to evaluate the effects of materials and contact design on fretting [238]. Some interesting conclusions were obtained:

- For standard gold-plated contacts, fretting corrosion was initiated only for normal forces higher than a given value (100 gf in the reported experiment).
- For standard gold-plated contacts, the magnitude of change in resistance decreased as displacement increased.
- Unpowered gold flash Pd Ni contacts did not remain stable and degradation was initiated.
- Powered gold flash Pd Ni contacts had catastrophic failures, which are tentatively attributed to the additional mechanisms of electrical erosion.

Procedures have been carried out under laboratory-controlled conditions to investigate both thermal and vibration fretting effects using environmental chambers and fretting tests. Both optical and visual inspections have been adopted to observe the movement at the contact interface. In situ measurement was made with a sensor designed to monitor the relative displacement [239]. The sensor was assembled into a connector sample, taking the place of the male component. When the interface experienced movement, the relative displacement of the contact point would cause a corresponding linear change of resistance measured across the male and female connection. The sensors were validated by a series of experiments and subsequently used in a field test to establish the relationships between the fretting effects with temperature, humidity and differential pressure (associated with temperature variation).

The harsh operating environment of the automotive application makes the semi-permanent connector susceptible to intermittent high contact resistance, which eventually leads to failure. Fretting corrosion is often the cause of such failures. However, laboratory testing of sample contact materials produced results that do not correlate with commercially tested connectors. A multicontact reliability model was developed in [240] to bring together the fundamental studies and studies conducted on commercially available connector terminals. It is based on fundamental studies of the single contact interfaces and applied to commercial multicontact terminals. The model takes into consideration, first, that a single contact interface may recover to low contact resistance after attaining a high value, and second, that a terminal consists of more than one contact interface. For the connector to fail, all contact interfaces have to be in the failed state at the same time.

Because connector contacts are exposed to air, dust is the main pollution in contact failure and seriously deteriorates the reliability of electronics and telecommunication systems. An investigation of the manner in which two solids touch when one is contaminated with dust was made in a historical paper [241]. The behaviour is discussed in terms of the probability that intimate contact will be established between the solids in any particular closing operation. The manner in which a dust particle can become trapped between the approaching bodies and thus prevent direct contact from occurring is considered and a simple physical model of the processes is proposed, depending on the nature of the contact and the number and size of the contaminating particles. It seems that the effectiveness of the contamination in preventing direct contact should fall off rapidly when the surface roughness is increased until the height of the irregularities is comparable with that of the particles. The effect has been verified experimentally. It has also been demonstrated that there is a sharp change in the probability of the occurrence of intimate contact if the area of contact between the solids is made comparable with the cross-sectional area of the particles.

Recently, based on a model from mechanics, Wang and Xu analysed the size of particles that would be pushed away and the failure conditions when rigid particles embed on a contact surface [242]. This is an elastic-plastic model of FE analysis, and is closer to real conditions. The behaviours of large-size and small-size particles and the influence of particle hardness were investigated. The calculated result of small-size particles presents a general hazardous size coefficient for different contact-surface morphologies; for large-size particles, it presents a hazardous size coefficient for complicated compositions of the dust; the effect of the dust shape is also discussed.

Housings of connectors are usually made of glass and mineral-filled thermoplastics (e.g. PET). When a connector must survive extended high temperatures, polyphenylene sulfide (PPS) may be used. PPS resists all solvents, but is affected by some amines and halogens. It is also more brittle than PET. Detailed examination of brass coaxial cable connectors which exhibited cracks in their outer housings was made [243]. FA indicated a brittle failure due to stress corrosion cracking. The research study confirmed the observed fracture mode; additional testing may be especially helpful in corrosion-related failures.

Intermittent continuity of contact fingers is one of the most difficult failures to identify in a high-volume production process. Proper determination of the application/end-use needs with qualification and control of the supplier base for raw cards will greatly reduce the occurrence of marginal plating thickness issues. X-ray fluorescence is a preferred method for measuring plating thickness as it saves time versus traditional metallurgical methods; it is also accurate and nondestructive. Several factors impact the accuracy and repeatability of XRF plating thickness measurements, but XRF can minimise errors in plating measurements [244].

Copper has a low resistance to corrosion, which can lead to electrical failure of connectors. For this reason, a layer of gold is often plated on the surfaces of connectors to seal off the base metal from direct exposure to the environment. As an economic practice, '*gold flashing*' (the gold layer is thinner than 0.25 μm) has been used to protect electrical contacts from corrosion. However, there is increasing evidence that gold flashing can be detrimental in applications calling for long-term reliability (synthesised in [245]). The gold-flash product performance varies from manufacturer to manufacturer as well as between product families. Putting aside basic limitations (such as durability), those products with gold flash which appear to perform in a satisfactory manner seem to be tied to the following interrelated variables, among others: (i) plating quality; (ii) normal force levels; (iii) contact geometry and configuration; (iv) surface conditions; and (v) contact sheltering techniques [238].

The most frequent FMos of PCB connectors are associated with *mating* and *unmating* of the connectors. The one-piece connector may sustain damage to the contact tabs during a mating cycle. As the connector is inserted or removed, extreme stresses on the tabs may destroy a contact connection. Another FMo is associated with insertion and withdrawal forces. Excessive force when inserting or withdrawing the connector can damage the connector contacts.

While pins are safe from bending in a mated connector pair, exposed pins are subject to bending with even gentle mishandling, and the initial damage is often too slight to notice. Unfortunately, the force required to join mating connector pairs is usually more than sufficient to flatten or crush a misaligned contact without any physical indication of damage. A single bent pin often causes both a broken circuit (since the pin is no longer in contact with the corresponding contact) and a short circuit (since the pin may make contact with an adjacent pin or the grounded connector body, or both). If this failure causes unacceptable system effects, the connector's contact assignments may have to be rearranged, or other connectors may be needed, in order to increase separation among critical signals. The shortcomings of the usual approach to bent-pin analysis are shown in [246], where some analysis rules that address these shortcomings and enable a large part of the work to be automated are also discussed.

5.2.4.3 Fibre-Optic Connectors

Fibre-optic (FO) technology is a relatively new process and the specific FMos and reliability levels have not been well documented yet. The dominant FM for FO is a fracture due to stress caused by proliferation of micro-cracks. Another FMo for cables occurs when hydrogen migrates into the core of the fibre, causing optical signal loss.

Basically, FO connector failures can arise during manufacture or because of environmental or handling factors. All the known FMos can be avoided with proper design, screening, testing and handling of connectors.

There are three causes of a broken fibre in the ferrule: (i) inserting a weak fibre into the ferrule; (ii) air voids in the epoxy; and (iii) uncontrolled epoxy expansion. To avoid these failures, companies should ask whoever builds their connectors whether they have looked at these FMos and what they are doing to prevent them [247].

Two other common FO connector failures involve fibre breaks caused by thermal changes [248]. Type I failures involve fibre buckling during cooling from the epoxy cure temperature and are related to the free length of fibre inside the ferrule/backbone assembly. Type II failures occur during the heating phase of thermal cycling and are caused by expansion of epoxy in the ferrule entry under certain conditions. These conditions lead to unexpected tension at the ferrule capillary entry. In both cases the failure probability is increased by fibre damage during the termination process.

5.2.4.4 Mobile Phone Connectors

A mobile phone connector mainly consists of a spring with a spherical head as a rider, contacting the PCB surface. During sliding or micro-motion, the surface plating of the spherical contact is mostly worn out. Contaminants including wear debris are usually accumulated on the plane surface and tend to form a tiny region with high resistance.

Since any contaminant at the contact surface is extremely small, it is very difficult to remove it for identification. The elements involved in the contaminant can be tested by X-ray energy spectroscopy (XES). In one study the composition of the contaminant was revealed as mainly Ni, Cu, S, O and Cl. By further study using electron spectroscopy for chemical analysis (ESCA), the compounds were identified as sulfates, chlorides and oxides of nickel and copper. Si and Ca are commonly found in dust [249].

In another experiment, looking at 95 failed connector contacts from 23 mobile phones, the contaminants at the contact areas were formed by various sizes of particles [250]. They all adhered together, and it seemed that the smaller ones accumulated the larger ones. The composition of contaminants was very complex, including dust particles containing mainly silicates (quartz, mica, feldspar) and calcium compounds (gypsum, calcite and lime), wear debris of surface materials, corrosion products (sulfates, chlorides and oxides of copper and nickel) and high concentrations of organics. Organic materials seemed to act as adhesives at different temperatures. Contaminants causing failures were usually located at or near the wear tracks. The most important function of micro-motion is to move the separated contaminants and accumulate them together at the contact; it also produces wear debris and destroys the surface metal layer. High contact resistance was found in several small regions of the contaminant with high enough thickness. Failure could occur within very short periods of time (three to four months), depending on the high-resistance region formed.

In [251], the special features causing contact failure are summarised. Particle accumulation is the main problem; organic material such as lactates from human sweat can act as an adhesive, sticking separate particles together and making them adhere on the contact surface; the chemical properties of dust cause serious local corrosion. The corrosion products may trap particles and firmly attach them to the contact surface; micro-motion frequently occurs at the contact interface. Hard particles can be embedded into the surface, and soft particles can be squeezed and inserted into the contact. Si compounds in dust play the most important role in forming high-resistance regions that lead to failure; deposition of particles depends on the amount of material, the static electricity attracting force and the force of gravity applied on the particles. Current dust tests can hardly reflect the serious contact failure.

5.2.5 Inductive Elements

Inductive elements, such as transformers, inductors, relays, solenoids and motors, store electrical energy in the form of a magnetic field.

In *transformers*, time-varying magnetic fields transfer the energy between an inductive element (winding) and other windings inductively linked to the fields. Transformers can be designed to change one voltage to another and to isolate circuits from DC flow. Inductive elements can experience catastrophic electrical failures. These devices rely on current that passes through small-diameter solid 'magnet' wire wound on a form or bobbin. When a wire breaks, the device fails, and engineers and failure analysts must determine what went wrong so that they can correct any manufacturing problems and prevent future failures. The analysis begins with the removal of coil-impregnation material, which adheres to a wire and often obscures the cause of failure. Then, to detect the presence of specific chemical elements and compounds, X-ray energy dispersive spectroscopy (XEDS), wet-chemistry analysis, atomic-emission spectroscopy or other analytical tools and techniques can be used.

Inductors are typically used in RF devices as frequency filters. Current inductors vary from planar inductors to multi-layer coil structures. Failure occurs whenever there is a disruption of the current flow through any part of the conductive element or if some condition interferes with the magnetic coupling between the elements of a transformer. Failure may also be caused by failure of the insulation on or between the coil windings (i.e. the inductive elements may be fused open by the flow of excessive current). Mechanical abrasion or the application of excessive electrical potential across the insulation may cause insulation failures.

ESD failure in an inductor can occur by the following: (i) inductor coil 'open' circuit, if a physical 'open' occurs in the inductor itself; (ii) inductor coil 'turn-to-turn' short, if a 'short' occurs between adjacent 'turns' in the inductor coil; (iii) inductor 'coil-to-coil' short (between two sides of an intertwined coil); and (iv) inductor resistance degradation (due to changes in the resistance).

If copper overheating is the main induction-coil FMo, a coil current reduction leads to reduced coil heat losses, which can appreciably lengthen coil life. In addition, a reduction in coil current results in reduced electromagnetic forces. If coil failure is related to stress-fatigue cracking, joint water leakage or copper banding, lower electromagnetic forces can increase coil life. The clamping area of the coil also contributes to short inductor life due to wear and contaminants, which can lead to excessive overheating and even arcing, and ultimately to premature coil failure. If arcing is primarily responsible for coil failure, the reduction in coil voltage due to the use of a flux concentrator can eliminate it and improve coil life. Degradation of the flux concentrator due to its overheating is the typical FMo. Premature coil failure can result if insufficient attention is paid to thermal expansion [252]. The factors related to premature induction-coil failure are: (i) process-related factors; (ii) coil copper-related factors; (iii) improper coil design, fabrication, maintenance and storage; and (iv) tooling and accessories.

Corrosion probably causes the most failures of wires in coils. The ends of corroded wires vary widely in appearance, but they usually have odd shapes and appear rough and dirty. Insulation may be split or cracked by the build-up of corrosion products between the wire and its insulation.

Metal fatigue, caused by cyclic bending of a magnet wire, produces a rough but relatively flat end when the wire fails. Insulation surrounding the broken end may be pulled away from the metal conductor. A nick, scratch or abrasion in the wire surface concentrates stress at this point and increases the susceptibility of the wire to tensile or bending stress. Intentionally nicking the wire surface with a sharp knife and then cyclically bending the wire until fatigue failure occurs produces a wire end bearing evidence of the nick [253].

On-chip planar inductors are constructed from interconnect technology; the inductors consist of conductive metal films, metal contacts, metal vias and ILDs. Planar-inductor ESD failure can occur through metal displacement, insulator cracking, via failure and cladding failure. A failure in an inductor can occur in the coil structure, the underpass structure or in connections to the input and output pad.

Coil geometry design can influence the failure level of inductor structures. Inductor-coil geometry influences the quality factor, the resistance and the coil current density within a given cross-section of the coil. With the introduction of MEMS devices to the RF CMOS world, new FMs can also occur in such structures [254].

Relays consist of inductive elements similar to those used in inductors to generate a magnetic field, which move one or more mechanical switch elements. As in other inductors, the coil windings may be damaged by the flow of excessive current or by the breakdown of the insulation between windings. The contacts on the switching elements may be damaged by excessive voltage or current causing spark generation. They may also fail through mechanical abrasion (wear).

Normally, failed relays show pitted or otherwise damaged switch contacts. In some cases, the contacts are welded together by excessive electrical stress applied across the contacts when switching inductive elements. When a charged inductive circuit is opened by the relay, the collapse of the magnetic field in the inductor generates a large voltage spike that may cause arcing across the relay switch contacts. This type of failure may be prevented or minimised by proper circuit design. Damaged contacts have also been observed in relays with gold contacts operating in a moistureless environment. A thin film of moisture provides lubrication for the gold; without moisture, the gold surfaces easily damage each other. A problem of this type can occur if a non-hermetic relay is used in a space application. Such relays usually have a small amount of moisture intentionally sealed inside.

5.2.6 Embedded Passive Components

In a typical PCB assembly, over 80% of the electronic components are passive components, such as resistors, inductors and capacitors, making up to 50% of the entire PCB area. By embedding the passive components within the substrate instead of mounting them on the surface, the embedded passives can reduce the system real estate, eliminate the need for surface-mounted discrete components, eliminate lead-based interconnects, enhance electrical performance and reliability, and potentially reduce the overall cost.

Embedded passives (also known as integral passives) have some potential advantages, such as:

- Significant reduction in overall system mass, volume and footprint by eliminating surface mounts.
- Improved electrical performance by eliminating leads and reducing parasitic inductance and capacitance.
- Increased design flexibility.
- Elimination of lead–base interconnects.
- Improved thermo-mechanical reliability by eliminating solder joints.
- Potential reduction of overall cost.

Most importantly, embedded passives are more reliable since they eliminate the two solder joints, which are the major failure location for discrete parts. However, it is difficult to confirm that the overall reliability is improved, mainly because not enough data has been accumulated to date. Use of new materials and processes may bring other concerns such as crack propagation, interface delamination and component instability. More layers may be necessary to accommodate embedded passives. This could mean integration of various new materials and processes that might cause significant thermo-mechanical stress due to CTE mismatch. Furthermore, unlike discrete components, in which defective parts can be replaced, rework is not a viable option for embedded passives. A single bad component can potentially lead to scrapping of the entire board.

The typical FMs of the embedded passive components are detailed in Table 5.18 [255].

Table 5.18 Typical FMs of embedded passive components

Passive component	Failure mode	Possible failure mechanisms
Resistors	Open circuit	Electrical overstress (EOS), leading to thermal overstress.
	Parameter drift	Application of high levels of stress, exposure to high humidity conditions and/or high-temperature operating environment.
Capacitors	High leakage	Rupture of oxide film in electrolytic capacitors due to application of high electric field.
		High temperature and faulty seal.
	Open circuit	Corrosion of the electrodes due to chemical action caused by contaminants and moisture.
	Short circuit	Dielectric breakdown due to application of high voltage beyond the rating.
	Parameter drift	Degradation of dielectric material due to exposure to humidity, high temperature and ageing.
Coils	Open circuit of coil wire	Thermal overstress caused by shorting of adjacent turns where insulation has been damaged during the winding process or due to a manufacturing-process fault.
		Nicks and kinks in the wire.
Transformers	Open circuit fault in primary and secondary windings	Excessive thermal stress caused by EOS, shorting of windings, as in the case of coils.
	Short circuit between primary and secondary	Poor isolation, low dielectric withstanding voltage.
	High levels of parasitics	Increasing leakage inductance and/or inter-winding capacitance due to faulty design and manufacturing technique.
	High levels of copper and eddy current losses	Poor design leading to high heat dissipation in the transformer and affects adjacent components.
Relays	Contact damage	Arcing induced; corrosion of contacts by ingress of moisture, flux, cleaning agents due to improper sealing; melting of contacts due to EOS.
	Coil damage	EOS.
	Damage to plastic body	Exposure to high temperature, for example during soldering, or internally generated heat due to EOS.
Printed circuit board (PCB)	Discoloration, delamination	Exposure to high temperature during soldering, heat dissipation of components on the board.
	Warping	Exposure to high temperature or faulty board design (e.g. insufficient thickness of the laminate, faulty layout and mounting of components on the board).

After [255].

5.3 Failure Modes and Mechanisms of Silicon Bi Technology

Over the last 50 years, silicon (Si) has been the most popular basic material for semiconductor components. Two features have led to this situation: (i) Si remains a semiconductor at higher temperatures than other materials (e.g. germanium) and (ii) SiO_2 easily be grown in a furnace and forms a better semiconductor/dielectric interface than any other material.

Si devices are produced on ultrapure Si wafers, which are doped with other elements to adjust their electrical response by allowing control of the number and charge of current carriers. A large variety of devices may be produced, such as diodes, transistors, ICs, microprocessors, solar cells and so on. There are two main groups of technologies for the fabrication of Si devices: the so-called *Bi technologies* and *MOS technologies*. The FMs of the Bi technologies will be detailed in this section, while Section 5.4 will list those of the MOS technologies. Because they have specific applications, we have chosen to treat those optoelectronic and photonic components which cannot be manufactured by either group of technologies separately, in Section 5.5.

The FMs of devices manufactured by Bi technologies are as follows:

- Si diodes, both power and small-signal (Section 5.3.1).
- Si Bi transistors, covering small-signal, medium-power and power transistors (Section 5.3.2).
- Si power devices, such as thyristors and insulated gate bipolar transistors, IGBTs (Section 5.3.3).
- Linear ICs (Section 5.3.4).

5.3.1 Silicon Diodes

Diodes are two-terminal components with a rectifying property: depending on the polarisation of the applied voltage, they allow an electric current to pass in one direction (called the forward-biased condition) and block it in the opposite one (the reverse-biased condition). The diode is the oldest semiconductor electronic component, invented in 1874 by the German physicist Ferdinand Braun. The first device, a point-contact diode (or 'cat's whisker diode'), used the rectifying properties of galena.[8] Later, a vacuum tube with two electrodes (a plate and a cathode) was used as a diode.

In this book, electronic components are grouped by basic material and manufacturing technology. Consequently, the diodes have been divided into three main groups:

- Si diodes, manufactured with planar-epitaxial or mesa technology.
- Optoelectronic diodes, with a large range of basic materials, including Si.
- Diodes manufactured on basic materials other than Si.

In this section, only the FMs of Si rectifier diodes are detailed; the other two groups are treated in Sections 5.5 and 5.6. The main types of Si diode are synthesised in Table 5.19.

Where the development phase is concerned, the FMs of semiconductor diodes can be divided into [1]:

- **Fabrication-induced FMs**, which may be:
  - **'Front-end' FMs**, produced by defects in the semiconductor material (point defects, dislocations, etc.) or defects induced by processing steps of die manufacturing, such as flaws in the thermally grown oxide or in the chemically deposited epitaxial layer (details were given in Section 5.1).
  - **'Back-end' FMs**, which are induced during assembling operations, such as dicing, die-attach, wire-bonding, encapsulation and so on (details were given in Section 5.1).
- **Operation-induced FMs**, which are event-dependent and directly related to either poor design (leading to EOS), careless handling of components (leading to static damages) or misuse during operation; consequently, they concern both the manufacturer and the user.

The typical FMs of the families of diodes mentioned in Table 5.19 are detailed below.

5.3.1.1 pn Junction Diodes

Parameters common to most single-junction devices (and indicative of their satisfactory operation) are forward voltage drop, reverse current and breakdown voltage.

[8] Galena is the natural mineral of lead sulfide.

Table 5.19 Main types of silicon diode

Diode type	Technology and characteristics
pn-junction diodes	Used as rectifiers. The vast majority of all diodes are *pn* diodes found in CMOS integrated circuits.
Varactor diodes	Varactor (varicap) diodes use the characteristics of *pn* junctions to changes their capacitance and series resistance as the applied bias is varied [256]. Employed as voltage-controlled capacitors, rather than rectifiers, they are commonly used in parametric amplifiers and oscillators, or in voltage-controlled oscillators.
Switching diodes	Normal *pn*-junction diodes, but with gold or platinum introduced as dopants, which act as recombination centres and provide the fast recombination of minority carriers. Compared with Schottky diodes, switching diodes are slower, but have lower current leakage.
Z diodes	Diodes operating in the breakdown zone, they conduct backwards at a precisely defined voltage, allowing them to be used as precision voltage references. In electrical circuits, Z diodes and switching diodes are connected in series and in opposite directions to balance the temperature coefficient to near zero. Two (equivalent) Z diodes in series and in reverse order, in the same package, constitute a transient absorber (TransZorb or Transorb), which fulfils the task of a TVS diode (see below).
Avalanche diodes	Designed to conduct in the reverse direction when the reverse-bias voltage exceeds the breakdown voltage. Electrically similar to Z diodes (the only practical difference is that the two types have temperature coefficients of opposite polarities), they break down using a different mechanism, the avalanche effect, without being destroyed.
Transient voltage-suppression (TVS) diodes	These are avalanche diodes designed specifically to protect other semiconductor devices from high-voltage transients. Their *pn* junctions have a much larger cross-sectional area than those of a normal diode, allowing them to conduct large currents to ground without sustaining damage.

The reverse current (I_R) of *pn* junctions is voltage-dependent. According to theory, I_R is proportionate with $V_R(1/n)$, where n = 2–3. However, experimentally it was found that beyond a certain voltage level (100–200 V), but well before the breakdown region, the variation is no longer a linear one, and an exponential increase of current with voltage occurs [257]. This behaviour was noticed for both Ge and Si *pn* junctions and is attributed to a significant leakage current flow at junction periphery. In the linear region, reverse current flow is uniformly distributed on the junction periphery, but at higher voltages current crowding causes local defects at the semiconductor – dielectric interface. These defects can be diminished or even avoided by appropriate passivation.

The packaging variant is a determinant for diode reliability. For surface-mounted glass diodes, a series of precautions must be taken to avoid failure risks [258].

5.3.1.2 Varactor Diodes

The characteristics of varactor diodes are low resistance, high capacitance change ratio and high reverse voltage, although the capacitance–voltage requirements differ in different applications [259].

There are two typical FMs of varactor diodes, both produced by misuse during operation [260]:

- Leakage increase when the high- frequency modulation circuit does not work or the modulation performance is damaged.
- Parameter drift when there is a distortion in the signal of the high-frequency modulation circuit.

5.3.1.3 Switching Diodes

Obviously, the switching diodes (also called fast-recovery diodes) are intended for switching applications. Some instabilities of the reverse current at high temperatures (typically higher than 125 °C) have been reported. Because the maximum operating junction temperature is 150 °C, the I–V reverse characteristics of switching diodes in the temperature range 100–150 °C were compared with those of standard-recovery diodes [261]. Only the switching diodes were instable, due to thermal runaway caused by the surface component flowing at the junction edge. FA performed on failed items led to the conclusion that damages at junction periphery were responsible for such instabilities.

5.3.1.4 Z Diodes

Si diodes that operate in the breakdown zone are called *Z diodes*. These diodes were originally called Zener diodes after the 'Zener effect' occurring in the operation domain, an effect discovered by Dr Clarence Melvin Zener (Southern Illinois University) in 1934 [262]. It later became clear however that this effect is decisive only for breakdown voltages smaller than about 5.5 V. At higher voltages, the parameter variation is governed by avalanche breakdown, as explained by K.G. McKay (Bell Telephone Laboratories, Murray Hill, New Jersey) in 1954 [263]. Dr Zener refused to give his name to a component which was not fully linked with the Zener effect, so the diodes are now called Z diodes.

The most common FMos and FMs of Z diodes are shown in Table 5.20.

Table 5.20 FMos and possible FMs of Z diodes

Failure mode	Possible failure mechanisms
Short circuit (breakdown)	Inefficient power dissipation produces excessive heat in junction area, which leads to thermal runaway. Migration of the dopant impurities can lead to short circuits and subsequently burnout of the *pn* junction [1]. If the diode is pressure-contact-mounted by a spring, the failure may occur in time, especially when the diode is subjected to variations in ambient temperature or in loading. Incipient failures of this type can be screened out to some extent by thermal cycling (−55 to +150 °C). The most usual form of damage is cracking. With a loose fragment, intermittent or permanent short circuits may be caused. If the crack runs across the active area, the actual breakdown voltage of the diode may be reduced by the direct flashover across the exposed *pn* junction.
I_R drift after extended operation	Contamination near the semiconductor junction, or under, into or on top of passivation layers. The use of Si nitride on to planar junctions (as a barrier against ionic contamination with lithium sodium and potassium) virtually eliminates this type of drift. Oxide passivation is bridged by purple plague.
I_R drift after damp heat test	Lack of hermeticity, allowing moisture to reach the semiconductor chip. Misalignment may result in a very small insulation path, where moisture or ion concentration can lead to high leakage.
Z_Z drift	Degradation of chip ohmic contacts. The glass package has a good hermeticity, and a good metallisation system, which guarantees ohmic contact integrity on all Z diodes. Strongly correlated with 1/f noise in the devices. The larger the initial 1/f noise of a diode, the earlier Z_Z drift occurs, being related to dislocations in the space-charge region of the *pn* junction. Based on results found, a *1/f* noise-screening approach is proposed for the high-reliability application of the devices [264].

5.3.1.5 Avalanche Diodes

The avalanche diode is designed to break down and conduct at a specified reverse-bias voltage. Avalanche breakdown is an electric-current multiplication, allowing a very large current to flow within insulating or semiconducting materials. This phenomenon occurs when the voltage applied across the insulating material is great enough to accelerate free electrons to the point where, when they strike atoms in the material, they can knock other electrons free. Generally, avalanche causes a catastrophic failure, but if the diode is designed to control the avalanche phenomenon, the avalanche caused by over-voltage can be tolerated and the device remains undamaged. Thus, avalanche-type diodes are often used in protecting circuits against transient high voltages that would otherwise damage them. Being optimised for the avalanche effect, avalanche diodes exhibit a small but significant voltage under breakdown conditions. The same effect is also found in Z diodes at high voltages, but an avalanche diode has a better surge protection than a simple Z diode and acts more like a gas discharge tube.

Since avalanche diodes are frequently used as lightning-surge protectors (due to their high-speed response and constant voltage-suppression characteristics), a transient-temperature response analysis was carried out to determine the relationship between temperature and failure under application of a lightning surge [265]. The non-steady-state thermal conduction equation was solved by the finite difference method, the nonlinear elements included in the analysis being: (i) the temperature dependencies of the thermal-conductivity thermal capacity and (ii) the breakdown voltage of the avalanche diode. The theoretical results agreed well with the measured data: it was found that the peak temperature in the avalanche diode under application of a standard lightning surge was near the half-value decay time and that the failure of the avalanche diode depended on the peak temperature. The critical-failure temperature was estimated to be about 490 °C.

The static breakdown behaviour of high-voltage diodes at elevated temperatures has also been investigated [266]. It was found that under isothermal and homogeneity conditions, the reverse-bias safe operation area (RBSOA) of power diodes is the same as the sustain-mode dynamic avalanche. After introducing current inhomogeneity, failure will occur at a reverse power density much less than the RBSOA predicted by the sustain-mode dynamic avalanche. The proposed FM of power diodes was based on these findings.

In the field of high-temperature modelling and simulation of semiconductor devices, most of the physical models available to date have been validated only to 400 K, in spite of the fact that local heating during the stress event can lead to local temperatures well in excess of this limit. The work in [267] deals with mobility and impact ionisation in Si at high temperature, focusing on the experimental determination of these physical parameters and the extension of the experimental temperature range. The mobility has been measured, using the Hall effect, up to 1000 K thanks to the use of Ti/TiN interconnections in combination with junction-free van der Pauw resistors, which are intrinsically immune to spurious thermal leakage currents. Hole and electron-impact ionisation have been determined as functions of electric field up to 673 and 613 K respectively, by measurement of the multiplication factor in Bi and static induction transistors.

The experimental and simulation studies performed so far have failed to identify the burnout breakdown mechanism for power diodes. Although the possibility of thermal breakdown has been suggested by preliminary results, not enough attention has been paid to the thermal effects. In [268], the effect of irradiation with 17-MeV C12 ions was simulated. When an avalanche diode was biased at 2700 V and above, a large increase in the current along with a temperature rise to the melting point of Si was observed, indicating diode burnout. The breakdown is attributed to the local heat generation due to impact-ionisation-generated charge. Local temperatures inside the diode increase due to the presence of high current and electric field following ion strike. Intrinsic carrier concentration, which is also a function of temperature, increases with temperature. As the intrinsic carrier concentration increases at high temperatures, more carriers become available, producing a larger current. A thermal feedback loop may be initiated, where an increase in current leads to more heating.

5.3.1.6 Transient Voltage-Suppression Diodes

A transient voltage-suppression (TVS) diode is an electronic component used to protect sensitive electronics from voltage spikes. It is also commonly referred to as a Transorb (after TransZorb, registered by General Semiconductor) or Transil (registered by ST Microelectronics). It can be used as a rectifier in the forward direction, like other avalanche diodes, but operates at higher peak currents (e.g. 1500 W of peak power, for a short time). If formed by two opposing avalanche diodes that are in series with one another and connected in parallel with the circuit to be protected, the TVS diode works bidirectionally.

TVS diodes are used to protect against very fast and often damaging voltage transients. They are commonly employed at voltage levels below 20 V, but series-connected TVS diodes may be used in high-voltage applications (even above 1000 V) [269].

5.3.2 *Bipolar Transistors*

There are two basic variants of transistors: *bipolar* transistors (operating with minority carriers) and *unipolar* transistors (operating with majority carriers). Each variant has strengths and weaknesses. In high-voltage applications, Bi transistors can obtain a sufficiently low voltage drop in the on state (due to the high levels of charge injection consisting of *both* electrons and holes into the thick portion of the device designed for high-voltage stand-off) and support a high voltage in the off state. Unipolar devices (in which only one type of charge carrier is responsible for conduction) are not considered useful for high-voltage applications because of their large voltage drop in the on state. They can however quickly remove a stored charge, making for an efficient transition from on to off state, so are suitable for fast turn-off. This is not possible in Bi devices, because of the long transit time of charge carriers in the thick stand-off region.

The ideal transistor would combine the low voltage drop of a high-voltage Bi device and the fast turn-off capabilities of a unipolar transistor. These features are especially important in power transistors. In 1984, the BiMOS was proposed, which combines the Bi process with CMOS circuits. BiMOS technology allows the configuration of potent and 'intelligent' devices suitable for microprocessor-based systems.

With Bi transistors, there are three main variants for manufacturing the die [1]:

- **Volt/base** (diffusion and epitaxy are used and a homogeneous base is obtained; suitable for linear IC, but also for small-signal transistors, including those used for switching applications).
- **Volt/collector** (diffusion and epitaxy are used and a homogeneous collector is obtained; suitable for medium-power transistors).
- **Volt/base and collector** (only epitaxy is used and homogeneous base and homogeneous collector are obtained; suitable for power transistors).

Basically, Bi transistors have four fundamental parameters – breakdown voltage, current gain, switching speed and dissipated power – but others may be significant in some specific applications.

For power devices manufactured by Bi Si technologies (e.g. Bi transistors, thyristors and IGBT), an important parameter is the *safe operating area* (SOA), which is defined as the voltage and current conditions needed for safe operation of the device. This is shown in the device datasheet as a graph having in abscissa the collector–emitter voltage (V_{CE}) and in ordinate the collector–emitter current (I_{CE}), the safe area being under the curve.

Failures of Bi transistors occur for many reasons: design flaws, poor-quality materials, manufacturing problems, improper conditions during transport or storage, overstress during operation and so on. An important primary cause of failure is an abnormal increase of the temperature, often spatially limited (so-called '*hot spot*'), which might be produced by an abnormal operation. In some cases, if no irreversible physical transformations occur, the transistor can regain its initial characteristics (for instance, when e-h pairs are formed by thermal agitation, leading to a thermal turn-on). If irreversible transformations occur, the transistor is damaged or destroyed (an example is the surface damage during soldering, which may lead to local melting of the eutectic Si/metal, irreversibly modifying the contact properties). Sometimes, some abnormal stresses, even limited and occurring once, lead to a progressive degradation (in normal operating conditions), building up to the failure of the component. This progressive degradation can occurs minutes or thousands of hours after the initial accident.

First, it is important to say that Bi transistors are affected by the common FMs detailed in Section 5.1: at *wafer level* (e.g. crystallographic defects, surface contamination, corrosion, microcracks, EM, diffusion defects, insulating oxide breakdown, etc.) and by *packaging* (e.g. bad solder joint, chip-mounting errors, use of improper materials for the contact area and connection wire, imperfect sealing, allowing contaminants and moisture into the case, etc.).

The specific FMos and possible FMs of Bi transistors are synthesised in Table 5.21.

New developments concerning some of the FMs mentioned in Table 5.20 are given below.

Second Breakdown (SB) is an irreversible phenomenon in Bi power transistors, inducing device failure while operating well within switching voltage and current conditions that should pose no threat. The phenomenon occurs as follows: the current concentrates in a single area of the base-emitter junction, causing local heating and destruction of the transistor. SB can occur with both forward and reverse base drive. Except at low collector–emitter voltages, the SB limit restricts the collector current more than the steady-state power dissipation of the device.

The first detailed characterisation of SB was given in [270]. Subsequently, serious attempts were made to understand this significant FM, which could occur well within the power dissipation limits of power transistors. On the basis of an observed delay time before the initiation of the breakdown, it was proposed that SB is related to a thermal mechanism. It was shown that the onset of SB cannot be predicted simply in terms of voltage and current, as had been the practice, but that it is important to characterise SB in terms of the energy dissipated in the transistor, and further, that the energy threshold (or delay time) is dependent on other factors such as ambient temperature and the biasing base current of the transistor. It was assumed [271] that the phenomenon observed during SB in a semiconductor is the occurrence of a current-controlled negative-resistance region. Once this event has occurred, the semiconductor may be modelled as solid-state plasma which undergoes a plasma pinch, as a consequence of its own self-magnetic field.

Note that it is possible to avoid failure after SB if appropriate measures are taken, which predominantly involve diversion or 'crowbarring' of the collector current [272]. For instance, a Z diode and resistor (cathode to collector) connected in parallel with the collector–emitter terminals may offer some protection against SB. Today, SB is well controlled through device design and the specification of SOAs, but contributions to understanding and avoiding SB are still being made.

Table 5.21 Specific FMos and possible FMs of bipolar transistors

Failure mode	Possible failure mechanisms
Drift of electrical parameters	Continuous *increase of leakage current* (I_{CBO}), often accompanied by a decrease of the *current gain* (h_{FE}), is a sure indication of a surface with impurities. Some items that have failed due to an increase in I_{CBO} may suddenly recover after a period of testing but in general the change is permanent and degraded items cannot be restored by stressing under any conditions.
Short circuit	Hot-spot phenomena, chip problems, a circuit defect or *second breakdown*.
Open circuit	A bad solder joint or a melted conductor due to an excessive current.
Combination of short circuit and open circuit	A melted conductor linked with the upper conducting layer.
Intermittent open circuit	A bad solder joint, especially when it occurs at high temperature or after thermal cycling; *thermal fatigue* may be involved.

The origin of a **leakage current** in several failed Bi transistors has been identified in [273] by complementary advanced FA techniques. After precise localisation of the failing area by PEM and optical beam-induced resistance-change investigations, a FIB technique was used to prepare thin lamellae adequate for TEM study. Characterisation of the related microstructure was performed by TEM and XDES nanobeam analyses. It was identified as Ti-W containing trickle-like residue located at the surface of the spacers. Current–voltage measurements can be related to such structural defects. The conduction mechanism involved was identified as the Poole–Frenkel effect.

The drift behaviour of the **current gain** of dedicated Bi transistors achieved by an automotive power wafer technology was studied as a function of stress time and qualification test [274]. In order to replace the end-of-life test requirement and to find a new approach to estimating the expected lifetime in the final device application, a power-law-fitting procedure was proposed.

Thermal fatigue is an FM of Si power transistors, mainly activated during thermal cycling. The phenomenon is caused by the mechanical stresses set up by the differential in the thermal expansions of the various materials used in the assembly and heat sink of the transistor.

The transistor heat sink plays an important role in heat dissipation. It is made from copper, steel or aluminum, materials with different dilatation coefficients to those of Si. Two variants can be used to obtain the link between the Si chip and the case:

- **'Soft' solder joint** with molybdenum, which has a dilatation coefficient close to that of Si and can swallow up the stress if the thickness of the molybdenum layer is well chosen. However, the stress induced by many thermal cycles is able to weaken the molybdenum–copper alloy and a crack may appear, leading to an increased thermal resistance.
- **'Hard' solder joint**, using lead, which can swallow up the stresses between the chip and the case, as the lead is modelled by plastic deformation. After deformation, recrystallisation restores the metal, acting better at higher solder temperature and at longer times. However, the formation of microscopic holes cannot be avoided. These holes lead to stress concentration; as soon as the twisting limit is reached locally, a crack appears at this concentration point, limiting the heat dissipation and modifying the thermal resistance.

The result of thermal fatigue may be cracking of the Si pellet, failure at the Si mounting interface or a slow degradation of the parameters. Generally, the phenomenon is linked to the mechanical stresses generated by the different dilatation coefficients, which influence the quality of the solder joints and of the metal–Si and passivant–Si joints. Crack propagation seems to be accelerated by the presence of voids in the joint region. The reactions between the Ni layers and Sn of the solder also affect the long-term reliability of the devices. Fractographs obtained from the failed devices reveal ductile fracture of the soft solder as well as creep or fatigue failure of the solder at the elevated temperature.

An important increase (25%) of the thermal resistance between the junction and the heat sink certifies a thermal fatigue. Usually, in all practical circuits, the power transistors undergo thermal stresses. In many applications, these stresses are very large and can lead to the physical destruction of the chip or the intermediate layers. The user may identify the ageing of the solder joints of a component in operation by the abnormal heating of the junction, leading to component failure.

Note that the behaviour at SB is a sensitive parameter for thermal fatigue. As soon as a microcrack is formed, the thermal resistance increases locally and the behaviour at SB will worsen. If a transistor does operate close to SB, it can suddenly be destroyed, without previous degradation of the connections by thermal fatigue.

5.3.3 Thyristors and Insulated-Gate Bipolar Transistors

The **thyristor** is an electronic component containing four semiconductor layers of alternating N- and P-type material, which form three *pn* junctions. Proposed in 1950 by William Shockley at Bell

Laboratories, the thyristor was developed in 1956 by the group led by Gordon Hall (General Electric) and commercialised in 1957 under the name silicon-controlled rectifier (SCR). It is used as a bistable switch, the conduction state being initiated by a current pulse applied on the gate, and conduction being forward-biased. Thyristors cannot be turned off by external control, only by disconnecting the forward bias. Thyristors are mainly employed where high currents and voltages are involved, and are often used to control alternating currents, where the change of polarity of the current causes the device to switch off automatically.

Thyristor reliability is strongly dependent on temperature. If the temperature produced during operation is not dissipated correctly (as designed), the thyristor will fail. An Al heat sink increases the surface used to dissipate the energy to air. Controllers with relatively small current capacities rely on natural convection. Higher-current-capacity controllers use a fan to force air past the fins in order to increase heat dissipation.

Occasionally, water-cooled heatsinks are used on SCR controllers with very high current ratings [275].

Catastrophic failure of thyristors results in a short circuit between anode and cathode. This phenomenon begins with an increase of the leakage current at low voltage. The recommended control method is to measure the resistance between anode and cathode, in both directions. If values lower than $10\,k\Omega$ are obtained, a failure risk has already developed.

Instabilities of the electrical characteristics of thyristors are induced by the leakage current at the junction edge, due to the non-uniform junction temperature [276]. The junction-edge current is governed by phenomena at the Si–dielectric passivation interface. It seems the glass-passivation method used today does not offer full control of these phenomena.

Other specific FMos of thyristors are:

- **Turn on di/dt**, in which the rate of rise of on-state current after triggering is higher than can be supported by the spreading speed of the active conduction area.
- **Forced commutation**, in which the transient peak reverse-recovery current causes such a high voltage drop in the subcathode region that it exceeds the reverse breakdown voltage of the gate cathode–diode junction.
- **Switch on dv/dt**, in which the thyristor is spuriously fired without a trigger from the gate when the rate of rise of voltage from anode to cathode is too great.

The **IGBT** is a minority-carrier device with high input impedance and large Bi current-carrying capability. The IGBT represents a compromise between Bi and MOS technologies: MOS input characteristics and Bi output characteristics are obtained, the IGBT being a voltage-controlled Bi device. An ptimised IGBT is available for both low conduction loss and low switching loss. The main advantages of the IGBT over a power MOSFET and a classical Bi junction transistor are:

- A very low on-state voltage drop due to conductivity modulation and superior on-state current density, making a smaller chip size possible and reducing costs.
- Low driving power and a simple drive circuit due to the input MOS gate structure. It is more easily controlled than current-controlled devices (thyristor, Bi transistor) in high-voltage and high-current applications.
- A wide SOA. It has superior current conduction capability compared to the Bi transistor and excellent forward- and reverse-blocking capabilities.

However, some disadvantages must be taken into account:

- Slower switching than power MOSFET (because there is a collector current tailing due to minority carriers).
- Possible latchup due to the internal *pnpn* thyristor structure.

The FM under short circuit just after turn-off has been investigated in [277], by considering the simulation of an FE 2D physically-based Trench IGBT model. The information collected at various simulation times allows us to examine device behaviour by considering device physics. Note that the decrease of the cathode junction potential induced by increase of the cathode temperature indicates that an assisted thermal phenomenon occurs, inducing the conduction enhancement of the parasitic *npn* component. This leads to the 'mode D failure' a few microseconds after turn-off under short circuit.

5.3.4 Bipolar Integrated Circuits

The first IC was reported by Jack Kilby and Robert Noyce in 1958, and the solid-state circuit based on an Si substrate was developed in 1959, when the planar technique became available. Immediately, the IC became the main driver of the semiconductor industry, with increasing complexity every year. In 1965 Gordon Moore (the founder of Intel) predicted a doubling of IC complexity every 18 months. This is the so-called 'Moore's law', which has proved to be true for more than 40 years.

But the reliability level of ICs has also constantly improved. For instance, between 1970 and 1997, the intrinsic reliability of a transistor from an IC improved by 2 orders of magnitude (i.e. the failure rate decreased from 10^{-6} to 10^{-8} hour^{-1}). It seems IC reliability increased even faster than predicted by Moore's law. The model for reliability growth is called 'Less's law', after the well-known phrase 'less is more'. Less's law predicts a tremendous decrease in an IC's failure rate: from 1000 failures in 10^9 devices $\times$ hours (or 1000 FITs) in 1970 to some single-digit number of FITs today. There is also a 'More than Moore' movement, which focuses on system integration rather than transistor density, and is aimed at revolutionary mega-function electronics.

Basically, there are two types of IC: (i) linear IC, which is an analogue device characterised by a theoretically infinite number of possible operating states and operating over a continuous range of input levels, and (ii) digital IC, which has a finite number of discrete input and output states. Generally, linear ICs are made by Bi technologies and digital ICs by MOS technologies. However, it is possible to use MOS technologies (particularly CMOS or BiCMOS) to achieve linear ICs. Linear ICs are employed in audio amplifiers, A/D (analogue-to-digital) converters, averaging amplifiers, differentiators, DC amplifiers, integrators, multivibrators, oscillators, audio filters and sweep generators.

In the remainder of this section, the typical FMs of Bi ICs will be detailed, while the FMs of MOS ICs will be treated in Section 5.4. In recent years ICs have been manufactured on semiconductor materials other than Si (e.g. monolithic microwave ICs, based on silicon-on-sapphire and gallium arsenide technologies), and the FMs of such ICs will be detailed in Section 5.6.

Generally, the FMs of Bi ICs are the same as those encountered in Bi transistors, with some specific differences. The most common type of FMo is the open circuit. Even if the Bi IC works at delivery, failure may be produced by a high-current density or by a thermal/mechanical shock. Most frequently, the damage will be induced by ultrasonic cleaning, a method used to remove the etching.

Another possible FMo is a short circuit, which may be produced by the causes mentioned in Section 5.1: (i) contamination (caused by insufficient cleaning, impurities on two semiconductor areas connected to different electrical potentials or metallic particles on the surface of the wafer); (ii) photolithography (photoresist defects, mask defects, etc.); (iii) metallisation (over-alloying of the surface metal with Si); and (iv) oxidation (oxide break, leading to short circuit between the surface metallisation and the substrate).

The degradation effects can be produced by the migration of ions (e.g. Na^+) in Si or by surface charges, which can produce surface inversion. ESD can be a cause of failure, but this arises mainly for MOS ICs.

Note that linear ICs are also used for many critical functions in spacecraft, particularly in instruments, interfaces between analogue and digital functions, and power control. The effects of space radiation on linear ICS are detailed in [278]: dose-rate effects; permanent damage in linear devices, caused by a combination of ionisation and displacement damage; and transient effects, due to high-energy cosmic rays or reaction products from high-energy protons that produce short-duration charge tracks.

Ionising radiation produces e-h pairs within the high-quality oxides that are used to isolate different regions of Bi transistors and circuits. Holes, which are less mobile than electrons, can be trapped at the interface between the Si surface of an active device and the SiO^2 region, altering the surface potential and increasing the recombination rate for minority carriers. The latter factor causes ionising radiation to change the gain of Bi transistors [278]. It has also been shown that the transistors used in typical Bi ICs are affected by dose rate, exhibiting significantly more damage at the low dose rates encountered in space than at the high dose rates typically used in laboratory testing [279]. This effect was unanticipated, and escaped discovery for many years.

Another possible stress factor is represented by high-energy protons, which may induce important displacement damage in electronics in many spacecraft. For the wide-base *pnp* transistors used in linear circuits this displacement damage has been identified [280]. Wide-base transistors require relatively long lifetimes in order for minority carriers to be transported through the base region. Because of this, they can be more heavily damaged by protons than by ionisation from gamma rays at equivalent total dose levels. Substrate transistors have narrower base regions than lateral *pnp* transistors, and are less affected by displacement damage.

5.4 Failure Modes and Mechanisms of MOS Technology

The field-effect transistor (FET) controls the conductivity of a channel of one type of charge carrier (n or p) in a semiconductor material using an electric field. Also called unipolar transistors (because unlike Bi transistors, FETs operates with single carriers), FETs were invented before Bi transistors, by Julius Lilenfield, in 1925 (the concept was later developed by Oskar Heil in 1934), but practical devices were manufactured only in 1952, after the first Bi transistor was achieved. In 1959, Ernesto Labate, Dawon Kahng and Martin M. (John) Atalla (Bell Laboratories) proposed a new type of FET, the so-called MOSFET, which soon became much more common than other types (e.g. junction field-effect transistors (JFETs), see below). This is why a MOS technology is now used to achieve unipolar devices.

5.4.1 Junction Field-Effect Transistors

The JFET is formed by a semiconductor channel doped with positive charge carriers (p-type) or negative charge carriers (n-type). At each end of the channel contacts form the source and drain. The gate surrounds the channel and is doped opposite it. Hence, a *pn* junction is formed at the interface. By applying a bias voltage to the gate, the channel is 'pinched', so that the electric current is impeded or switched off completely.

The JFET has some important advantages, such as: linearity, high input impedance, negative temperature coefficient for the drain current (preventing SB and protecting against short circuit when the device is placed at the output of an amplifier).

The resistance to radiation of a JFET used at the input stage of an operational amplifier was studied in [278]. Although the circuit is not designed to be hardened, older data shows that the device continues to operate when it is tested at high dose rates up to 1 Mrad (Si), and the manufacturer's data sheet states that the part has excellent radiation hardness. However, the input offset voltage changes drastically when the tests are carried out at low dose rates. The changes become more and more severe as the dose rate is decreased, and the practical failure level is about 10 krad (Si) when dose-rate effects are taken into account. Note that these effects are permanent, and do not anneal after irradiation.

5.4.2 MOS Transistors

In MOS technology, two structural configurations are regularly used: vertical structures (the current flows vertically across the chip) and horizontal structures (the current always flows at the surface). There are also integrated devices, in which the current can flow both vertically and horizontally. Examples of vertical devices are the VMOS transistor (with the V shape etched in Si) and the VDMOS transistor (vertical double-diffused, which allows a higher voltage to be obtained than in VMOS transistors). The horizontal devices are basically double-diffused lateral MOS transistors (LDMOS) [281] with highly integrated gates and source geometry.

The power MOSFET is a majority carrier device that has the advantage of avoiding the secondary breakdown effect (common for Bi power transistors). This is a high-voltage transistor that conducts large amounts of current when turned on; it is potentially attractive for switching applications. One example of a power MOSFET is the LDMOS, which is used at high voltages and currents (e.g. 20 V and 10 A/mm^2).

In 1984, a new technology for power transistors was proposed: BiMOS, which combines the Bi process with CMOS circuits. The CMOS technology begins with n-type substrates, unlike Bi technology, which begins with p-doped substrates.

A comprehensive study of the reliability of power MOSFETs, and their FMos and FMs at various stress profiles for futuristic application requirements, is given in [282]. It was found that the reliability of power MOSFETs is limited by packaging issues, such as thermo-mechanical wear-out of the wire in the bond wires and the soft solder of the die attach. Investigation of the failed devices revealed that bond wires were either elongated or broken at the loop. At lower temperature swing, bond-wire lift-off from the die and cracks at the heel of the bond wire were found. One of the most important factors influencing the lifetime of the bond wires is the operation of the device above the glass-transition temperature of epoxy. Above the glass-transition temperature, the CTE of the epoxy increases drastically, thereby exerting an additional shear force on the bond wires.

In [283], power MOSFET FMs are reviewed and discussed, with emphasis on the parasitic *npn* Bi transistor, a structure inherent to n-channel MOSFET, which was identified as playing a key role in the burnout mechanism. The drain–substrate junction is reverse-biased in n-channel MOSFETs. When an energetic particle passes through the junction, the generated e-h pairs separate under the influence of the electric field, producing current. The current across the reverse-biased junction may turn on the parasitic *npn* transistor. The resulting large currents and high voltages may generate high power, leading to burnout of the MOSFET. The MOSFET-specific FMs reviewed in [283] result from the high value of dU_{ds}/dt (U_{ds} is the ON-voltage), the slow reverse recovery of the MOSFET body diode and the single-event breakdown due to inadequate voltage derating of the MOSFET. A condition for the parasitic Bi junction transistor turn-on is derived and used to discuss the failure.

A DfR approach to optimisation of the performance of a power MOSFET (more specifically, an RF LDMOSFET) is discussed in [284]. Constraints introduced by the demands of performance from the structure on its fabrication process are described, followed by an explanation of some of the difficulties associated with the experimental determination of accelerated ageing factors for RF power devices. Device reliability is discussed in terms of three aspects: hot carriers, packaging and EM. The physics of degradation due to HCI and its impact on power gain and linearity are evaluated. Drift-region engineering with a particular emphasis on design for reliability is carried out for four designs, keeping process conditions identical for all structures.

Health management is a new concept, based on understanding how and why a device fails. An FMo is defined as what the user of the device will see when it fails or what the device will exhibit when it fails. There are many FMos that are specific to MOSFET operation, in addition to those that are applicable to all semiconductor devices. Specific FMos include those within the MOSFET device itself, and those that occur due to packaging. Many underlying mechanisms can contribute the same FMo. A failure precursor is an event or series of events that is indicative of an impending failure. Precursor parameters can be identified based on factors that are crucial to safety, that are likely to cause catastrophic failures, that are essential for mission success or that can result in long downtimes [285].

Electrically, the MOSFET device can degrade over time. When this happens, the typical FMos are an open circuit, a short circuit and operation outside the range of application specifications. In opens and shorts, the output of the device is severely changed and the result is an electrical failure. Operating outside the range of specifications is usually attributed to electrical parameter degradation. There are numerous modes associated with going out of specification, but most fall into the categories of loss of gate control, increased leakage current or a change in the on-resistance [$R_{ds(ON)}$]. The complexity of the MOSFET device and packaging lends itself to a number of possible mechanisms leading to failure. Note the comprehensive review paper of William Tonti (IBM) [286], which review some of the early MOSFET reliability issues and solutions that have allowed this industry to flourish over the last 25 years. A discussion of how these past issues and new advances affect power versus performance is the framework and motivation of this paper.

Generally, the FMs of MOS transistors are those already described in Section 5.1, where the process-induced FMs were detailed. Some of the FMs, such as those produced by crystallographic defects or EM and hot-carrier effects, can be avoided by using specific foundry design rules. The stress induced by certain fabrication steps may lead to latent damage: contamination with mobile ions (mostly sodium ones) will render transistor characteristics unstable and encourage early DB. Process-induced oxide charging, caused by injection of charge into gate oxides during certain ion-etching processes, will reduce the lifetime and cause some transistor degradation similar to hot-carrier effects. Metal-stress migration, which is caused by a large thermal coefficient of expansion difference between metal interconnect and inter-level dielectrics, can lead to voiding of metal lines similar to the damage caused by EM. In Table 5.22, all possible FMs are gathered.

Five FMs (marked in bold in Table 5.22) are considered to be specific to MOS technology. The two FMs that may be encountered in MOS transistors are detailed in this section: radiation-induced soft errors and NBTI. The other three FMs specific to MOS technology (latch-up, ESD and plasma-charging damage (PCD)), which are linked mainly to MOS ICs, will be presented in Section 5.4.3.

5.4.2.1 Radiation-Induced Soft Errors

MOS devices are largely used for space applications, which is an environment with high radiation risk. Moreover, the new technology trends (i.e. smaller feature sizes, lower voltage levels, higher operating frequencies, etc.) also cause an increase in the **soft-error** failures induced by radiation. Consequently, MOS devices resistant to radiation are needed. Radiation produces two fundamental phenomena:

- **Lattice displacement**, caused by neutrons, protons, alpha particles, heavy ions and very high energy gamma photons. This is especially important in Bi devices. The arrangement of the atoms

Table 5.22 Possible FMs of MOS technology

Process	Failure mechanism	See Section
Semiconductor	Crystallographic defects	5.1.1.1
	Latch-up	5.4.3
	Radiation-induced soft errors	This section
Passivation	Hot-carrier effects, dielectric (gate-oxide) breakdown, interface state generation	5.1.1.2
	Electrostatic discharge	5.4.3
	Negative bias temperature instability	This section
	Plasma-charging damage	5.4.3
Metallisation	Electromigration, hillocks, voids, corrosion, delamination	5.1.1.3
Packaging	Wire bonds, purple plague	5.1.2

in the lattice is changed, creating damage, and increasing the number of recombination centres and depleting the minority carriers. It is interesting to note that higher doses over short time cause partial healing of the damaged lattice, leading to a lower degree of damage than the same doses delivered in low intensity over a long time.

- **Ionisation effects**, caused by charged particles with energy too low to cause lattice effects. These affect MOS devices. The effect is usually a transient one, creating soft errors, but can lead to destruction of the device if these errors trigger other damage mechanisms (e.g. a latch-up). Holes are gradually accumulated in the oxide layer of MOSFET, worsening performance, until the device fails when the dose gets high enough. The effects can vary wildly depending on the type of radiation, total dose and the radiation flux, combination of types of radiation, and even the kind of device load (operating frequency, operating voltage, actual state of the transistor during the instant it is struck by the particle), which makes thorough testing difficult and time-consuming, and requires a lot of test samples.

One of the FMs induced by radiation in MOS devices is the single-event burnout (SEB), caused by the heavy ion component of galactic cosmic rays and solar flares found in space environments: an ion traverses the transistor structure through the source and can induce a current flow that turns on the parasitic *npn* transistor below the source, thereby initiating forward-biased SB. This leads to device destruction, if sufficient short-circuit energy is available. Experimental data gathered for avionics systems at 40 000 feet shows that the SEB failure rate for 400 V MOS devices derated 75% is 9×10^{-5} failures per device-day; the rate is 2×10^{-3} failures per device-day for similarly derated 500 V devices. Failure rates at ground level are about 1/300 of this. Avionics and ground-level SEB failures are a consideration only in higher-voltage devices, but for 400 and 500 V devices, derating to minimise SEB failures should be considered in avionic and ground-level applications [287].

Another FM is the single-event gate rupture (SEGR), sometimes called single-event gate damage (SEGD): a heavy ion traverses the transistor through the gate, but avoids the *p*-regions, and can generate a plasma filament through the *n*-epi layer that applies the drain potential to the gate oxide, damaging (increased gate leakage) or rupturing the gate oxide insulation (device destruction). When not destructive, the same mechanisms affect memory and logic circuits and are generally known as single-event upsets (SEUs) and soft error rates (SERs).

A first experimental determination of the SEB sensitive area in a power MOSFET irradiated with a high-LET heavy-ion microbeam is given in [288]. A spectroscopy technique was used to perform coincident measurements of the charges collected in both source and drain junctions together, using a nondestructive technique (current limitation). The resulting charge-collection images were related to the physical structure of the individual cells. These experimental data revealed the complex three-dimensional behaviour of a real structure, which cannot easily be simulated using available tools. As the drain voltage increased, the onset of burnout was reached, characterised by a sudden change in the charge-collection image. 'Hot spots' were observed where the collected charge reached its maximum value. Those spots, due to burnout-triggering events, corresponded to areas where the Si was degraded through thermal effects along a single ion track. This direct observation of SEB-sensitive areas has applications in device hardening, by modifying doping profiles or layout of the cells, and in code calibration and device simulation.

It was demonstrated that the super-junction power metal-oxide semiconductor field-effect transistor (SJ-MOSFET, which adds *n* and *p* pillars in what would be the body material in a conventional device, lowering the on-resistance) is much less sensitive to SEB and SEGR than the standard power MOSFET. The pillars have the added benefit of reducing SEB and SEGR rates in the device by lowering field gradients [289].

An investigation into the failure rate of power MOSFETs (vertical DMOSFETs) with room-temperature junctions has been performed at sea level, covering a range of drain voltages up to 110% of device rating [290]. Phenomenally high failures rates (over 10% per week, several orders of magnitude higher than the expected Arrhenius model rate, with a maximum near room temperature) were recorded using a test arrangement in which the devices were configured to block continuous forward DC voltage. The data exceeded the estimated sea-level SEB and SEGR described in space and avionics radiation research papers, at over 80% voltage stress.

5.4.2.2 Negative-Bias Temperature Instability

NBTI is the key reliability issue for *p*-channel MOS devices stressed with negative gate voltages at high temperature. The FMo is an increase of the threshold voltage, accompanied by a decrease in drain current and transconductance. It seems that NBTI degradation is due to generation of interface traps, which are unsaturated Si dangling bonds [1]. The phenomenon is exacerbated by the very small distances involved in the deep submicron processes. The FM follows the reaction diffusion model, based on the generation of interface traps, which is a hole-induced electro-chemical reaction at the $Si-SiO_2$ interface. Initially, the degradation is controlled by the reaction rate, but with time the phenomenon becomes diffusion-limited, being nonsaturated (increasing continuously with the applied stress). However, until recently, there were difficulties and confusions in exploring the NBTI mechanism, as several aspects were poorly understood [291]. The reaction–diffusion of hydrogen species interacting with Si dangling-bond-induced interface traps at the $Si-SiO_2$ interface and the trapping/detrapping of oxide traps were proposed as explanations for NBTI degradation and recovery [292].

In the last 10 years, increased attention has been given to NBTI, because: (i) the effective operating oxide electric field and temperature are increased, which induces the NBTI in operational conditions and (ii) the nitrogen that is introduced to the thin oxide layer to reduce gate leakage current and to suppress boron penetration is obviously enhancing the NBTI. The effects of interface states and oxide positive charges on NBTI were studied in [293] and the importance of stress temperature and measurement speed was emphasised. Two major conclusions were given: (i) interface-state generation is negligible compared with the oxide positive-charge formation under device operational conditions and (ii) the positive charge formed by NBTI is not always of the same type. An example of the first is given by trapped holes that are tunnelling holes trapped by oxide, which are insensitive to stress temperature. An example of the second is generated positive charge, formed by breaking the bonds and accelerated by temperature. In the reported case, the NBTI was dominated by positive charge, and the trapped-hole characteristics observed in the experiments were explained by physical properties.

In 2008, a study of NBTI degradation in *p*MOSFETs with an ultrathin SiON gate oxide was reported. Two NBTI components were proposed: a slow component due to interface trap degradation and a fast component contributed by trapping/detrapping of oxide charge. This assumption was verified by rigorous experiments [294].

5.4.2.3 Strained MOSFET

In recent years, a method for improving transistor performance in smaller CMOS devices was proposed, by enhancing carrier mobility using a mechanical stress, the so-called strained-Si MOSFET. For this purpose, the technique of inducing stress through a tensile or compressive nitride capping layer was promoted as a necessity for future nanoscale-device and circuit design.

In [295], the impact of strain on the threshold voltage of short-channel (sub-100 nm) nanoscale strained-Si/SiGe MOSFETs was studied, by solving the 2D Poisson equation in the strained-Si thin film and analysing the dependence of threshold voltage on various device parameters, such as strain (Ge mole fraction in SiGe), gate length, junction depth and so on. The proposed model is useful in the design and characterisation of high-performance strained-Si/SiGe nanoscale MOSFETs when short-channel effects have to be included.

For the same strained-Si MOSFET, it has been reported that deuterium (D) annealing can be used instead of hydrogen (H) to improve hot-carrier reliability, and lifetime improvement correlates with the

D incorporation at the SiO_2–Si interface [296]. D replaces the existing H at the interface, resulting in a strong kinetic isotope effect that enhances the reliability lifetime. Obviously, the Si–D bond is more resistant to hot-electron excitation than the Si–H bond. In addition, the high-pressure annealing process not only improves device characteristics but makes the subsequent annealing time short, because a high-pressure system can make a high-concentration environment of D gas. NMOS and PMOS devices have been studied. With annealing, tensile-stressed NMOS devices showed a more reduced charge-pumping current, an increased hot-carrier lifetime and a reduced $1/f$ noise power compared to control NMOS devices. The PMOS devices showed improved results too, but the compressive-stressed PMOS devices showed an increased Vth spreading, a reduced NBTI lifetime and an increased normalised drain current noise power.

5.4.3 MOS Integrated Circuits

MOS technology is well-suited for IC fabrication, being simpler and cheaper than the Bi process. Only one diffusion step is required to form both the source and the drain regions, while in the Bi process two to four diffusion steps are required. The crossover between components of MOS ICs is diffused simultaneously with the drain and source and there is no need to isolate regions between MOS transistors, because each source and drain region is isolated from the others by the *pn* junctions formed within the P-type substrate. Moreover, the MOS IC typically occupies a smaller space than Bi ICs (only 5% of the surface required by an epitaxial double-diffused transistor in a conventional IC), which is extremely important for modern ICs. The main drawback of MOS ICs is a smaller operating speed than Bi ICs. Therefore, MOS ICs are not suitable for ultra-high-speed applications. However, because of their advantages of low cost, low power consumption and high packing density, MOS ICs find wide application in LSI and VLSI chips, such as calculator chips, memory chips and microprocessors.

Many possible methods for improving MOS IC reliability have been proposed. With more mature and powerful computer techniques, most aspects of modern chip design are modelled and simulated before designs are committed to Si. However, the lack of effective SPICE compact models capable of reliability simulation and characterisation in truly dynamic circuit-simulation environments remains a problem. This result is attributable to the complexity of MOSFET failure physics and the uncertainty of failure statistics. By introducing a set of accelerated-lifetime and SPICE compact models of the most important intrinsic FMs in advanced MOSFETs, some new concepts in reliability modelling are given in [297]. This comprehensive paper (with 94 references) addresses various FMs, such as HCI, TDDB and NBTI, integrated into a broader environment with a holistic method for circuit-reliability design. Based on the new accelerated-lifetime and SPICE compact models of these wear-out mechanisms, a simple but effective SPICE reliability simulation approach was developed and demonstrated by ageing simulations of an SRAM design with reliability-model parameters extrapolated from a commercial 90 nm technology to help designers accurately predict circuit reliability and failure rate from a system point of view.

The typical FMs of MOS technology have already been detailed in Section 5.4.2. However, the MOS ICs have some specific FMs of their own, which will be presented below: ESD, latch-up, plasma-charge damage and leakage currents. Details specific to MOS ICs will also be given about other FMs previously presented, such as parametric failures, soft errors and EM. Finally, technological improvements aimed at mitigating the action of these FMs will be shown.

5.4.3.1 Electrostatic Discharge

ESD is '*the sudden and momentary electric current that flows between two objects at different electrical potentials caused by direct contact or induced by an electrostatic field*' [298]. ESD is a subset of the broad spectrum of EOS, which includes lightning and electromagnetic pulse (EMP). EOS commonly refers to events other than ESD that encompass timescales in the microsecond and millisecond ranges, compared to the 100 ns range associated with ESD [299].

In the electronics industry, ESD describes momentary unwanted currents that can cause damage to electronic equipment. For ICs, ESD represents a serious issue, because the used materials are permanently damaged when subjected to high voltages. It was reported that ESD is responsible for close to 10% of all failures in Si ICs [300].

The charged device model (CDM), proposed in 1970 by a group from Bell Laboratories, has been associated with mechanical handling of ICs as a reason for failure. This model was simple but effective, and allowed many designers (at many locations, due to Bell's willingness to talk and write about it) to improve their semiconductor components. Bell continued its work on CDM in the late 1980s and early 1990s in the development of a machine that evolved into the commercial testers of today [301].

The mains ESD issues in IC fabrication are [299]:

- The static charge generation (initiated in the wafer fabrication area) needs to be suppressed, using prevention methods such as antistatic coating of the materials and air ionisers aimed at neutralising charges.
- Damage caused by human handling must be reduced by using wrist straps to ground accumulated charges and shielded bags to carry individual wafers.
- Static control and awareness are needed to avoid ESD in the semiconductor-manufacturing environment.
- Protection circuits are implemented within the IC chip. With effective protection circuits in place, the packaged device can be handled safely from device characterisation through to device application.
- Antistatic precautions are needed during the wire-bonding and assembly phases.

After years of experimental work in ESD protection, many guidelines for designing robust ESD devices and test methods were proposed. Moreover, the Electrostatic Discharge Association organises an annual 'EOS/ESD Symposium', which deals with all areas of ESD. A. Amerasekera and C. Duvvury gathered many of the valuable papers from this symposium and other contributions together in a single book [299]. Being the most informative text on the subject today, a short presentation of the book's contents may be beneficial to the reader. First, the details of the ESD phenomenon, introducing the 'charge' and 'discharge' effects, are presented, followed by a discussion of the various test models and methods: the human body model (HBM, to represent human handling), the machine model (to emulate machine contact) and the CDM (to determine the effects of field-induced charging of the packaged IC). The next chapter is devoted to the physics of FMs and to the operation of the semiconductor protection devices under the high-current short-duration ESD pulses. To illustrate the transistor phenomena and the design techniques discussed in the earlier chapters, the main FMos observed in advanced Si ICs are discussed, together with case studies related to the effects of design and layout on ESD performance. This analysis involves a thorough stress methodology for characterisation and a full study of the FMos. Several actual case studies are presented, indicating the common and more unusual ESD problems. A brief summary of the FA techniques useful for ESD and the post-stress failure criteria are reviewed. In the final part of the book, the principal aspects related to process effects, such as the impact of LDD junctions or silicide diffusions on ESD performance are discussed. A review of the device modelling techniques based on the high-current behaviour of the protection circuits is then given.

5.4.3.2 Latch-Up

Latch-up is a particular type of short circuit that can occur in an improperly designed MOS IC: the inadvertent creation of a low-impedance path between the power supply of the circuit and ground, triggering a parasitic structure (equivalent of an SCR) that acts as a *pnp* and an *npn* transistor stacked next to each other, disrupting proper functioning of the component and leading to its destruction, due to over-current. During a latch-up, when one of the transistors is conducting, the other begins conducting too, both being in saturation for as long as the structure is forward-biased and some current flows through them. Some possible causes of latch-up are:

- A supply voltage higher than the absolute maximum rating, which may be a transient spike in the power supply, leading to a breakdown of some internal junction.
- Ionising radiation.
- Cable discharge events (CDEs).
- External noise.

The sensitivity of devices to latch-up triggered by short-duration pulses is an often overlooked root cause for severe field failures. Standard JEDEC 17 tests applying quasi-static voltages and currents often fail to identify these problems. In addition, wide pulses may cause thermal damage in the device before it triggers. In [302], a new analytical test technique is introduced on the basis of very fast square pulses. The method allows the in situ monitoring of the voltages and currents during triggering and helps to gain fundamental insights into the underlying mechanisms.

One of the technological solutions for avoiding latch-up is a shallow trench, which is a layer of insulating oxide that surrounds both the NMOS and the PMOS transistors and aims to break the parasitic SCR structure between them. Other possible solutions are: deep trench, retrograde wells, connecting implants, subcollectors, heavily-doped buried layers and buried grids. A radical solution is to use a silicon-on-insulator (SOI) technology, because SOI devices are inherently latch-up-resistant.

The latch-up problem was essentially solved for CMOS circuits. However, the aggressive scaling of CMOS, SOI and BiCMOS technologies led to greater numbers of transistors in a given die size. Moreover, with the introduction of triple-well bulk CMOS technologies, new *npn* and *pnp* transistors have been formed that will need to be considered beyond the classical ones, formed in dual-well bulk CMOS technology. In [303], evaluations of several types of CMOS device after nondestructive latch-up are reported, revealing structural changes in interconnects due to localised ejection of part of the metallisation due to melting. This creates localised voids within interconnects that reduce the cross-section by 1–2 orders of magnitude in the damaged region. These effects must be considered when testing devices for damage from latch-up, as well as when establishing limits for current detection and shutdown as a means of latch-up protection. For a variety of CMOS devices, high current-density conditions during single-event latch-up (SEL) may produce noncatastrophic interconnect damage from melting. Because this type of structural damage is permanent and significantly reduces interconnect cross-sections in the damaged area, it raises a concern about vulnerability to future device failure due to EM or additional SEL events.

A latch-up phenomenon occurred in a high-voltage IC product during a latch-up test and was identified within the diodes used for ESD protection [304]. A parasitic *npn* Bi device formed by ESD protection diodes was trigger-activated and produced a large current, resulting in EOS failure. This was verified by electrical measurement from TLP and curve-tracer as well as by physical FA. Failure originated from activation of a parasitic *npn* Bi device formed by power-ground diodes. Corresponding layout solutions were proposed and this anomalous latch-up failure was solved successfully. Therefore, ESD protection diodes should be laid carefully for true latch-up-robust design.

5.4.3.3 Plasma-Charging Damage

During plasma processing of MOS ICs, it is possible that an unintended high level of tunnelling current will flow through the gate oxide. This phenomenon is called plasma-charging damage. The current stress degrades the transistor properties and shortens the gate-oxide breakdown lifetime [305]. Unfortunately, it is more likely in modern devices, because advanced ultrathin gate oxide greatly reduces the voltage needed to induce significant tunnelling.

Currently, there is a common perception in the industry that PCD is no longer a serious concern for ultrathin gate oxide in advanced CMOS technology (in multiple gate-oxide technology, damage is still a concern for the thicker oxides). The basic argument is that the ultrathin gate oxide is so leaky that all of the charging current flows right through it without leaving any damage. However, that does not mean that all plasma-charging has become negligible. Some serious charging damage in ultrathin gate-oxide cases has been reported in the literature. In [305] it is shown that the absence of detectable damage does not mean that charging damage is not severe enough to cause a gate oxide to fail a reliability specification. The combination of advanced plasma and ultrathin gate oxide makes the detection of PCD extremely difficult. There is currently no viable method for handling this problem. In [305], soft breakdown (SBD) was used as a failure criterion for simplicity. Since plasma damage tends to produce very soft breakdown, its impact on the product is further reduced.

5.4.3.4 Leakage Currents in MOS ICs

In MOS ICs, leakage has two components [306]:

- **Subthreshold leakage**, consisting of source–drain currents when the transistor is supposed to be nonconducting. These currents flow through the substrate of the transistors due to effects near the active regions which heavily depend on the length of the transistor gate.
- **Gate-oxide leakage**, from currents that tunnel through the very thin oxide layer between gate and source, drain or bulk.

Clearly, both types of leakage depend on the device size, and also on the voltages at the terminals. Further, by altering the doping of the substrate, the threshold voltage, *Vth*, can be changed, enabling the design of low-leakage transistors with higher *Vth* values. High *Vth* has weaker drive and will deteriorate the speed of the circuitry, however.

The **drain leakage current** is induced by the high electric field between gate and drain. This sub-breakdown leakage current is called gate-induced leakage (GIDL) current and needs to be minimised. It has been found that, for the present bias conditions, the GIDL current is caused by the electron band-to-band tunnelling in the reverse-biased channel-to-drain *pn* junction. A band-to-band tunnelling current equation in the reverse-biased *pn* junction was used to fit the measured GIDL currents for various channel-doping levels and bias conditions, including the dependence of the GIDL current on body bias and the lateral electric field [307]. The effects of channel width and temperature on the GIDL current were characterised and discussed for 45 nm state-of-the-art CMOS technology. The observed GIDL current was found to increase in MOSFET devices with higher channel doping levels.

The first irreversible increase in **gate leakage current** is considered an oxide failure criterion, the first sign that the oxide will breakdown. So, any damage to gate-oxide integrity may lead eventually to oxide breakdown. The reliability of gate oxides depends on the electric stresses applied, a phenomenon that is cumulative in time. The amount of lifetime being consumed by a certain electric stress increases with increasing voltage drop across the oxide (which has the largest influence), increasing temperature and increasing active gate-oxide area [308]. With decreasing gate-oxide thickness below about 1.3 nm, the time range of further leakage-current increase becomes significantly large compared to the time-to-formation of the breakdown path itself. New DfR rules are proposed in [308]. It is strongly recommended that any overshoot events of the voltage drop across the gate oxide between any terminal and the gate be identified. The new definition of gate-oxide failure criterion tolerates gate-leakage levels well above the direct tunnelling current of the fresh gate oxide as well as an additional source of noise caused by such a progressive SBD spot in the gate oxide.

5.4.3.5 Parametric Failures of ICs

Parametric failures are mostly speed-related failures, which are detected in the field, when customers use the parts at temperatures or power-supply voltages different from those employed in the production tests. There are two major categories of parametric failure:

- Those occurring in defect-free or intrinsic material, when the statistical distribution of the speed-related IC parameters is skewed such that transistor and signal interconnects work against each other; a longer-channel transistor driving a line with higher resistance may fail while a single-parameter deviation may not cause failure.
- Those occurring in parts with subtle defects, such as soft defects that damage IC performance, but perhaps only under a certain set of temperature, power supply or radiation conditions.

Examples of parametric failures are resistive vias, metal slivers and gate-oxide shorts in ultrathin oxides. Unique tests are required to detect these forms of failure.

In [309], the pervasive forms of parametric failures are presented and some present and forthcoming test techniques to address this form of IC failure, which is difficult to detect, are discussed. As individual transistor and metal interconnection parameters vary widely within a die, die-to-die and lot-to-lot, the exact transistor drive (speed) prediction is difficult to make. The transistor and metal interconnection parameters that can significantly alter the circuit speed include: channel-length variation, random-doping variation, threshold voltage and diffusion resistance, channel-width variation, effective gate-oxide-thickness variation, via and contact resistance, and metal interconnection uniformity – metal width, thickness, spacing, granularity and current density. Multiparameter test approaches can detect certain defect-related failures, but there is no systematic approach at this time to detect failures due to statistical variation. Multiple VDD–temperature corner testing is one approach to exposing intrinsic parameter failures. The traditional stuck-at-fault, delay fault, functional and IDDQ tests are insufficient in their original simplicity.

5.4.3.6 Soft Errors of ICs

Radiation-induced soft errors in MOS technology have been discussed in Section 5.4.2.1. However, there are some specific elements related to MOS ICs to be discussed below.

Novel techniques are needed to mitigate the SER of ICs, and in some cases a combination of different mitigation techniques may be required for significant performance improvements. In [310], an autonomous single-event transient (SET) pulse-width characterisation technique is used to quantify the effect of hardened-by-design structures, such as guard rings, in reducing the collected charge and SER of ICs. Experimental results obtained for a 0.35 μm technology show a reduced SET pulse width for devices with guard rings. Heavy-ion test results indicate a reduction of approximately 30% in the event cross-section (the number of SET pulses with a given width to the total fluency) and in the maximum SET pulse width. However, it seems that the described method for single-event mitigation has to be used in combination with other mitigation techniques to improve reliability.

Another possible mitigation technique is associated with charge sharing. In [311] this method was used for a 90 nm dual-interlocked-cell latch. In this case, the use of guard rings shows no noticeable effect on upset cross-section, but the use of nodal spacing as a mitigation technique led to an order-of-magnitude decrease on upset cross-section as compared to a conventional layout. Simulation results show that the angle and direction of the incident ion strongly influence the amount of charge collected by multiple nodes.

5.4.3.7 Electromigration at IC

EM of interconnects is one of the most common causes of IC failure, with a rate that is exponentially dependent on temperature. Based on this model, the maximum tolerable operating temperature for an IC is calculated, with resulting cooling costs. The typical EM models assume a uniform, typically worst-case, temperature. In [312], a model that accounts for temporal and spatial variations in temperature is proposed. This approach allows a more accurate prediction of EM lifetime. For example, for a fixed target lifetime, intermittent higher operating temperatures and performance can be tolerated if compensated by lower temperatures at other times during the product's lifetime.

For a particular spatial-gradient pattern, by assuming a uniform average temperature along the interconnect, the expected lifetime is overestimated by 30% compared to the actual case, and using a uniform maximum temperature underestimates expected lifetime by 80%. This means that the typical approach of worst-case analysis will yield drastically lower temperature specifications than necessary, which again will require an unnecessarily expensive cooling solution or unnecessary performance sacrifices. Also, the same modelling approach applies to temperature-related gate-oxide breakdown, another common cause of IC failure.

5.4.3.8 High-k Dielectrics

The decrease of the dielectric film thickness to an oxide-equivalent value of 2 nm and below in MOS ICs requires replacement of SiO_2 gate oxides, due to too-high leakage currents. A material with higher permittivity is needed, a so-called high-k dielectric: oxynitrides or oxide/nitride stacks, metal-oxides or compounds, such as metal-silicates [313]. After almost a decade of intense research, the family of hafnium-oxide-based materials HfO_2, $HfSi_xO_y$, HfO_xN_y and $HfSi_xO_yN_z$ has emerged as a leading candidate to replace SiO_2 gate dielectrics in advanced CMOS applications [314].

For the new high-k MOS devices, the physics of failure (PoF) and traditional reliability testing techniques must be re-examined for ultrathin gate oxides that exhibit excessive tunnelling currents and SBD. Electrical and reliability characterisation methodologies need to be developed and enhanced to address issues associated with both ultrathin SiO_2 and alternate dielectrics including large leakage currents, quantum effects and thickness-dependent properties.

Basically, high-k dielectrics have been proposed to replace SiO_2 in order to reduce gate-leakage current. However, among the reliability concerns of high-k dielectrics, hot-carrier effects may represent one of the major limitations. In [315] the results of a study into whether hot carrier-induced degradation might be accompanied by electron-trapping in the bulk of the high-k film due to the high density of structural defects in the high-k dielectrics are shown. This bulk electron-trapping, which is not observed in SiO_2 dielectrics, can significantly affect transistor parameters and therefore complicates evaluation of the hot-carrier degradation properties of the high-k gate stacks.

An early challenge for the fabrication of this new MOS device was to establish a reliability model for the high-k gate dielectrics. A generated subordinate carrier injection (GSCI) model is proposed in [316], a model that had already been used for quantitative prediction of the breakdown lifetime. With this model, the gradual increase of gate-leakage current under stress has been clarified as the multiple events of the SBD in addition to the large initial leakage current. The proposed multiple-SBD-based model points out the possibility that the first observed breakdown is not the actual first breakdown, so the mechanism of gradual increase must be taken into account for the evaluation of the breakdown statistic. The proposed new method offers precise evaluation of the breakdown statistics without detecting the first breakdown, concealed in large and noisy leakage currents.

A method for passivating the interface states of high-k gate dielectric, proposed in [317], is based on a new high-pressure (up to 100 atm), pure (100%) hydrogen annealing system. The effect of high-pressure hydrogen and deuterium post-metal anneal on the electrical and reliability characteristics of high-k (hafnium) *n*MOSFET was investigated. Compared with a forming gas annealed sample, the MOSFET annealed in high-pressure deuterium ambient exhibits excellent performance and reliability, which can be attributed to the significant improvement of interfacial oxide quality, reduction of fast trap sites in the high-k layer and the heavy mass effect of deuterium. By optimising the process parameters, device performance and reliability characteristics were improved.

In the last few years, it has become clear that mere replacement of the gate insulator (GI), with no concurrent change of electrode material (currently heavily doped polysilicon), may not be sufficient for device scaling. Polysilicon gate electrodes are known to suffer from a polysilicon depletion effect, which cannot be ignored for sub-2 nm gate stacks. Therefore, research on dual-work-function metal-gate electrodes is gaining momentum, since conventional gate stacks are approaching a limit to scaling as a means of improving performance for nano-CMOS (i.e. sub-65 nm) technologies. A comprehensive paper [314], with 145 references, is reviewing current understanding of advanced metal-gate/high-k stacks from the perspective of integrating both basic materials and devices. Because reliability is an important factor, especially for long-term device operation, some reliability aspects of advanced gate stacks are also covered.

The quality of MOSFET gate stacks where high-k materials are implemented as gate dielectrics is discussed in [318]. Drain- and gate-current noises are analysed in order to obtain information about the defect content of the gate stack. The overall quality of the gate stack is evaluated, depending on the kind of high-k material, the interfacial layer thickness, the kind of gate-electrode material, the strain engineering and the substrate type. The issues to be addressed in order to achieve improved quality of the gate stack from a $1/f$ noise point of view are identified.

One of the most serious issues for high-k gate-dielectric reliability is NBTI, which cannot be described by the hydrogen R&D model alone. In [319], the application of a technique to separate bulk hole-trap effects from interface state degradation in NBTI is discussed, with the aim of understanding hole-trap behaviour, including MOSFET fabrication process dependence. As a result of this study, it seems rapid thermal annealing (RTA) is effective for reducing pre-existing hole traps, while nitrogen incorporation is effective in the thermal deactivation of hole traps. To suppress hole traps following NBTI lifetime improvement, nitrogen incorporation in the high-k films and gate-first processes is desirable for reliable high-k/metal gate-stack pMOSFET fabrication.

5.4.3.9 Shallow Trench Isolation

In recent years, shallow trench isolation (STI) has been used as an isolation scheme for ultra-large-scale integration (ULSI) applications. The devices manufactured by this procedure have a major reliability risk, represented by dislocations and oxidation-induced stacking fault, which increases junction leakage current through trench isolation. In [320], the mechanism of parasitic STI MOSFET formation is analysed and the temperature dependence of the leakage current is investigated. The abnormal temperature dependence of the parasitic drain current with floating body is proposed for use as a faulty STI indicator.

The results of a study into the effects of STI-induced mechanical stress on hot-carrier-induced degradation of *n*/*p*MOSFETs are reported in [321]. It was found that mechanical stress increased with source/drain (S/D) area reduction and had no impact on the hot-carrier degradation for *n*/*p*MOSFETs with large channel width. However, hot carrier became a significant failure risk when channel width was reduced. The results also showed that hot-carrier degradation phenomena due to STI-induced mechanical stress cannot be explained by the piezoresistance effect. The effects of longitudinal and transverse mechanical stress distribution at different regions seem to be the cause. In addition, the effect of a tensile pocket at the STI edge on device degradation by hot carriers cannot be neglected.

5.4.4 Memories

Memories are devices used to store digital information, using paper, magnetic material, a semiconductor or a passive component as their basic material. In this section we will focus on semiconductor memories, which are based on IC technology, manufactured by both Bi technologies (e.g. standard, TTL, Schottky TTL, ECL, I^2L, etc.) and MOS technologies (e.g. PMOS, NMOS, CMOS, HMOS,

VMOS, CCD, SOS/SOI, etc.). These are the most used today. Some reliability issues of MOS memories are discussed below.

Tunnelling through the GI is a potential reliability issue. In theory, the tunnelling limit is around 3 nm. In practice, oxide quality and defect density have resulted in thicker oxides, but improved processing already allows insulators at close to the tunnelling limit.

The *maximum allowable field* in the depletion regions and the GI represents another reliability issue. If the fields go too high, hot-electron effects, punch-through or breakdown may result. As dimensions are reduced, the supply voltage cannot be made arbitrarily low. Even for a well-designed device with good characteristics, subthreshold conduction will limit how low the threshold voltage can be reduced. Thermal energy allows some fraction of the carriers in the Si to surmount the barrier which the gate electrode creates as the device is turned off. In good devices, the current decreases by an order of magnitude for every 80–90 mV reduction in gate voltage, in the subthreshold region. Another challenge with device fields is to control where the field lines terminate. Device threshold voltage can vary due to short channel effects, which arise when the device threshold voltage depends upon the source drain spacing and drain voltage.

Soft errors are important reliability risks. As device dimensions and supply voltage are reduced, the amount of charge involved in a switching or retentive operation is correspondingly reduced. Soft errors (see Sections 5.4.2.1 and 5.4.3.6) can occur when minority carriers cross a *pn* junction into a node in sufficient quantity to upset the state of the node. DRAM memories are most sensitive, because they involve storing small amounts of charge in the memory cell. Memory can handle such errors through error-correction codes, and logic can use parity to detect and retry. One source of minority carriers is ionising radiation such as α-particles or cosmic rays.

One method for accurately analysing single-event vulnerability of SRAM cells leads to precisely calculated SERs [322]. The critical charge is used as a measure to determine whether a memory cell can be upset, and most hardening techniques are based on that single value. Accurately estimating the statistical spread of the critical charge helps define proper design margins and account for SEUs occurring outside the expected median value. The spread in critical charge required for an upset due to statistical variations in threshold voltage in the 130 and 90 nm technologies is quantified.

5.4.5 Microprocessors

A microprocessor is an IC incorporating all the functions of the central processing unit (CPU) of a computer. Invented in the early 1970s (the first product was Intel 4004, launched in November 1971), the microprocessor rapidly became the only option for modern computers. The first variant was a 4-bit microprocessor, but since the early 2000s a 64-bit architecture (first developed in the early 1990s) has been used in personal computers (PCs).

Microprocessors have the same FMs as standard devices manufactured using MOS or Bi technologies, but due to their larger chip areas, the incidence of defects inherent in the semiconductor materials (e.g. pinholes in the oxides, micro-cracks in the metallisation, etc.) is higher [1]. Oxide rupture, interruption of Al lines or wire-bond failures may induce catastrophic failures or failures related to the interrelationship between software and hardware. The functional defects lead to soft failures, which can be eliminated by adjustment. As passivated Al(Cu) lines become narrower, the metal exhibits increasingly elastic behaviour with higher stress levels, a combination of stress characteristics which favours void formation. Stress relaxation in Al(Cu) films and lines has been measured by bending beam and X-ray diffraction methods [323].

A microprocessor may be considered as a system consisting of a series of separate units, each one contributing to the overall reliability of the device [324]. The functionality of the whole device can be tested by means of a suitable programme, but such techniques cannot guarantee that every functional unit in the device is exercised. The common practice of some microprocessor manufacturers is to perform functional tests that verify the correct performance of the functions specified in the machine's instruction set. Separate tests are done to verify data transfer and storage, data manipulation (arithmetic and so on), register decoding, and instruction decoding and control. These functional tests are then evaluated in terms of their fault-detection effectiveness. For example, the test engineer may pass a single 1 bit through the microprocessor to verify that the bus lines and register cells have no single fault causing the 1 to be changed permanently to (or 'stuck at') 0; similarly, the passage of a 0 may be used to find a stuck-at-1 fault. A combination of 0s and 1s may be used to verify that no pairs of bus lines or register cells have stuck-at faults [324].

For microprocessors, burn-in coupled with a good functional test programme is generally accepted as the most effective screen: surface-related defects are detected using a static burn-in, whereas dynamic burn-in should weed out defects in MOS memory cells due to weak oxides. The best screen is considered to be a succession of several burn-in programmes, which may prove to be uneconomical. Burn-in has greatly reduced the failure rate in the field, since it identifies and allows elimination of early failures, or infant mortalities, during the first 1000 hours of operation. After burn-in, functional tests may be performed to check storage and readout, address patterns, timing voltages and signal margins. An approach used by some memory manufacturers is to plug a number of RAMs into one board in the test chamber; signals from a pattern generator are distributed to a bank of amplifiers which drive the RAMs. Comparator circuits compare the RAM outputs with the expected outputs; every bit in the test pattern is compared at strobe time with expected data. When an error occurs, the comparator sets a flag bit in the memory location corresponding to the failed RAM.

5.4.6 Silicon-on-Insulator Technology

SOI technology refers to the use of a layered silicon-insulator-silicon substrate in place of conventional Si substrates in semiconductor manufacture, in order to reduce parasitic device capacitance and thereby improve performance. SOI technology was invented for the fabrication of radiation-hard MOS devices. The upper Si layer containing the active device region is fully isolated from the inactive substrate by a buried oxide, which differs from a thermal oxide [1]: it is Si-rich, which implies high densities of electrons traps and trap centres. Larger than the thermal-oxide interface at the gate, but small enough not to adversely affect the circuit performance, the buried oxide is more subject to degradation than the gate oxide; its defects may jeopardise, via coupling effects, the performance of CMOS circuits. However, the use of the buried oxide to isolate the active device region has outstanding merits and results in substantial theoretical advantages [325] over bulk Si: improved speed and current drivability, higher integration density, attenuated short-channel effects, lower power consumption and elimination of substrate-related parasitic effects. Another strong argument in favour of this technology is the inexpensive control of the interface degradation induced by HCI, which is a key challenge for Si technology. The first high-speed transistor manufactured with SOI technology and able to be used in microprocessors was announced by IBM in August 1998 [326]. Today, a very serious opportunity is offered to SOI by the aggressive development of ultradense, deep-submicron CMOS circuits operating at low voltage. The subsisting obstacle is the credibility of SOI when competing with bulk Si, which is still extremely efficient.

SOI architectures emerged from the idea of isolating the active device overlay from the detrimental influence of Si substrate. In ULSI circuits the advantages of thin-film SOI device structure over its bulk Si counterparts include high carrier mobility, sharp subthreshold slope, ameliorated short-channel effects, reduced hot-carrier degradation and increased drain current. In a sub-quarter micron regime, the ultra-thin-film CMOS/SIMOX (Separation by IMplantation of OXygen)[9] technology is very promising for future low-voltage ULSIs. Recently, the development of fully-depleted (FD) SOI devices fabricated on ultrathin SOI films has provoked great interest in SIMOX and bonded SOI wafers

[9] SIMOX is a method of manufacturing SOI devices consisting in an oxygen ion beam implantation process followed by high-temperature annealing to create a buried SiO_2 layer.

for high-speed ULSI applications. This is mainly due to the reduction of parasitic capacitances, bodies-charging effects, threshold-voltage shifting, latch-up effects, short-channel effects and higher drive-current ability [1].

The aggressive scaling of MOSFETs led to the idea of replacing the -Si gate in future CMOS devices with metal gate electrodes, in order to achieve an equivalent oxide thickness <1 nm for semiconductor industry requirements. One possible solution is a fully silicided (FUSI) metal gate [327], with the variant NiSi FUSI gates, which seems to be an appropriate gate technology for *n*/*p*MOSFETs [328].

Moreover, as the gate length of CMOSFET goes below 100 nm, increased channel doping is required to suppress short-channel effects that will cause higher ionised impurity scattering and further result in the degradation of carrier mobility. A strained technology has been proposed [329] to improve device performance, particularly for deep-submicrometre MOSFET design. This involves an extra process step, a high-strained contact etch stop layer (CESL) that can generate higher strain to induce great carrier mobility enhancement in the channel even at large vertical electric fields. This is a very useful method for improving device performance by combining CESL with FUSI gate technology.

In [330] the impact of the CESL technique on FUSI gate SOI CMOSFET performance and voltage-stressing-induced (including HC and bias instability) device degradation was investigated. Related noise analyses, as well as charge-pumping techniques, were employed in the inspection of strain-induced defects that accelerate device degradation after long-time HC voltage stressing and/or bias instability voltage stressing.

The thermal management of high-performance devices is an important design issue that must be discussed if long-term reliability is under question. In [331], the self-heating effect in high-performance sub-0.18 μm bulk and SOI CMOS circuits is analysed using fast transient quasi-DC thermal simulations. The impact of the self-heating effect and technology scaling on the metallisation lifetime and the gate oxide time-to-breakdown reduction are also investigated. Based on simulation results, an optimised clock-driver design is proposed. The proposed layout reduces the hot-spot temperature by 15 °C and 7 °C in 0.09 μm SOI and bulk CMOS technologies, respectively.

5.5 Failure Modes and Mechanisms of Optoelectronic and Photonic Technologies

Optoelectronics is a discipline focused on studying and developing electronic devices that emit, detect and control light (visible light and invisible radiation[10]). Basically, optoelectronic components are devices which use electrical-to-optical or optical-to-electrical transducers. Today, optoelectronics is considered a subfield of photonics, together with electro-optics, optomechanics, quantum electronics and quantum optics. Note the difference between optoelectronics and electro-optics: the latter is the study of all interactions between light and electric fields, including those not linked with electronic components.

The main families of optoelectronic and photonic components are described in Table 5.23.

The typical FMs of these components are detailed below. To accelerate or stimulate the FM of optoelectronic components, accelerated lifetime tests can be performed at excess temperature, humidity, voltage, pressure, vibration and so on. Such failures may be due to mechanical fatigue, corrosion, chemical reactions, diffusion and charge migration within individual electronics chips, among other things. The FMs under the applied stresses are kept consistent with those under normal operation; only the time scale is different [332].

5.5.1 Light-Emitting Diodes

The phenomenon of electroluminescence was discovered in 1907, by H.J. Round (Marconi Labs). The first light-emitting diodes (LEDs) were reported 20 years later, but the real interest in the study of

[10] Invisible radiation includes gamma rays, X-rays, ultraviolet and infrared rays.

Table 5.23 Main types of optoelectronic and photonic component

Component family	Basic material	Technology and characteristics
Light-emitting diodes (LEDs)	AlGaAs, AlGaP, AlGaInP, GaAsP, GaP, GaN, InGaN, and so on	Semiconductor diodes that emit incoherent narrow-spectrum light (electroluminescence) when electrically biased in the forward direction of the *pn* junction. An LED is usually a small-area light source, often with optics added to the chip to shape its radiation pattern, and is used in small indicator lights on electronic devices and increasingly in higher-power applications such as flashlights and area lighting. The colour of the emitted light depends on the composition and condition of the semiconducting material used, and can be infrared, visible or ultraviolet.
Photodiodes	Si, Ge, GaAs, InP, CdTe	Diodes (usually avalanche or PIN type) that use the optical charge-carrier-generation effect, packaged in materials that allow light to pass. They are used in photometry or optical communications. Multiple photodiodes may be packaged in a single device, either as a linear array or as a two-dimensional array.
Phototransistors	Si, GaAs	Designed specifically to take advantage of light sensitivity (e.g. NPN bipolar transistor with an exposed base region). The light strikes the base (instead of voltage being applied to the base, as in transistors) and the phototransistor amplifies variations within it. The phototransistor has an advantage over the photodiode in that the resulting photocurrent is amplified through the normal current-gain function of a transistor.
Optocouplers	LED + photodetector	Obtained by optically coupling (in the same package, but keeping them electrically isolated) an LED with a photodetector.
Photonic displays	LCD, PDP	A liquid-crystal display (LCD) uses the light-modulating properties of liquid crystals (LCs). Each screen element (pixel) of an LCD is an electrode (made by a transparent conductor called indium tin oxide, ITO) in contact with the LC material and can align the LC molecules in a particular direction. A plasma display panel (PDP) is made of many tiny cells between two panels of glass, which are filled with a mixture of noble gases (usually neon and xenon). The gas in the cells is electrically turned into a plasma, which emits ultraviolet light that excites phosphorus to emit visible light.
Solar cells	Si, CdTe	Convert the energy of sunlight directly into electricity, based on the photovoltaic effect.

LEDs only began in the early 1950s. In 1962, it was observed that the light's wavelength could be shifted from IR to the visible spectrum by changing the chemical composition of GaAs to GaAsP. A few years later the first commercial LEDs appeared. Recently, an organic light-emitting diode (OLED) was proposed [333].

Generally, LED failure is a gradual process, produced by various types of degradation [334]:

- The presence of crystal defects (dislocations or precipitation of host atoms) may affect the radiative recombination of injected carriers that emits light in the active region. On high electrical injections, the chemical components can electromigrate into the other regions. The structural changes generate crystalline defects such as dislocations and point defects, which act as nonradiative centres, hindering the natural radiative decay and producing more heat within the active layer.

- Electrode degradation arises mainly due to metal diffusion on to the inner region, or so-called outer diffusion of semiconductor material. Diffusion increases as the injected current and ambient temperature increase. In the same area, current-crowding in high-power LEDs is another source of reliability risk. The solution is an optimised electrode design allowing vertical electrical current flow. Some electrodes, such as transparent conducting oxide (indium tin oxide, ITO) and reflective metals (silver), have several problems, including EM and thermal instabilities.
- Reverse discharge of EOS and ESD is a significant reliability issue for LEDs. The solution might be to use a Z diode or Schottky barrier to achieve specific ratings of ESD classification. Most commercial InGaN/GaN LEDs are grown on sapphire substrate, which has no electrical conduction. This leads to more residual electrical charges in the device, which make it more susceptible to EOS/ESD damage.
- Thermal runaway can be initiated if the voids exist in the solder used for bonding the LED to the heat sink of the substrate. An insufficient thermal path is created and hot spots result, eventually leading to thermal runaway and failure. Voids can occur because of poor processing conditions, metal diffusion at the interface (i.e. Kirkendall voiding) or EM. When a sufficiently high current density is available in the metal, vacancies and metal ions will migrate towards opposite poles, leading to void formation (vacancies), crystals, hillocks and whiskers.
- A mismatched CTE between bonded parts and the bonding solder introduces stresses during temperature cycling in the manufacturing process, which can cause delamination between the attachments. When power devices undergo cycling stress, for example, the performances of hard-soldered and soft-soldered devices can differ. Thermal fatigue is observable in soft solder, while hard solder is stable against thermal cycle stressing.
- Package-related failures can occur in encapsulant (due to thermomechanical stress from high temperature or because the epoxy resin reaches its glass-transition temperature or package cracking at very low temperatures) or in wires (bond breakage or detachment and die-attach strength loss are due to overheated epoxy and may cause a delamination between the chip and the epoxy). Another FM is produced by mechanical stress from lead wires as they can generate open circuits inside the device. Inappropriate pressure, position and direction applied to lead-wire soldering can add stress at normal operating temperatures, as can leads bent too close to the body of the LED. Two concurrent phenomena take place on the contact layers and solder during stress operation. One dominates at temperatures lower than +250 °C and tends to improve the series resistance value; the other dominates at temperatures higher than +250 °C and tends to worsen the series resistance value. This implies that the stresses without heat sink have a much faster degradation than the stresses performed at lower temperatures, because the active-layer efficiency decreases appreciably as the temperature increases and negative effects such as indium segregation can be activated. Both stresses are caused by semiconductor chip degradation; this degradation can be correlated with the creation of deep levels in the semiconductor, or with the worsening of the active-layer rectifying function [335].
- The transparent epoxy or silicone gel may cause another FM, linked to the light-transmission efficiency. A model incorporating some ideas from Monte Carlo ray tracing into the context of radiometry was proposed for predicting failure risks [336].
- Specific defects have been created in double-heterostructure AlGaAsGaAs commercial LEDs by neutron irradiation. Using controlled neutron energy, only one FM can be activated. Defects are located in the side of the chip and increase the leakage current, driven by the well-known Pool–Frenkel effect with E_C-E_T = 130 meV electron-trap energy level. The maximal amplitude of optical spectrum also reveals a drop of about 20% associated with the rise of leakage current [337].

In the late 1980s, the OLEDs were proposed as a more economic option, as they can be made on transparent and flexible substrates and can be mass produced by printing techniques. It seems OLEDs will become the dominant choice for future-generation flexible and flat-panel displays. However, stability, degradation in device luminance and high operating voltages are the major problems hindering the rapid commercialisation of OLEDs [338].

Relatively recently, the *p*-type doping in ZnO allowed blue and ultraviolet (UV) LEDs to be obtained, which can be an alternative to those based on III-nitrides. The device structure has been optimised and a reduced point defect density was achieved in the material [339].

5.5.2 Photodiodes

The main FMs of photodiodes are those of avalanche and PIN diodes and will be discussed in Section 5.6. Some specific reliability issues are detailed below.

Generally, avalanche photodiodes (APDs) are encapsulated in nonhermetic packages, in order to achieve lower-cost optoelectronic modules (transmitters and receivers). Experiments (operation in humid ambient as a function of temperature, RH and voltage) were carried out on InP-based APDs and a moisture-related FM was identified, related to corrosion of the passivating SiN and the underlying semiconductor, eventually producing a short circuit. Based on the voltage- and RH-dependence of the degradation rate, a model for the failure process was developed. The mechanism involves RH-dependent transport of charge across the surface of the SiN and Poole–Frenkel current flow through the SiN [340].

High-speed APDs used in harsh radiation environments suffer from background SETs due to the generation of high-energy heavy-ion secondary recoils and nuclear reactions. These transients degrade the bit error rate of an optical receiver, introducing spurious noise. The spatial dependence has been examined using a high-energy focused ion microbeam and the transient ion-beam-induced-current technique allowed measurement of the SET data collected on an InP InGaAs APD device [341].

Fault-tolerant redundancy in an active pixel sensor (APS) is obtained by splitting the photodiode and readout transistors into two parallel operating devices, while keeping a common row select transistor. This creates a redundant APS that is self-correcting for most common faults. Simulations suggest that, by combining hardware fault-tolerance capability with software correction, APS arrays could be made virtually immune to defects. Based on this, a fault-tolerant photodiode APS was designed and fabricated using a CMOS 0.18 μm process [342].

The degradation mode and reliability of planar waveguide photodiodes (WGPDs) for the optical hybrid module for subscriber systems were investigated in [343]. From electroluminescence topography observations and current–voltage characteristics, it seems the degradation mode for early failure is caused by the concentration of electric field or microplasma at the edge of the *pn* junction. From optical-beam-induced current images, it is clear that wear-out degradation, which governs the lifetime of WGPDs, occurs at the *pn* junction perimeter on a cleaved facet.

The failure risks of the planar InAlAs APD without guard ring were studied in [344]. The identified FMos were: (i) an increase in the dark current (generated in the upper side of the absorbing layer); (ii) an decrease in the breakdown voltage (because the depletion width in the absorbing layer shrinks from the region side); and (iii) short circuit. Their thermal activation energies were 0.96, 1.30 and 0.93 eV, respectively, and their estimated mean times to failures at 85 °C were 25, 100 and 22 million hours. Any degradation mode on the surfaced *pn* junction did not appear in spite of the 10 000-hour ageing test at high temperature.

The post-annealing effect on the dark current of the InGaAs WGPDs, which are developed for 40 Gbps optical receiver applications, was experimentally investigated in [345]. The dark current was significantly decreased and the breakdown voltage was slightly increased after annealing at 250 and 300 °C, but both were almost constant after annealing at 200 °C. It seems the post-annealing is more effective for the dark current improvement than the conventional curing process. The degradation mechanism for WGPDs can be explained by the formation of a leakage current path by ionic impurities in the passivation layer on the exposed *pn* junction. Nevertheless, it can be concluded that the WGPD test structures exhibit sufficient reliability for practical 40 Gbps optical-receiver applications [346].

The accelerated life testing of AlGaAs/GaAs multiple quantum well (MQW) APDs has shown an increase in dark current results in a reduction of the APD signal-to-noise ratio and breakdown voltage. Based on investigations with SEM (the EBIC method), it seems the FM is produced by ionic impurities or contamination in the passivation layer at the junction perimeter [347].

For a PIN photodiode, a failure was obtained as the current was increased under a high voltage and temperature, beyond a threshold current. Experimental data suggested that the junction failure occured due to the crystal breaking at the end facet as a result of thermal heat or energetic carriers. As long as the current was limited below the threshold current, no such failure events were noticed [348].

5.5.3 Phototransistors

A primary source of failure risk in phototransistors is ionic contamination. Damage in phototransistors is often accelerated due to relatively high operating temperatures resulting from the greater power dissipation of these devices. To achieve maximum reliability it is particularly important to operate phototransistors under derated voltage, power and temperature conditions. Accelerated testing at elevated temperatures has indicated an Arrhenius temperature dependence of failure rate with equivalent activation energy of about 0.7 eV [349].

The behaviour of phototransistors in special environmental conditions has been investigated. For phototransistors exposed to proton irradiations and in space applications, the same abnormal fluctuations of phototransistor collector current were noticed, the FM being in both cases produced by the mobile charges located in the photobase passivation layer [350].

Recently, organic phototransistors were proposed as one of several viable candidates for use in optical transducers, because such devices require both precise photo-conducting properties and high transistor performance. Organic phototransistors (e g. organic field effect phototransistors, OFETs) combine light-detection and signal-amplification properties in a single device without the noise increment associated with APDs.

5.5.4 Optocouplers

The main optocoupler ageing problem is current transfer ratio (CTR) degradation, which is caused by the reduction of the total photon flux emitted from the LED [351]. A diagnostic method using reverse-recovery time was proposed for the damage in the internal LED [352]. It depends only on electrical measurements, and can be used as a first-order parameter to determine LED properties when the LED technology of the optocoupler is unknown.

The noise properties of an optocoupler, especially in a low frequency range, can substantially affect operation of electronic systems. A recent paper [353] has shown that the intensity of LED noise and of the optical channel is negligibly low and does not influence the output noise of the optocoupler, which depends on noise sources existing in the phototransistor. An analysis of optocouplers with phototransistors using low-frequency noise measurements [354] confirmed this assumption, showing that the corner frequency in output noise is caused by the frequency dependence of dynamic current gain of the phototransistor under conditions of open base circuit.

A significant improvement of optocoupler reliability is achieved by using a thin ITO film deposited over the passivation. When this film is electrically connected to ground potential through contacts, it acts as a shield to avoid inversion failures by sinking any charge build-up to ground. In order to perform a full electrical FA on these optocoupler detector chips, the ITO layer must be removed. The technique for removing this film using an argon gas etch technique is discussed in [355]. Electrical FA of the die continues at this point, and a subsequent buffered oxide etch (BOE) removes the remaining passivation, leaving the exposed metallisation and oxide completely intact.

FMs in plastic-packaged optocouplers have been investigated in [356]. The high failure rate of the optocouplers, specifically the reverse-bias leakage current of the LEDs, was found to be moisture-related. Using SEM coupled with XEDS, a significant concentration of copper was detected on the surface of failed LEDs, possibly migrated from the copper lead frame. By applying infrared photoemission microscopy it was found that bandgap emission was associated with the reverse leakage path in the area of the *pn* junction of the LEDs.

Optocouplers are frequently used in space, where the environment mainly consists of protons and electrons that cause permanent damage due to the large number of interactions that result from exposure to relatively high fluencies of these particles. There are also galactic cosmic rays (relatively few in number compared to protons and electrons), the effect arising from the interaction of a single particle with the device (such as an SEU). Different types of damage mechanism have their origin in different particles, but the typical FMs are those already discussed for LEDs, photodiodes and phototransistors. Space failures have occurred due to two different mechanisms: displacement damage from high-energy protons, which produces permanent degradation, and transient upsets from heavy ions or protons. Transient-upset effects are generally important only for optocouplers with high-gain amplifiers, and are expected to be of secondary importance for these devices compared to displacement degradation [357]. Oxide defects play an important role in breakdown from heavy ions and breakdown occurs more readily when an ion strike occurs close to a defect site and the minority carrier lifetime reduces the CTR of optocouplers [358].

5.5.5 Photonic Displays

5.5.5.1 Liquid-Crystal Displays

Liquid-crystal properties were discovered around 1900. In 1972, the first LCD was commercially produced. At the end of 2007, for the first time, LCD televisions surpassed CRT units in worldwide sales.

The main reliability risk of twisted nematic LCDs is linked to misalignment issues. A life test is reported in [359], showing that misalignment life is longer with a lower voltage, a smaller duty ratio and a higher frequency. These results are highly appropriate for models with alternating-current electrolysis of the liquid-crystal material. In addition, electrostatic capacity and applied-voltage properties were used to study the relationship of liquid-crystal dielectric anisotropy and the coupling coefficient with temporal misalignment changes. The coupling coefficient decreases with misalignment failure.

Other FMs of greatest concern for LCDs, especially when used at high altitudes, are failure of the seal, battery leakage and lack of sufficient air cushion for the hard drive. Initial tests indicated no immediate failure of an LCD at 35 000 ft [360]. This should indicate limited issues during early use. The LCD seal will most likely leak after repeated pressure cycles; however, the number of cycles required for failure is unknown and could exceed the expected lifetime. While the probably of battery leakage is unknown, this should be seriously considered as a design limitation due to the potential for fire and explosion.

A new method linking experimental and theoretical approaches to precisely analyse the failure of LCD panels under mechanical drop is proposed in [361]. Generally, the thickness of a glass plate for an LCD panel is less than 0.7 mm, and thus requires sufficient strength to withstand mechanical shock. The fracture toughness (K_{IC}) of panel glass obtained experimentally is compared with the stress-intensity factor (K_I) calculated using the maximum stress at weak points via an FE drop simulation in order to evaluate crack growth on the panel. It has been demonstrated that this approach is simple but effective in determining the specification of components at the design level, including the roughness of glass-panel edges. Furthermore, important information related to the shockproof design of an LCD panel can be gathered from a dynamic simulation.

ITO is typically used for the conducting layer in display technologies. However, the process temperatures required to obtain low sheet resistance and high optical throughput properties for ITO on glass are incompatible with plastic substrates. Therefore, lower-temperature processes have to be developed for ITO in order for it to be considered for flexible display applications. Although ITO has excellent sheet resistance and optical properties, it does have one shortcoming in the flexible display realm: when ITO is deposited on a polymeric substrate, it can crack (buckle) under tensile (compressive) strain. In a flexible display application, ITO cracking can cause catastrophic failure. The mechanics of ITO on polymeric substrates are becoming better understood in flexible display applications [362].

In [363], 90 nm thick conducting ITO layers on polymer substrates are used to study the failure behaviour of brittle layers. In a two-point bending test the resistance of uniform ITO layers and narrow ITO lines (10–300 mm width) are determined as a function of the applied tensile strain. At a certain strain, the thin layer will crack and the resistance will strongly increase. Early stages of crack development are studied in the fragmentation test. Different FMs of thin brittle layers on the electrical conductivity of ITO layers are shown. When the strain in the ITO layer increases, stable cracks of a limited length are initiated at defects in the layer (crack initiation). At the critical strain, the crack is no longer stable and it will propagate over the whole width of the layer (crack propagation).The results show a wide failure strain distribution of the narrow ITO lines. This is determined by crack initiation, which is determined in turn by the distribution of defects. The uniform layers show a narrow failure strain distribution. This is determined by the well-defined crack-propagation strain.

5.5.5.2 Flat-Panel Displays

The LCD flat-panel display technology was invented at RCA Corp.'s David Sarnoff Research Center in 1964. Recent progress in flat-panel displays has been based on TFT technology, which has improved image quality (e.g. addressability, contrast). TFT LCD is one type of active-matrix LCD, though all LCD screens are based on TFT active-matrix addressing.

In [364], an LCD integrated with a gate driver and a source driver using amorphous In-Ga-Zn-oxide TFTs was reported. Using bottom-gate bottom-contact (BGBC) TFT, superior characteristics could be obtained with little variation, even when the thickness of the GI film was reduced due to etching of S/D wiring, which is a typical process for the BGBC TFT. Moreover, a favourable on-state current was obtained even when an In-Ga-Zn-oxide layer was formed over the S/D electrode. Since the upper portion of the In-Ga-Zn-oxide layer is not etched, the BGBC structure is predicted to be effective in thinning the In-Ga-Zn-oxide layer in the future. Upon evaluation, we found that the prototyped liquid-crystal panel integrated with the gate and source drivers using the TFTs with improved characteristics had stable drive [364].

In [365], the locations of process-induced defects in hydrogenated amorphous silicon TFTs, which are used as elements of active-matrix LCDs, were investigated by combining FIB techniques with X-TEM. The FIB technique is applied to TFT FA problems, which require considerable localised etching without inducing mechanical stress or damage at fragile failure locations. TFT defects such as pinholes and portions of the multilayer damaged by mechanical stress were characterised. A dramatic improvement brought about by the FIB technique is the increase in temporal efficiency of sample preparations. X-TEM observations also led to identification of the fault and analysis of its cause, which in turn led to a marked yield improvement.

The electrical characteristics of short-channel p-type excimer laser-annealed polycrystalline silicon TFTs, under high gate and drain bias stress, were investigated in [366]. The threshold voltage of short-channel TFTs was significantly shifted in the negative direction, an effect that may be attributed to interface-state generation near the source junction and deep-trap state creation near the drain junction between the poly-Si film and the GI layer. The effects of high gate and drain bias stress are related to hot-hole-induced donor-like interface-state generation. The transfer characteristics of the forward and reverse modes after the high gate and drain bias stress also indicate that the interface-state generation at the GI–channel interface occurred near the source junction region. To identify the existence of positively trapped charges on the interface between the poly-Si thin film and the GI in the short-channel TFTs, C–V measurement was employed. The variation in the C–V characteristics with an applied frequency, before and after high gate and drain bias stress, indicates whether the main origin of the dominant degradation mechanism in the short-channel poly-Si TFTs under the high gate and drain bias stress is fixed traps or trap states.

5.5.5.3 Plasma Display Panels

The plasma display panel (PDP) has a unique light-producing mechanism: the electric field of the addressing electrode in the back panel triggers the gas to become a plasma state, which is used to produce light, with a technique very similar to that used for fluorescent tubes.

One of the problems with PDPs is their relatively short life as a result of degraded strength. The heat dissipated due to low power efficiency leads not only to a decrease in the display quality but to structural failure attributable to a critical thermal gradient. Thermally induced stress is the major FM of early structural failure in PDPs [367].

The mechanism for Y-scan buffer IC failure of PDPs was analysed in [368] using the real-time electrical stress analysis (RTESA) method. This method shows the phenomenon, in which electrical stress comes into an IC under device operating condition, as a photoemission profile, connecting a photoemission spectrum analyser (PSA) with an external electrical stress source and bias network. In other words, this method easily finds the soft damage level, damage injection path and damage mechanism by showing the electrical stress-effect procedure as a photoemission profile. The equipment used in the analysis was the HP4155C semiconductor parameter, the PSA, an optical microscope and an SEM.

5.5.6 Solar Cells

Solar cells come under the heading ‘photovoltaics’ (PVs), a field of technology and research related to directly converting sunlight into energy. The term ‘photovoltaic cell’ is used to describing a device that converts a source of light other than sunlight into energy.

The degradation mechanisms of solar cells may involve either a gradual reduction in the output power of a PV module over time or an overall reduction in power due to failure of an individual solar cell in the module. The PV modules do not have moving parts, so the reliability is largely determined by the stability and resistance to corrosion of the materials from which they are constructed. There are several FMos and degradation mechanisms which may reduce the power output or cause the module to fail. Nearly all of these mechanisms are related to water ingress or temperature stress.

The **sudden failures** of solar cells that are part of PV modules could be: (i) short-circuited cells, a common FMo for thin-film cells since top and rear contacts are much closer together and stand more chance of being shorted together by pin holes or regions of corroded or damaged cell material or (ii) open-circuited cells, another common FMo, although redundant contact points plus ‘interconnect-bus bars’ allow the cell to continue functioning. Cell cracking can be caused by thermal stress, hail or damage during processing and assembly, resulting in ‘latent cracks’, which are not detectable on manufacturing inspection, but appear sometime later.

PV modules can fail by:

- **Gradual degradation**, caused by: (i) increases in R_S due to decreased adherence of contacts or corrosion (usually caused by water vapour); (ii) decreases in R_{SH} due to metal migration through the *pn* junction; or (iii) antireflection coating deterioration. In [369], the degradation in field-aged modules is reported, including degradation of packaging materials, adhesional loss, degradation of interconnects, degradation due to moisture intrusion and semiconductor device degradation.
- **Open circuit** of the module or of the interconnections (produced by the fatigue due to cyclic thermal stress and wind-loading [370]).

- **Short circuit** of the module, which may be the result of manufacturing defects occuring due to insulation degradation with weathering, resulting in delamination, cracking or electrochemical corrosion.
- **Delamination of the module**, usually caused by reductions in bond strength, either environmentally induced by moisture, or by photothermal ageing and stress, which is caused by differential thermal and humidity expansion.
- **Failure of the encapsulant**, produced by the degradation of the material (e.g. degradation of the ethylene vinyl acetate (EVA) encapsulant including its delamination, or even degradation of the glass cover [371]).

Many technologies have been proposed for solar cells. The dye-sensitised solar cell (DSSC), which is based on a photo-electrochemical mechanism, resembling the photosynthesis in plant leaves, seems to be a good solution [372]. The dye adsorbed to a nanostructured TiO_2 film has been studied with UV-VIS and IR spectroscopy following exposure to visual and ultraviolet radiation, increased temperature, air, electrolyte and water in the electrolyte. The identified FM was that sensitive material is lost from the dye together with the electrolyte [373]. In another experiment conducted under continuous simulated sunlight illumination, the cells showed fast degradation under full-spectrum sunlight illumination, but rather good stability when the UV part of the illumination was removed. XPS measurements showed evidence that TiO_2 could oxidise CuI in the presence of UV light. The photo-degradation mechanism of the cells was discussed in [374]. The long-term stability of the solid-state DSSC was found to be improved under simulated sunlight by coating the TiO_2 porous electrode with an ultrathin MgO layer, which was able to block the photo-oxidative activity of the TiO_2.

The concept of three-dimensional PVs was explored computationally using a genetic algorithm to optimise the energy production during a day for arbitrarily shaped 3D solar cells confined to a given area footprint and total volume [375]. It seems the optimal 3D structures are not simple box-like shapes, and that design attributes such as reflectivity can be optimised using three-dimensionality.

In [376], the failure risks of PV devices comprising a bilayer heterojunction formed between poly(3-carboxythiophene-2,5-diyl-co-thiophene-2,5-diyl) (P3CT) and buckminsterfullerene (C_{60}) sandwiched between ITO and Al electrodes were elucidated by TOF-SIMS analysis in conjunction with isotopic labelling using $^{18}O_2$ after a total testing time of 13 000 hours. The cell was subjected to continuous illumination with an incident light intensity of $1000\,W\,m^{-2}$ (AM1·5) at $72 \pm 2\,°C$ under a vacuum of $<10^{-6}$ mBar. It was found that the oxygen diffuses into the device through the Al electrode in between the Al grains and through microscopic holes in the Al electrode. Once inside the device, the oxygen diffuses in the lateral and vertical plane until the counter electrode is reached. C_{60} was found to be susceptible to the incorporation of ^{18}O but P3CT was not under the conditions in question. The other prominent degradation pathway was found to be the diffusion of electrode materials into the device. Both electrode materials diffuse through the entire device to the counter-electrode.

5.6 Failure Modes and Mechanisms of Non-Silicon Technologies

Generally, semiconductor materials other than silicon are used for two families of electronic components: (i) optoelectronic and photonic devices (discussed in Section 5.5) and (ii) power devices (to be discussed in this section). Even the first power semiconductor device, launched in 1952, was a germanium power diode (invented by R.N. Hall of General Electric), with a voltage capability of 200 V and a current rating of 35 A. Later, power devices were also fabricated using silicon as a starting material (as discussed in Sections 5.3 and 5.4).

The FMs and FMos of three main categories of power device fabricated using semiconductor materials other than silicon will be discussed below: (i) diodes, (ii) transistors and (iii) other devices.

5.6.1 Diodes

In this section, the FMs of power diodes manufactured with semiconductors other than silicon will be discussed. Table 5.24 shows the main families of power diode and the basic semiconductor material used in each case. Details about the FMs of each of the diode types in Table 5.24 are presented below.

5.6.1.1 Tunnel Diodes

The tunnel diode was invented by Leo Esaki (Tokyo Tsushin Kogyo, later with Sony) in August 1957. In 1973 Esaki received the Nobel Prize in Physics for discovering the electron-tunnelling effect used in these diodes. They are capable of very fast operation into the microwave frequency domain and have a heavily doped *pn* junction only some 10 nm wide. The basic material is usually germanium, but tunnel diodes can also be made by gallium arsenide or even silicon materials.

In [377], the effect of electric stress on the characteristics of $Al/SiO_2/p^+$-Si MOS tunnel diodes ($d_{ox} = 2.5-3$ nm) is studied. Along with the gradual current changes, superimposed by the SBD-related

Table 5.24 Main types of non-silicon diode

Diode type	Basic material	Technology and characteristics
Tunnel diodes	Ge, Si, GaAs, InP	Diodes based on the electron-tunnelling effect, with a negative-resistance region of operation caused by quantum tunnelling, which allows amplification of signals and very simple bistable circuits. Note the high resistance of these diodes to nuclear radiation.
Gunn diodes	GaAs, InP	Diodes similar to tunnel diodes, but with a region of negative differential resistance. They are used in high-frequency microwave oscillators, because, when appropriately biased, dipole domains form and travel across the diode.
IMPATT diodes	SiC	IMPATT diodes (IMPact ionisation Avalanche Transit-Time) are high-power diodes, with high breakdown voltages, used in high-frequency electronics and microwave devices (frequencies between about 3 and 100 GHz or more). A major issue of IMPATT diodes is the high level of phase noise they generate due to the statistical nature of the avalanche process. They are used in microwave generators for many applications.
Schottky diodes	SiC	Diodes built from a metal-to-semiconductor contact, with a lower forward voltage drop than *pn*-junction diodes. They can be used in voltage-clamping applications and prevention of transistor saturation, but also as (relatively) low-loss rectifiers. Being a majority carrier device, the Schottky diode does not suffer from minority-carrier storage problems, which slow down many other diodes. They may thus have a faster 'reverse recovery' than *pn*-junction diodes and a high switching speed (due to the junction capacitance, which is lower than for *pn* diodes), and are used in RF devices (mixers, detectors, etc.).
PIN diodes	GaAs, Si	Diodes with a central undoped or intrinsic layer, forming a *p*-type/intrinsic/*n*-type structure, operating as variable resistors at RF and microwave frequencies. They are used as radio-frequency switches, attenuators, large-volume ionising-radiation detectors and photodetectors. They can also be used in power electronics, as their central layer can withstand high voltages.

steps, a nontrivial abrupt decrease in current is also revealed during constant voltage stress. The latter occurs predominately under high bias and may be considered an unusual appearance of the same SBD events. In case of substantial spatial oxide thickness deviation, this effect is important even if it occurs within a small area.

The results of an analysis of the degradation mechanisms that occur in resonant tunnelling diodes made of both the arsenides and the antimonides is furnished in [378]. Microstructure analysis shows that Al0.6Ga0.4As/GaAs resonant tunnelling diodes grown at different growth temperatures offer the same crystalline quality. Thermal stress and TLPs were used to degrade the resonant tunnelling diodes. Only the TLP method led to device failure. The appearance of a parallel conductance is attributed to local thermal degradation at the interface, whereas the shift of the resonance voltage to higher values is attributed to the degradation of the ohmic contacts.

5.6.1.2 Gunn Diodes

Discovered by J. Gunn in 1963, the 'Gunn effect' is linked to microwave current instabilities in bulk N-type GaAs. A 'negative resistance' phenomenon results when the electrons in N-type GaAs traverse from a high-mobility to a lower-mobility valley, thus producing a lower net electron velocity. This 'negative resistance' phenomenon continues until the breakdown voltage causes diode failure. A transferred electron device (TED) was created, later becoming the Gunn diode, which produces oscillations using the negative-resistance property of bulk GaAs.

Gunn diodes are fabricated from a single piece of n-type silicon, having two connecting regions: the top and bottom areas, heavily doped to give $N+$ material. The device is mounted on a conducting base, to which a wire connection is made, acting also as a heat-sink for the generated heat. The connection to the other terminal of the diode is made via a gold connection deposited on the top surface. Gold is required for its relative stability and high conductivity. The centre area of the device, which is the active region, has a thickness of around 10 microns (its thickness will vary as it is one of the major frequency-determining elements). Note that the device consists only of n-type material; there is no pn junction and in fact it is not a true diode, operating on totally different principles.

The Gunn diode is used as transferred electron oscillator by employing the negative-resistance property of bulk GaAs for generation of microwave power with a relatively low efficiency – about 2–5%. Consequently, considerable power is dissipated. That is why adequate heat-sinking must be provided in order for the diode to operate properly. The diode is prone to oscillating at low frequencies with the lead inductance from bias circuit connections. A Gunn diode made from gallium nitride can reach 3 THz operation [1].

An initial paper on the FMs of Gunn diodes was published in 1967 [379], reporting observations on the electrical properties and FMs of planar Gunn diodes. A conducting channel of Sn(In) is proved to be responsible for the final catastrophic failure. Excess heat is generated at the anode, where the breakdown is observed to start. The mechanisms causing the breakdown (hole injection and combined ion drift and diffusion of impurities) are discussed.

Generally, the **thermal issues** seem to be the major failure risk. Two basic FMos of Gunn diodes – the thermal mode and breakdown in a strong-field domain – are identified and discussed. The malfunctions and dynamics of Gunn diode overheating are analysed and the effects of impact ionisation in the strong-field domain are examined. Conditions under which impact ionisation is a more serious problem than overheating and thermal breakdown are formulated.

Another significant failure risk factor for Gunn diodes is **metallisation damage**. An experiment performed on two types of Gunn diode, which were submitted to a biased life test at elevated temperature (70 °C), show a difference in burn-out percentage between the two [380]. Based on an investigation with a volume of 8.5×10^5 device hours, it was concluded that deterioration of the metallisation is the main FM for both types of diode, but the fraction failed of n^+–n–n^+-type diodes is significantly lower than that of n–n^+-type diodes.

5.6.1.3 IMPATT Diodes

The impact ionisation avalanche transit-time (IMPATT) diode is a *pn*-junction diode with a depletion region adjacent to the junction, allowing the drift of electrons and holes, which is biased beyond the avalanche breakdown voltage. The basic material is usually silicon carbide (SiC), which allows high breakdown field. The negative characteristic is produced by a combination of impact avalanche breakdown and charge-carrier transit-time effects. A major drawback is the high level of phase noise, resulting from the statistical nature of the avalanche process. Nevertheless these diodes make excellent microwave generators for many applications. The IMPATT diode family includes many different junctions and metal semiconductor devices (Si, GaAs, InSb, SiC, etc.). Avalanche breakdown occurs when the electric field across the diode is high enough for the charge carriers (holes or electrons) to create e-h pairs [1].

The basic advantage of IMPATT diodes is their high values of junction temperature and reverse bias, higher than those of other semiconductor devices. Consequently, any device with surface contamination is eliminated, because it can cause excessive leakage currents and may diminish the performance of the device. The major degradation mechanism of GaAs IMPATT diodes is the **interdiffusion of Au** through Pt, forming metallic spikes which extend into the GaAs, forming a short circuit of the junction either in the bulk or at the metal–GaAs interface. In addition, since more than 90% of the DC input power can be dissipated in the high-field region [38], the attendant rise in junction temperature can result in concomitant **increase in leakage current**. Excessive surface leakage current or metallisation overhang in a defective diode can lead to early failure, even under normally safe operating conditions.

Other current failure risks are **bonding** and **metallisation**, which are generally responsible for a high percentage of semiconductor failures. Screening tests (thermal resistance and HTRB) are used to remove the devices with weak die attach, metal contact or bonding. The maximum operation temperature of IMPATT diodes is mainly limited by the reliability of the assembly and interconnection technology for the required power/temperature cycling. Known weak points are bond wires and solder interconnection technologies; it is necessary to improve these technologies or to replace them with better technologies in order to meet the requirements of harsh temperature environments [381].

Another failure cause is the **diffusion of the contact metal into the semiconductor material**. The main methods for eliminating this FM are: (i) choice of metals used in the contacting system; (ii) careful control while applying the metals; and (iii) control of junction temperature. For any given metallisation system, the diffusion of the contact metal into the semiconductor is an electrochemical process, to be described by the Arrhenius equation, with an activation energy of 1.8 eV [1].

The results of an experimental study on IMPATT diodes are reported in [382]. Constant-stress (temperature) operating tests have been conducted on silicon Ka-band diodes. Statistics of failure fit a log-normal distribution except near the early-failure tail; deviations are ascribed to manufacturing-process non-uniformities. Analysis of failed devices shows a single predominant FM: gold contamination of the *pn*-junction area, observable at junction operating temperatures between +230 and +350 °C.

Tuning-induced burnout can easily be avoided once the circumstances that result in this failure are understood. One possible mechanism which might be responsible for diode burnout under this condition has been described in [383]. It is suggested that the low-frequency negative resistance induced by large RF modulation could lead to a transversely non-uniform current density within the diode.

Diode failures are a limiting factor for the reliability of power circuits. One failure reason is **dynamic avalanche**, three degrees of which can be distinguished; some designs are rugged up to the third degree. Design modifications for improving the dynamic ruggedness and suitable test conditions are proposed in [384].

5.6.1.4 Schottky Diodes

The Schottky diode is formed by a Schottky barrier (named after the German physicist Walter H. Schottky), which is a potential barrier formed at a metal–semiconductor junction with rectifying characteristics. The Schottky diodes are ideal for power-supply switching operations due to two main parameters of their metal–semiconductor junction which are better than those of standard diodes: a smaller forward-bias voltage drop of only 0.15–0.45 V (compared to 0.5–0.7 V) and a much shorter switching time. In fact, the most important difference between junction diode and Schottky diode is reverse recovery time, when the diode switches from nonconducting to conducting state and vice versa. Where in a *pn* diode the reverse recovery time can be in the order of hundreds of nanoseconds, and less than 100 ns for fast diodes, for Schottky diodes the switching time is ~100 ps for the small-signal diodes, and up to tens of nanoseconds for special high-capacity power diodes. The Schottky diodes essentially switch instantly, with only slight capacitive loading. The drawback of the Schottky diode is a much higher reverse-bias leakage current rating, but because *pn* recombination is not a factor in switching delay time, only capacitance affects the reverse-switching time.

Silicon carbide (SiC) was proposed as the basic material for Schottky diodes, after trialling several types of semiconductor, by Siemens Semiconductors (now Infineon) in 2001, and is preferred today by almost all manufacturers. SiC has a high thermal conductivity, and temperature has little influence on its switching and thermal characteristics. SiC Schottky diodes have about 40 times lower reverse leakage current compared to silicon Schottky diodes and are available in 300 and 600 V variants. With special packaging, it is possible to have operating junction temperatures even higher than 200 °C. Moreover, SiC is unique among compound semiconductors in that its native oxide is SiO_2, the same oxide as silicon [1]. Recent advances coupled with its proven track record in the field have made the SiC Schottky diode an excellent choice for any high-power application requiring fast switching and near-zero reverse recovery losses [385].

Material defects in SiC are the cause of many reliability issues in SiC devices, such as reduced critical electric field, higher leakage currents during reverse-bias operation and degradation in the on-state performance of Bi devices. Detailed experiments conducted on SiC devices [386] have shown that the on-state voltage drop on these devices increases with time when they are kept in the forward-biased mode for an appreciable length of time.

Contact migration (another form of Al migration; however, the physical process governing the movement of Al atoms is different from that of EM) is a particular problem of Schottky diodes. All diodes must be designed within the specifications required to prevent an EM of the metallisation.

ESD is a possible FM of Schottky diodes during EOS conditions. They are sensitive to static potentials, and can be destroyed by a permanent breakdown across the reverse-biased junction. The usual silicon-device forward-surge current failure $n_i(T)$- mechanism should be preceded by a $\mu(T)$- (i.e. mobility versus temperature) mechanism. [387] reports experimental results supporting this theory. In addition, *pn*-junction voltage-drop measurements interpreted by electro-thermal simulations indicate that nondestructive instant temperatures may reach 1600–1800 K in SiC. As capacitance is not expected to increase by more than 5% at 85 °C, the increased risk of charge build-up and ESD is considered to be a small one [388].

Elevated temperature combined with repeated power cycling could drive **fatigue at the die attach**. Basically, Schottky diode failure as a result of increased temperature is almost entirely dependent on proper heat dissipation of the diode through its heat sink and can be mitigated by the placement of the diodes away from other heat-generating devices. Assembly failure due to altered operating characteristics of the diode (i.e. greater reverse current) is more likely than direct failure of the diodes themselves.

The **wear-out FM** may particularly arise when the Schottky diodes are used on the output of power supplies, where failure is normally due to a single reverse-current effect. Reliability studies have been published on SiC Schottky diodes that indicate a lifetime greater than 50 years. Since the majority of failures are caused by EOS, long-term degradation at the die level is not expected to be an issue [388].

Cosmic radiation has been identified as a decisive factor in power-device reliability. Energetic neutrons create ionising recoils within the semiconductor substrate, with many leading to device burnout. Radiation hardness intrinsic to the SiC material used to be assumed, but as experimental data was scarce, reliability problems due to radiation-induced device failure could not be ruled out.

Recent accelerated testing results indicate that cosmic radiation will indeed affect the reliability of SiC power devices, as is the case with their silicon counterparts, but the problem can be contained very effectively by good device design [389].

The Schottky diode has a relatively low **sensitivity to a radiation environment**. However, ionising radiation (e.g. low-energy electrons, gamma rays and X-rays) produces surface effects, with a build-up of a positive space charge in the oxide and an increase in the surface velocity of planar devices. These surface effects are responsible for the degradation of the reverse current and excess forward current when the Schottky barrier diode is exposed to this environment. The primary effect of neutron radiation is bulk damage, which includes carrier removal and decrease in bulk lifetime. The series resistance of the diode increases because of the carrier removal, while the decrease in bulk lifetime results in an increase in forward and reverse current.

A set of complex radiation experiments on various types of Schottky diode led to some interesting results [390]:

- The exposure of two different Schottky diodes (gold-silicon and chromium-protactinium-platinum-silicon) to a total gamma dose of 10^8 rad resulted in increases of the reverse current that varied from a nominal change to almost 4 orders of magnitude. Reactor irradiation to a neutron fluency of 10^{14} n/cm^2 resulted in negligible changes in the characteristics of these same devices. Annealing experiments following the irradiation of the above devices resulted in significant annealing at 150 °C for units experiencing large changes in reverse current, while those having small increases required a temperature of 300 °C.
- Various physical configurations of Al-Si Schottky barrier diodes were irradiated with low-energy electrons (15–20 keV) to a dose of 10^9 rad at room temperature with their leads shorted. These configurations included a *pn* guard-ring diode, an overlap diode and a non-overlap diode. The *pn* guard-ring diode experienced no degradation in reverse current below a dose of 10^8 rad (SiO_2) and an increase of approximately 1 order of magnitude to 1.0 nA at 10^9 rad (SiO_2). The overlap diode was more sensitive: the reverse current increased an order of magnitude at 10^8 rad (SiO_2) and approached another order of magnitude increase at $\sim 5 \times 10^8$ rad (SiO_2) for a reverse current of 100 nA. The non-overlap diode was even more sensitive to the radiation environment.
- Platinum-silicon (Pt-Si) devices and Al-Si devices were tested to an electron of 10^8 rad (SiO_2), and a large increase in reverse current was noticed. The Pt-Si devices were considerably more sensitive to the radiation than the Al-Si devices and to the degradation of their forward characteristic, in which they experienced increases in forward current at low voltages. The forward characteristic of the Al-Si units was insensitive to the low-energy electron dose of 10^8 rad (SiO_2), with any excess current from the irradiation being masked by the normal thermionic emission current of these devices. The effect of the gate voltage upon the forward and reverse currents of the gate-controlled diodes was shifted by exposure to the low-energy electron environment in such a manner as to require a more negative gate voltage to obtain a change in current similar to that found prior to irradiation.
- Neutron irradiation to fluencies of 10^{15} n/cm^2 ($E > 0.11$ MeV) also resulted in large amounts of excess current at low forward voltages in the Pt-Si devices due to recombination in the space-charge region. The normal thermionic emission current of the Al-Si devices again completely masked any increase that might occur in these units. An increase in the diode series resistance from carrier removal was evident in both the Pt-Si and Al-Si devices.

The most serious FM in a radiation environment is SEB, which is a widely recognised problem for space applications, but may also affect devices in terrestrial applications. In [391], the current state of knowledge of power-diode vulnerability to SEB is reviewed, based on both experimental and simulation results. It is shown that present models are limited by the lack of detailed descriptions of thermal processes that lead to physical failure.

Power diodes may undergo destructive failures when they are struck by high-energy particles during the off state (high reverse-bias voltage). In [392], this FM is described using a coupled electro-thermal model. The specific case of a 3500 V diode is considered and it is shown that the temperatures reached when high voltages are applied are sufficient to cause damage to the constituent materials of the diode. The voltages at which failure occurs (e.g. 2700 V for a 17-MeV carbon ion) are consistent with previously reported data. The simulation results indicate that catastrophic failures result from local heating caused by avalanche multiplication of ion-generated carriers.

5.6.1.5 PIN Diodes

A PIN diode has a wide, lightly doped region (near intrinsic semiconductor), sandwiched between a p-type semiconductor region and an n-type semiconductor region, both heavily doped in order to be used as ohmic contacts. The PIN diode was invented by Jun-ichi Nishizawa in 1950. PIN diode design depends on the intended application (RF switches, attenuators and photodetectors): by increasing the dimensions of the intrinsic region (and its stored charge), the diode is made to act like a resistor at lower frequencies; this adversely affects the time needed to turn off the diode and its shunt capacitance.

The main FM of PIN diodes is **avalanche breakdown**. The experiment reported in [393] aimed to analyse the breakdown failure points in the 4H-SiC PiN using the EBIC mode of SEM. The failure points were determined by emission microscopy. The basal plane dislocations around the failure point and at measured temperatures below 200 K were observed and the dark spots were identified by EBIC. However, in the X-ray topography image, no spots were found around the dislocations, leading to the conclusion that these spots originated from the metal contamination. The electric field was multiplied due to a permittivity change, and this multiplication caused the avalanche breakdown.

5.6.2 Transistors

5.6.2.1 Heterojunction Bipolar Transistors

The heterojunction bipolar transistors (HBTs) are a relatively new type of Bi transistor, but this technology has already become a major player in wireless-communication, power-amplifier, mixer and frequency-synthesiser applications. The HBTs extend the advantages of silicon Bi transistors to significantly higher frequencies, being used in military applications requiring a high current drive, high transconductance, high voltage-handling capability, low-noise oscillator and uniform threshold voltage.

The original idea of HBTs was to use different semiconductor materials for the emitter and base regions, creating a heterojunction. This structure limits the injection of holes from the base into the emitter region, since the potential barrier in the valence band is higher than in the conduction band. Unlike Bi junction transistors, HBTs allows a high doping density to be used in the base, reducing the base resistance while maintaining gain. The efficiency of the devices is measured by the Kroemer factor.[11] A large range of materials can be used for the substrate, such as silicon, gallium arsenide and indium phosphide. The epitaxial layer can be manufactured by a silicon/silicon-germanium alloy. Emerging HBT technologies allow the integration of a large quantity of high-performance RF circuits and high-speed digital circuits on a single chip [394].

An investigation of the high-voltage/high-current mixed-mode (M-M) **stress-induced damage** mechanisms of *pnp* SiGe HBTs is presented in [395]. Different accelerated stress methods, including M-M stress, reverse emitter-base (EB) stress and forward-collector-plus-reverse-EB stress, were applied to *pnp* SiGe HBTs from a state-of-the-art complementary-SiGe BiCMOS process technology platform. The operative damage mechanism of the M-M stress method was identified. Experimental evidence of collector-current change due to the M-M stress and experimental proof of the type of hot carrier (electrons versus holes) responsible for the observed M-M stress damage were presented.

[11] Herbert Kroemer (from the University of California, Santa Barbara) received a Nobel Prize for his work in this field in 2000.

The main reliability concern for SiGe HBTs is the robustness of the base–emitter junction to **hot carriers** which introduce current-gain degradation. This degradation is not likely to become a significant limit for future devices scaled for high-speed performance [394]. Another possible FM of GaAs HBTs is the sudden and catastrophic DC current-gain degradation. In [396] this FM is explained by the recombination-enhanced defect reaction (REDR) process, which may cause long-term current-gain degradation.

Parametric degradation in HBTs is described to help explain overall reliability and relationship to device junction ageing, and to aid in predicting parameter drift. The electrical parameter involved is the leakage current, which is a key device parameter for HBTs. The transistor beta ageing is directly proportional to the fractional change in the base–emitter leakage current. A model is presented in [397], describing the degradation that links ageing to junction temperature-dependent leakage-current mechanisms. Another multiparameter model of h_{FE} time-to-fail for M-M stress conditions for high-performance NPN HBTs is shown in [398]. This model is tested over an extensive range of parametric space, covering multiple orders of magnitude in V_{CB}-I_E M-M stress.

The relationship between operating voltage constraints and reliability issues in SiGe HBTs is discussed in [399], examining **breakdown-related instabilities** as they relate to technology generation, device geometry, bias configuration and current density. Practical design and reliability implications of breakdown are explored through careful measurement and simulation using quasi-3D compact models. Overall, SiGe HBTs biased in common-base configuration and driven by constant emitter current are shown to possess higher stable operating voltage limits than those biased in common-emitter configuration and driven by constant base current. However, biasing beyond the determined SOA may significantly increases **hot-carrier degradation** at higher currents, and must be carefully considered from a circuit-design perspective.

A theoretical thermo-electro-feedback model has been developed for the thermal design of high-power GaAlAs/GaAs HBTs. The power-handling capability, **thermal instability**, junction temperature and current distributions of HBTs with multiple emitter fingers have been numerically studied. The calculated results indicate that power HBTs on Si substrates (or with Si as the collector) have excellent potential power performance and reliability. The power-handling capability on Si is 3.5 and 2.7 times as large as that on GaAs and InP substrates, respectively. The peak junction temperature and temperature difference on the chip decrease in comparison to the commonly used Si homostructure power transistor with the same geometry and power dissipation [400].

The RF (6 GHz) power performance characteristics of SiGe power HBTs at cryogenic (77 K) and high-operation temperatures (chuck temperature +120 °C, junction temperature up to +160 °C) are discussed in [401]. Without specific device optimisations for cryogenic operation, the power SiGe HBTs exhibit excellent large-signal characteristics at 77 K. Compared with room-temperature operation, similar power gain and output power were obtained when the devices were operated at cryogenic temperature. The SiGe power HBTs also operate well at high junction temperature with reasonable power gain and **output power degradations**. The modelling of the SiGe power HBTs under high-operation temperature indicates significant increases of base resistance (R_B) and emitter resistance (R_E) that account for the degradation of power performance of these devices.

The reliability of another variant of high-voltage HBT, manufactured by an InGaP/GaAs technology for 28 V operation, is discussed in [402]. Long-term reliability tests performed at 28 V bias voltage, 5.2 kA/cm^2 current density and 310 °C junction temperature resulted in zero device failures after >3600 hours of stress. WLR tests using extreme current-stress conditions were used to force transistors to degrade in a short time for quick checking of transistor integrity and process reliability. Transistor lifetimes were generally higher than those in conventional low-voltage HBTs at similar junction temperatures. Combined with the previously reported ruggedness and linearity performance, it seems the high voltage InGaP/GaAs HBT is a mature technology for use in the 28 V high-linearity power amplification.

5.6.2.2 High-Electron-Mobility Transistors

The high-electron-mobility transistors (HEMTs) are FETs with their channel formed by a heterero-junction between two materials, instead of a doped region as is generally the case for a MOSFET. The two materials used for a heterojunction must have the same lattice constant and the most used combination is GaAs and AlGaAs. The classical devices are doped with impurities to generate mobile electrons, which are slowed down through collisions with the dopant impurities. HEMTs avoid this through the use of high-mobility electrons generated using the heterojunction of a highly-doped wide-bandgap n-type donor–supply layer (e.g. AlGaAs) and a non-doped narrow-bandgap channel layer with no dopant impurities (e.g. GaAs). Other combinations of materials could be used, depending on the application of the device: if more indium is incorporated, better high-frequency performance is obtained, and if GaN is used, the high-power performance is improved.

HEMTs are also known as heterostructure field effect transistors (HFETs) or modulation-doped field effect transistors (MODFETs). They were invented by Takashi Mimura (Fujitsu), but a group led by Ray Dingle from Bell Laboratories was also involved.

There are two variants of HEMT manufactured by two materials with different lattice constants:

- Pseudomorphic high-electron-mobility transistors (pHEMTs), achieved using an extremely thin layer of one of the materials – so thin that the crystal lattice simply stretches to fit the other material.
- Metamorphic high-electron-mobility transistors (mHEMTs), containing a buffer layer made of AlInAs, with the indium concentration graded so that it can match the lattice constant of both the GaAs substrate and the GaInAs channel; the indium concentration in the channel can be optimised for different applications: low concentration provides low noise and high concentration gives high gain.

HFETs fabricated from nitride semiconductors utilising the AlGaN/GaN structure have high electron-output power, an order of magnitude higher than the power available from GaAs- and InP-based devices. However, the nitride devices demonstrate a reliability problem, where the DC current and RF output power continually decrease as a function of time [403]. The decrease of the DC current and RF output power over time is attributed to **gate tunnelling** that is determined by the magnitude of electric field at the gate edge.

A detailed physical investigation of trapping effects in GaAs power HFETs is presented in [404]. The **hot-carrier** degradation of the drain current and the **impact-ionisation-dominated reverse gate current** are the two identified failure risks. Thoroughly consistent results show that: (i) the dominating effect is given by the traps at the source–gate recess surface and (ii) as far as the hot-carrier degradation is concerned, only a simultaneous increase of the trap density at the drain gate recess surface and at the channel buffer interface (again at the drain side of the channel) is able to account for the simultaneous decrease in the drain current and increase in the impact-ionisation-dominated reverse gate current.

The AlGaN/GaN HEMTs are used for operation at higher voltage with higher power density and at lower power consumption for next-generation high-speed wireless communication systems, operating at +200 °C. It has been found that the current degradation has relationships with the isolation structure. In [405], the differences in current degradations between two isolation structures (implantation-isolated and mesa-isolated device) are studied and some interpretations concerning the source of these degradations are discussed. The current degradation of the implantation-isolated structure device depends on channel temperature, drain bias and on-state drain current. At off state no degradation occurs. From these results, one can confirm that these degradations are caused by **hot carriers**. The mesa-isolation structured device prevents current degradation through the side-gate effect.

One of the main challenges in HEMT fabrication consists in finding an optimal layered structure to enhance reliability. A significant reduction in the saturation drain current is achieved by increasing the temperature, caused by the decrease in the saturation carrier velocity in the HEMT. Devices suffer from pronounced degradation when thin substrates reduce the thermal impedance due to the heat-spreading effect. It has been demonstrated that the temperature increase for similar devices on sapphire and SiC substrates at the same added power is about twice as high as that on sapphire-based substrates. To overcome the low thermal conductivity of sapphire, FC integration with a high-thermal-conductivity AlN substrate can be used. However, this approach demands additional process steps, such as metal contact-pad deposition and FC bonding by simultaneously applying heat and pressure to the carrier substrate and the HEMT, followed by a thermal reflow of the solder material. The alternative to FC technology in improving the performance of the devices is a high-thermal conductivity SiC substrate.

In [406], the transport and noise properties of AlGaN/GaN heterostructures are studied with respect to reliability. Heterostructures grown on sapphire and SiC substrates show an opposite dependence of self-heating on the buffer-layer thickness. An improvement in the transport properties in AlGaN/GaN heterostructures is noticed after treatment with small doses of **gamma irradiation**, which is irreversible with time. The results suggest that the improvements in the transport characteristics can be achieved by using gamma-quanta irradiation in processing technology.

5.6.3 Integrated Circuits

5.6.3.1 Monolithic Microwave Integrated Circuits

Today, GaAs monolithic microwave integrated-circuits (MMICs) are largely used in communications satellites technology in most communications payload subsystems. Multiple-scanning beam antenna systems also use GaAs MMICs to increase functional capability, to reduce volume, weight and cost, and to greatly improve system reliability. MMIC technology, including gigabit-rate GaAs digital ICs, offers substantial advantages in power consumption and weight over silicon technologies for high-throughput, on-board baseband processor systems [407].

In [408], the reliability of GaAs MMIC for power amplifiers is discussed, with the aim of determining some meaningful derating rules (in terms of temperature, current, voltage, power) applied to microwave circuits. The main degradation mechanisms of GaAs FETs are reviewed and a methodology for the evaluation of MMIC reliability (by high storage temperature and DC life-test results) based on test-vehicle definition (technological characterisation vehicle and dynamic evaluation circuit) and the well-suited life-test conditions is presented. The power-amplifier application is validated through life-test under dynamic electrical stress. A new degradation mechanism linked to the **multiplication ionisation phenomenon of carrier** by impact ionisation is brought to the fore. This effect is translated by a reduction of the output power and has a release threshold opposed to the cumulative phenomena activated by temperature (for example, **metal diffusion**). This explains why the operation of a power amplifier in a safe domain can be evaluated quickly.

5.7 Failure Modes and Mechanisms of Hybrid Technology

Hybrid technology aims to manufacture hybrid integrated circuits (HICs), which are miniaturised electronic circuits constructed from individual devices, such as active and passive electronic components, bonded to a substrate. The word hybrid means that this technique sits between a complete integration (monolithic ICs) and a combination of discrete elements placed on a PCB.

The basic principle is the same as for monolithic ICs; the difference between the two types of device is how they are constructed and manufactured. The advantage of HICs is the potential for including components that could not be integrated in monolithic ICs, such as capacitors of large value, wound components and so on.

On the other hand, HICs can be much more reliable than the corresponding circuits formed by distinct components on PCBs, due to the smaller number of soldering points, the more stable substrate, the greater resistance under mechanical stresses and the replacement of several cases by one single case.

Today, special elements in miniature form are available, carefully encapsulated, measured and selected, whose terminals can be *reflowed*. As well as transistors and ICs, tantalum and ceramic capacitors and high-frequency inductors are also available, all of them isolated and taking the desired form. The partitioned substrate is heated for a short time over the tinning temperature, until the solder becomes fluid (*reflow*). In this manner a very great number of reliable soldering points are made, in the shortest possible time, and through the fluid soldering surface a supplementary self-centring takes place.

There are two main techniques for manufacturing HICs: thin film and thick film, the difference being not so much the thickness of the film as the method of obtaining the conducting layers. These techniques are not rivals, but complementary to each other.

The hermetically sealed ceramic chip carrier package and the plastic small-outline (SO) IC package are both now widely used in hybrid microelectronics. These packages are currently being applied along with other discrete surface-mount devices to printed wiring board (PWB)[12] technology in order to increase circuit density and reduce weight. Two soldering techniques for the attachment of the chip carrier and SOIC packages to epoxy glass PWBs are evaluated in [409]: (i) by solder cream printing or pre-tinning of the circuit board and subsequent attachment by solder reflow and (ii) by a novel jet-soldering technique. The thermally induced expansion mismatch between epoxy glass and ceramic chip carriers and the strain-induced fatigue this causes in the solder joints are now well documented, and results have been presented for the effects of temperature-cycling ceramic chip carriers soldered on to PWBs. Various PWB materials have been assessed, including FR4, elastomer-coated FR4, polyimide kevlar and epoxy glass-laminated copper-clad invar. The reliability of these assemblies is discussed in terms of the appearance of micro-cracks in the solder fillets and the occurrence of electrical discontinuity in the solder joints during temperature cycling.

HICs are usually custom-built and nonstandard. Reliability considerations require knowledge of their technology, complexity and quality of manufacture, as well as their operating environment. Active hybrid microcircuits can be significantly more reliable than the equivalent assembly of discrete components for the following main reasons: (i) they reduce the number of electrical joints; (ii) they use a more stable substrate material; (iii) they have a greater resistance to mechanical stresses; and (iv) they replace a number of packages by a single one capable of providing a dry gas environment for all circuit elements. They can provide a wide range of electronic functions or subsystems on a single substrate, utilising the inherent reliability of semiconductor devices. Thus there is a wide range of complexity and reliability. A major reliability problem for a manufacturer of HICs is the difficulty of controlling the quality of add-on chip components bought externally. The manufacturer must also assure themself of the reliability of their wire-bonding over the full temperature range quoted. Thermal considerations of the overall circuit packaging are important, in order that hot spots be minimised.

5.7.1 Thin-Film Hybrid Circuits

The main generic bottlenecks identified in thin-film R&D are: (i) lack of understanding of adhesion mechanisms between thin-film/substrates and between different layers; (ii) lack of software for modelling and simulation of thin-film formulation and performance, resulting in long development times and expensive trial-and-error approaches; (iii) lack of equipment for meaningful and quick characterisation; and (iv) lack of understanding of industrial environments that require integration into existing production processes [410].

In [411], a case history of an FA performed on hybrid devices that exhibited electrical failures during thermal cycle with bias is presented. Electrical testing revealed resistive shorts between V_{CC} and ground bond pads on ICs within the package. A dendritic growth containing gold and copper extended from V_{CC} to ground across the surface of the passivation. Crystalline-like corrosion products were observed on the bond pads. The FM is shown to be an electrochemical corrosion reaction within the hybrid. Hybrid devices can contain reactive chemical species from outgassing of organic

[12] Another name for PCB.

adhesives, residual cleaning solvents and other process-related materials. These are present as adsorbed species and as gases within the package cavity. The constituents in the package ambient can react with the surface of wire bonds, bond pads and substrate metallisation to form dendritic growth and other corrosion products. The copper migration through reaction of the hybrid package atmosphere and copper oxide present on the surface of the thick-film gold-substrate metallisation was studied. Corrective actions are needed to alleviate the condition that led to the observed failures.

Aside from the problem of outgassing from organic die-attach materials, component-attach failures could be another cause of failure. A wide variety of substrate metallisation materials and combinations of materials are available to the hybrid designer. Several of these have been shown to be inherently unreliable. Others have been shown to be unreliable under certain conditions or when used in conjunction with some other material. There are, however, several FMs which will affect most materials. Lack of adhesion to the substrate, cracks, corrosion caused by various contaminants and shorts caused by particles are the most prevalent of these mechanisms. Particle-induced shorts are about the same in this respect.

The performance of a hybrid case-based reasoning (CBR) method that integrates a multi-layer perception (MLP) neural network with CBR for the automatic identification of FMs is analysed in [412]. The trained MLP neural network provides the basis for obtaining attribute weights, whereas CBR serves as a classifier to identify the FM. Different parameters of the hybrid methods are varied to study their effects. The results indicate that better performance can be (but is not always) achieved by the proposed hybrid method than by using conventional CBR or MLP neural networks alone.

One possible way to extend Moore's law [413] beyond the limits of transistor scaling is to obtain the equivalent circuit functionality using fewer devices or components; that is, to get more computing per transistor on a chip. One proposal for achieving this end was the hybrid CMOL (CMOS/molecule) architecture of Strukov and Lihkarev [414], which was modified by Snider and Williams [415] to improve its manufacturability and separate its routing and computing functions. This was called field-programmable nanowire interconnect (FPNI), which rather than relentlessly shrinking transistor sizes, separated the logic elements from the data-routing network by lifting the configuration bits, routing switches and associated components out of the CMOS layer and making them a part of the interconnect. Memristor[13] crossbars [416] can be fabricated directly above the CMOS circuits, and serve as the reconfigurable data-routing network. A 2D array of vias provides electrical connectivity between the CMOS and the memristor layer. Memristors are ideal for this field-programmable gate array (FPGA)-like application because a single device is capable of realising functions that require several transistors in a CMOS circuit, namely a configuration-bit flip-flop and associated data-routing multiplexer.

A successful implementation of the first memristor–CMOS HICs with demonstrated FPGA-like functionality was reported in [417]. The titanium dioxide memristor crossbars were integrated on top of a CMOS substrate using nanoimprint lithography (NIL) and processes that did not disrupt the CMOS circuitry in the substrate. This seemed to be the first demonstration of NIL on an active CMOS substrate that was fabricated in a commercial semiconductor fabrication facility. The successful integration shows that memristors and the enabling NIL technology are compatible with a standard logic-type CMOS process.

Another hybrid nano/CMOS reconfigurable architecture, named NATURE, consists of CMOS reconfigurable logic and interconnections and CNT-based nonvolatile on-chip configuration memory. Compared to existing CMOS-based FPGAs, NATURE increases logic density by more than an order of magnitude and offers cycle-by-cycle runtime reconfiguration capability, and is also compatible with mainstream photolithography fabrication techniques [418].

[13] The fourth element that completes the trinity of fundamental passive electronic components (the resistor, the capacitor and the inductor) is the memory resistor, or memristor (predicted by Chua in 1971 and realised only in 2008, by researchers from Hewlett-Packard laboratories).

Devices and circuits based on ultrathin nanograin polysilicon wire (polySiNW) dedicated to room-temperature-operated hybrid CMOS-'nano' ICs are developed and evaluated in [419]. The proposed polySiNW device is an FET in which the transistor channel is a nanograin polysilicon wire, and its operation (programmation) is controlled by a silicon-buried gate bias. The nanograin material is expected to offer interesting properties for single-electron memory applications. The second main objective of the research was to realise and qualify a CMOS-polySiNW hybrid circuit, which offers novel functionalities and/or outperforming characteristics compared with state-of-the-art CMOS. The original hybridisation approach allows new functionalities (using polySiNW original characteristics) to be brought out, as well as much higher current levels (provided by CMOS high-current drive) than traditional nanoelectronic devices.

5.7.2 Thick-Film Hybrid Circuits

The most significant factor influencing the future of thick-film hybrid (TFH) circuits comes from developments in new materials. The advent of the monolithic IC was largely responsible for pushing thick-film technology aside, and it has remained in the shadow of silicon technology since the early 1960s. Another reason for TFH technology falling out of favour is the poor performance of the early thick-film resistors. They were unstable and possessed many other undesirable characteristics. Modern ruthenium-based resistor compositions constitute one of the greatest assets of the technology, possessing low temperature coefficients, excellent stability and high reliability at a very low cost. The future of thick-film sensors looks set to expand into areas that could not have been contemplated in the early days of TFH technology. The creativity of the sensor designer will probably be the major factor influencing the boundaries within which the technology can be applied [420].

An advantage of the TFH is the possibility of obtaining (with the aid of various pastes) very different values of resistance (in practice, from $10\,\Omega$ to $10\,M\Omega$) in the same circuit. The bond wires on a TFH, which are protected by a glob top of epoxy resin, may be examined using X-ray radiography to look for breaks. A hybrid with several ICs and resistor networks was found to have failed as a result of the corrosion of the resistor tracks. In [421], FA on several hybrid devices is reported. The analysis was initiated because of electrical testing failures – a voltage divider compared the outputs of two thick-film resistors. Analysis of the hybrid failures revealed a copper-dendritic growth propagating across the laser trims in the thick-film resistor.

A method of in situ electromagnetic field (EMF) measurement, leakage current measurement and impedance spectroscopy was proposed for detecting spontaneous and forced blistering in thick-film multilayers during formation at high temperatures [422]. The occurrence of high-temperature shorts in Ag-dielectric-Ag multilayers under DC-bias is also detectable.

The reliability of adhesive FC attachments on TFH substrate was examined in [423]. Studied bumped test chips were attached to the ceramic TFH substrates using anisotropically conductive film (ACF) and nonconductive film (NCF). The reliability of the assemblies was assessed with thermal cycling and constant temperature/humidity testing. Based on the results, both the tested adhesives are well suited for FC attachments on TFH, although the NCF was substantially more susceptible to the bonding pressure than the ACF. The ACF adhesive samples were observed to outlast NCF ones with all used bonding pressures. Differences in reliability were due to NCF's low glass-transition temperature (T_g) and higher CTE values, which caused faster relaxation of the contact force, particularly in the temperature-cycling test.

5.8 Failure Modes and Mechanisms of Microsystem Technologies

Microsystems are miniaturised devices which perform non-electronic functions, such as sensing and actuation, fabricated by IC-compatible batch-processing techniques [424]. They integrate electrical components (e.g. capacitors, piezoresistors), mechanical components (e.g. cantilevers, microswitches), optical components (e.g. micromirrors) and fluidic components (e g. flow sensors). Initial synonyms for microsystems were 'MEMS', used mainly in the United States, and 'micromachines', in Japan. A typical microsystem contains, on the same chip, a microsensor, a microactuator (a mechanical component) and the necessary electronics, so one may say that a microsystem has 'eyes' (microsensor), 'arms' (microactuator) and a 'brain' (electronics) [425].

In recent years, microsystem technology (MST) has developed towards micro-optical electromechanical systems (MOEMSs), biological micro-electro mechanical systems (BioMEMSs, which handle cells or microbeads through fluidic streams, magnetic and electric fields, thermal gradients and so on) and chemical microsystems (sample pretreatment, separation and detection are built on microchips, this domain being known as microfluidics or lab-on-a-chip). There are *biomimetic microsystems* (built on the basic principles of living matter) and *intelligent microsystems* (with advanced intelligence capabilities, known also as smart systems) [426].

Many disciplines are involved in MST, employing hybrid terms such as mechatronics, chemtronics and bioinformatics, but the term *microtechnology* (technology of micro fabrication) seems to be the most appropriate [1]. In Europe, microtechnology includes both microelectronics (the 'classical' devices) and MST, with silicon still being the basic material. The major markets for MEMS are optical communications, bio technology, the car industry and RF applications. Many different MEMS sensors and actuators have also been implemented.

MST is used to create microsystems and nanodevices (due to be developed in the near future to become nanosystems). In this section, details about typical FMs of microsystems and nanodevices are presented.

5.8.1 Microsystems

Over the last few years, considerable effort has gone into the study of FMs in microsystems. Although still very incomplete, our knowledge of the reliability issues relevant to microsystems is growing. For example, an overview of the FMs commonly encountered in microsystems is provided in [427]. It focuses on the reliability issues of microscale devices, but briefly touches on the macroscopic counterparts of some. The paper discusses generic structures used in MEMS and their FMs, such as: stiction, creep, fatigue, brittle fatigue in silicon, wear, dielectric charging, breakdown, contamination and packaging. The FMs identified in various MEMS components and technologies (bulk and surface micromachining process technologies) are gathered in [428].

The reliability of MEMS-based systems might be increased by the use of a PoF methodology. This methodology, based on FA, necessitates the development of corresponding models that can be used for reliability evaluation [429]. A good knowledge of the effects of environmental influences on MEMS (e.g. temperature variations and irradiation) is required is order to develop behavioural models, which have a good ratio between simulation time and accuracy.

The well-known FMs of IC technology are supplemented by mechanisms specific to microsystems. In Tables 5.25 and 5.26, the most important FMs of microsystems are synthesised, grouped according to the source of failure risks: the material and the design and fabrication.

Stiction is an informal expression for 'static friction', describing the phenomenon that makes two parallel plates pressing against each other stick together. Some threshold of force is needed in order to overcome this static cohesion. The van der Waals forces are one possible cause of stiction. These are attractive or repulsive forces between molecules (other than covalent bonds or electrostatic interactions of ions with one another or with neutral molecules) and include: (i) forces between permanent dipoles; (ii) forces between permanent and induced dipoles; and (iii) forces between instantaneous-induced and induced dipoles. The hydrogen bond might also be responsible for stiction: this is the attractive force of the hydrogen between an electronegative atom of one molecule and an electronegative atom of another molecule. Other possible causes of stiction are the electrostatic forces and solid bridging.

If it arises during operation, stiction may lead to ESD, causing arcing between electrode surfaces and, eventually, microwelding. Humidity is a detrimental factor, changing surface properties

Table 5.25 FMs that depend on material quality

Failure mechanisms	Recommended corrective actions
Silicon crystal irregularities – after etching (deep reactive-ion etching, DRIE, a highly anisotropic etch process developed for MEMS, which creates deep, steep-sided holes and trenches in wafers) can create protuberances that may obstruct the moving elements of the microsystem, leading to failure. High-resolution X-ray diffraction methods such as the rocking-curve method and reciprocal space-mapping can monitor crystalline imperfection in single-crystal silicon devices.	Thermal annealing improves the crystal quality, removing some crystal irregularities.
Surface roughness – coupled with the limiting capacity of some microfabrication process such as DRIE for deep-trench etching in silicon substrates can introduce significant fitting problems during assembly.	Optimisation of DRIE may diminish the failure risks.
Mismatched CTEs between layers of thin films of dissimilar materials – may induce significant failure risks, especially after long-term thermal cycling.	Careful choice of the materials in contact.
Variations in material parameters (e.g. resistivity change in materials due to thermally induced resistance during operation).	Careful choice of material.
Internal stresses – This is a temperature-dependent process.	Temperature changes must be minimised by adequate process design.
Ageing – associated with polymers and plastics, which harden with time, resulting in a continuous change of material characteristics, leading to malfunction of the devices (e.g. malfunction of pressure microsensors using polymer protection coatings to silicon die).	Careful choice of material.
Clogging or build-up of material in strategic active regions – polymers and plastics continue to release gases after being sealed in packages. This is detrimental in microfluidics, making the device inoperable by obstructing or restricting the fluid flow.	Careful choice of material.

and favouring stiction between surfaces. It is possible to avoid stiction by using surface-assembly monolayers or by designing low-energy surfaces.

Another type of adhesion can arise between the biological molecular layer and the substrate (bio-adhesion); reduction of friction and wear of biological layers, biocompatibility and bio-fouling are also important in diminishing failure risks [430].

Fundamental studies of hot-switched DC (gold versus gold) and capacitive (gold versus silicon nitride) MEMS RF switch contacts were conducted in a controlled-air environment at MEMS-scale forces using a micro/nanoadhesion apparatus as a switch simulator [431]. Stiction increased during cycling at low current (1–10 μA) and was linked to the creation of smooth contact surfaces, increased van der Waals interaction and chemical bonding. Surface roughening by nanowire formation (which also caused shorting) prevented adhesion at high current (1–10 mA). Ageing of the contacts in air led to hydrocarbon adsorption and less adhesion. Studies of capacitive switches demonstrated that excessive adhesion leading to stiction was the primary FM and that both mechanical and electrical effects were contributing factors:

- The mechanical effect is stiction growth with cycling due to surface smoothening, which allows increased van der Waals interaction and chemical bonding.
- The electrical effect on stiction is due to electrostatic force associated with trapped parasitic charge in the dielectric, and was only observed after operating the switch at 40 V bias and above.

Table 5.26 FMs that depend on design and fabrication processes

Failure mechanisms	Recommended corrective actions
FMs specific to the *release of the suspended parts* of the microsystem (membranes, beams, etc.): (i) a *partial release* (meaning the suspended part is not totally released from the surrounding material), initiated by insufficient etching, oxide residuals that prevent adequate etching, slow etching rate due to inadequate solution or redepositions of etched materials; and (ii) *breakage of the part* due to mechanical rupture.	Use control and monitoring points on the manufacturing flow.
Some FMs are induced by *etching operation*: (i) *buckling* is the deformation induced by thermal strain, caused by residual stresses and observed in MEMS during etching of the underlying sacrificial layer; and (ii) *residual stresses*, which induce plastic deformation/displacement by relaxation.	Use control and monitoring points on the manufacturing flow.
TDDB, explained by a percolation model. An electric field applied to an oxide film causes the injection of holes into the oxide film to occur on the anode side, which consequently causes traps to appear in the oxide film. As the number of traps increases, an electric current through the traps is observed as a stress-induced leakage current (SILC) due to hopping or tunnelling. It has been reported that if the number of traps continues to increase and the traps connect between the gate electrode and the Si substrate, the connection carries a high current that causes the gate-oxide film to break down.	The level of the traps in an oxide film strongly influences TDDB, and it is necessary to characterise the oxide film quality using accelerated tests and feed the results into design rules. It is also important to use SiO_2 film, which does not easily produce defects, and to develop a method of forming an oxide film thereby.
Particle contamination (produced by environment pollution and alteration) may mechanically obstruct device motion, resulting in electrical short-out. Contamination can occur in packaging and during storage; for example, a particulate dust that lands on one of the electrodes of a comb drive can cause catastrophic failure.	Sources of particle contamination, such as residues left after fabrication and wear-induced debris or environmental contaminations must be eliminated.
Electromigration, which is specific to IC technology, may also arise for microsystems. It is caused at high-current densities by the gradual displacement of metals atoms, causing a change of conductor dimensions and, eventually, highly resistive spots and failure due to destruction of the conductor at these spots.	Appropriate design rules may diminish or eliminate such failure risks.
Stiction is typical for solid objects that are in contact during operation.	Stiction is detailed in the main text.
FMs related to the presence of *mechanical movement*, which introduces new classes of reliability issue that are not found in traditional devices, include cycled mechanical deformations and steady-state vibrations, which introduce new stress mechanisms on the structural parts of these devices.	Avoid mechanical relaxation of residual material stress, plastic deformations under large signal regime, creep formations and fatigue.

Moreover, the two effects are additive; however, the electrical effect was not present until the surfaces were worn smooth by cycling. Surface smoothening increases the electric field in the dielectric, which results in trapped charges, alterations in electrostatic force and higher adhesion.

An extensive study on stiction, reported in [432], has identified the possible causes: capillary effects due to changed environment conditions, electrostatic charge accumulation or redistribution within dielectric layers, and microwelding of metals due to DC or RF power. Electrical ohmic contacts occurring between two metallic surfaces can also suffer from stability problems due to cycling, resulting in changes in ohmic contact resistance. This might be the effect of surface contaminations, material transfer and erosion, or of surface changes due to absorption or oxidation.

The microsystem **packaging** is an important source of failure risks. Heat-transfer analysis and thermal management become more complex when packing different functional components into a tight space. The miniaturisation also raises issues such as coupling between system configurations and the overall heat dissipation to the environment. The configuration of the system shell becomes important for the heat dissipation from the system to the environment [433]. Heat-spreading in a thin space is one of the most important modes of heat transfer in microsystems. In general, strategies for heat transfer in a microsystem can be presented as: first, diffuse heat as rapidly as possible from the heat source; second, maximise the heat dissipation from the system shell to the environment.

There are basically three levels of packaging strategy that may be adaptable for MEMS packaging. These are: (i) die-level packaging, involving the passivation and isolation of the delicate and fragile devices, which are diced and wire-bonded; (ii) device-level packaging, involving connection of the power supply, signal and interconnection lines; and (iii) system-level packaging, integrating the microsystem with signal-conditioning circuitry or application-specific integrated circuits (ASICs) for custom applications. Today, the major barriers in MEMS packaging technology can be attributed to a lack of information and standards of materials and a shortage of cross-disciplinary knowledge and experience in the electrical, mechanical, RF, optics, materials, processing, analysis and software fields. Packaging design standards should be unified. Apart from certain types of pressure and inertial sensors used by the automotive industry, most MEMS devices are custom-built. A standardised design and packaging methodology is virtually impossible at this time because of the lack of data in these areas. However, the joint efforts of industry and academic and research institutions could develop sets of standards for the design of microsystems. The thin-film mechanics, including constitutive relations of thin-film materials used in the finite-element method (FEM) and other numerical analysis systems, need to be thoroughly investigated [434].

An example of the study of MEMS packaging is given in [435], where the FM of a thin-film nitride MEMS package is studied using an integrated test structure. The cause of the failure is investigated by advanced characterisation techniques and accelerated tests on the packaging material. The conclusion of this research was that PECVD silicon nitride is a proper sealing material for thin-film packaging due to its good sealing property. However, outgassing of this material, at elevated temperature, remains the main concern for reliability.

Operational and environmental stresses are responsible for a series of FMs, which are detailed below.

Mechanical or thermal fatigue may arise during cycling loading below the fracture stress (yield) of a material, being induced by the formation of surface micro-cracks and localised plastic deformations. *Mechanical fatigue* is the progressive damage of the material, leading to failure when the accumulation of defects becomes critical. The fatigue effect is evident in structures suffering from oscillations at hinges or elastic suspensions. Many sensors and actuators operating using thermal actuations are subject to thermal fatigue, depending on relevant temperature gradients and structural strain levels, resulting in thermal cycling, high temperature or creep. The structural integrity of many highly stressed components is crucial in avoiding their fracture collapse; the mechanical strength is the parameter used to define the material robustness that must be maximised; for example, micro-needles for bio-MEMS or thermal posts for micro-heat exchangers. When surfaces come in contact with one another the adhesion properties must be controlled in order to avoid contact failures [436]. Thermal stressing and relaxation caused by thermal variations can create material delamination and *thermal fatigue* in cantilevers. In large temperature changes, as experienced in outer space, bimetallic beams will also experience warping due to mismatched CTEs [430].

Vibration can cause failure either by inducing **surface adhesion** or by **fracture** of the device support structure. Long-term vibration will also contribute to **fatigue**. For example, MEMS microrelays (cantilevers[14] or optical switches) can cause the reversal of maximum stress in the structure, from tension to compression and vice versa. The amplitudes of these alternating stresses may be significantly intensified in the areas where stress-raisers are located. The device will be subjected to cyclic stresses with large magnitudes and this may lead to structural failure due to fatigue of the material. Vibration-induced fatigue is more likely in MEMS using polymers and plastics [437].

Fracture occurs when the load on a microdevice is greater than the strength of the material. Fracture is a serious reliability concern: for brittle materials, fracture immediately leads to catastrophic failures; for less brittle materials, repeated loading over a long period of time causes fatigue, which will also lead to the breaking or fracturing of the device. In principle, this FMo is relatively easy to observe and simple to predict if the material properties of thin films are known. Many microsystems operate near their thermal dissipation limit, where hot spots may cause failures, particularly in weak structures (diaphragms or cantilevers) [430].

Brittle fracture is a typical FM for digital micromirrors[15] used as optical switches and scanning devices, because in these applications, the beams supporting the micromirror are twisted and deformed to a large extent. In the experiment reported in [438], specimens suitable for microtesting were designed and an experimental procedure for pure bending and a combined (torsion and bending) loading test were performed on the samples. Then a fracture criterion based on the experimental data and a general safety design criterion (based on Bayesian reliability analysis) for microstructure under torsional loading were proposed.

Creep is an FM that arises after a long period of operation, being produced by the time-dependent mass transfer through glide and diffusion mechanisms introduced by high stress and stress gradients. In [439] a new methodology which uses a commercial nanoindenter coupling with electrical measurement on test vehicles specially designed to investigate the micro-contact reliability is presented. This study examines the response of gold contacts with $5\,\mu m^2$ square bumps under various levels of current flowing through contact asperities. A rise in contact temperature is observed, leading to shifts of the mechanical properties of the contact material, modifications of the contact topology and a diminution of the time-dependence creep effect. The data provides a better understanding of microscale contact physics, especially FMs due to the heating of the contact on MEMS switches.

Mechanical shock is a direct transfer of mechanical energy across the device. This can cause material **fracture**, **electrical shorting** or **stiction**.

Ambient pressure may affect MEMS characteristics. For example, in the case of thermally actuated MEMS, the presence of vacuum has a detrimental effect because of the reduced heat dissipation from the heater, which in turn increases the thermal resistance of the heater, resulting in thermal runaway [1].

As for IC devices, the radiation traps charged particles in dielectric layers, creating a permanent electric field. This field will change resonant characteristics and alter the output of many sensors. High levels of Z radiation[16] can lead to fracturing by creating massive disorder within the crystal lattice and severe performance degradation. Accelerated ageing induces some specific FMs, which are synthesised in [440]. Temperature and humidity are used to age vapour-deposited SAM-coated electrostatic-actuated MEMS devices. A critical factor in the long-term reliability of surfaces in contact with one another during storage is the stability of monolayer coupling agents applied during processing to reduce adhesion, which leads to stiction. These coatings are popular processing aids because they can be applied at the back end of the manufacturing line and thus have no impact on the fabrication process. The coatings are typically one molecule thick and as such do not modify the stress state

[14] A cantilever is a beam supported on only one end. Cantilever construction allows for overhanging structures without external bracing.

[15] The digital micromirror device (DMD) developed by Texas Instruments (TIs) before 1990, is a precise light switch. A mirror is rotated by electrostatic attraction produced by voltage differences developed across an air gap between the mirror and the underlying memory cell. Incident light is reflected and modulated by the tilt mirror.

[16] Z radiation is radiation emitted by elements with a large number of protons (the Z number) in their nucleus (e.g. heavy metals).

of the polycrystalline silicon layers. The adsorbed films are also self-limiting in thickness and can penetrate through the liquid or vapour phase to coat deeply hidden interfaces. The FM is linked to contacted surfaces and is dependent on both temperature and humidity, with longer life at both lower temperatures and lower humidity levels. A nice tutorial is presented in [441], which reviews some of the reliability aspects of various types of MEMS, such as:

- **Contact-related effects:** contact degradation, surface film formation, material property changes. This is the primary FM for MEMS switches; the two primary FMos are high contact resistance and contact stiction. Both of these FMos originate from changes at the contact surface and can be caused by mechanical and thermal/electrical stresses.
- **Actuator-related effects:** charge-trapping in dielectrics, flexure fatigue or fracture, stress changes in structural elements. Dielectric charge-trapping is the primary FM for membrane-style capacitive switches, where the use of insulating materials can lead to trapping of electrical charges in the interposed dielectric layer. This FMo is in general less severe in the ohmic contact switches during normal operation, but can become a big issue when the ohmic switches operate in an ionising environment such as space.
- **Environment/packaging-related effects:** humidity effects, outgassing and residues from die-attach and lid-attach processes. Packaging and environment are critical elements in life-cycle reliability for both switch types, due to the surface-dominated nature of MEMS devices. Humidity may also have a strong effect due to the scaling of capillarity forces at small dimensions. ESD events can also impair the reliability of microgap-based devices.
- **Radiation-related effects:** radiofrequency micro-electro-mechanical system (RF-MEMS) switch performances can be strongly impacted by ionising radiation. In particular, both the actuation voltage and the S-parameters are impaired. There are three main possible effects of radiation testing on MEMS: (i) single-event effects; (ii) total-ionising-dose effects; and (iii) non-ionising energy loss.

5.8.2 Nanosystems

Nanosystems are produced by nanotechnologies. There are two basic ways of developing nanotechnologies:

- **The top-down approach:** the dimensions of structures that are designed at micrometre scale are shrunk, and new devices are obtained.
- **The bottom-up approach:** nanoscale-structured blocks are assembled and synthesised by chemical processes or inspired by biology, and complex architectures are built, with new electronic or optical properties. The self-assembly of biological systems (e.g. DNA molecules) may be used to control the organisation of some species, such as carbon nanotubes (CNTs), leading in the future to the ability to grow new ICs instead of having to shrink the existing ones.

For the moment, developing controlled assembly strategies for integrating nanostructures and nanoarchitectures in electronic devices and circuits remains a long-term challenge. In the medium term, a hybrid strategy, combining the two approaches – top-down and bottom-up – seems to be the right solution.

The products of nanotechnologies are the nanomaterials and nanodevices. The *nanomaterials* are materials restructured at nano-level, with physical and chemical properties that are seldom different (improved) compared to the same properties at micro-level. Nanomaterials may enter the composition of *nanodevices*, which can be either the 'classical' type of electronic device (ICs, transistors, etc.) or electro-mechanical nanosystems, called nano-electro-mechanical-systems (by analogy with MEMS, which are at micro-level). The typical FMs of nanomaterials and nanodevices are presented below.

5.8.2.1 Nanostructured Materials

The nanomaterials are nano-objects with at least one dimension smaller than 100 nm. Some materials have been commercialised for many years without being known as nanomaterials, such as: nanoparticles of black carbon, silica precipitate, silica gels and carbonates. The second category of nanomaterials, those nanostructured as nanomaterials by design, is the object of this section. Some examples include nanoparticles (aluminum oxide, colloidal silica, zinc oxide), CNTs, quantum dots (QDs), nanowires and nonporous materials. In many cases (if not in all), the properties of a nanoscale material are different from those of the equivalent micro- or macroscale material.

Moreover, the nanostructure properties can be modulated under the action of some stress factors. This has been an important research goal in recent years. For instance, for *QDs* of InGaAS (with diameters of 20 nm) processed by a MEMS technique (being included in a air-bridge with a thickness of 0.11 μm), an external stress can be used to modulate the electronic states: by applying an electrostatic force, the number of carriers, the energy levels and the spin configuration can be manipulated [442].

Some environmental stresses can be detrimental. These form the object of reliability as a discipline. Some common environmental stresses acting on nanostructured materials are presented below.

CNTs have properties which make them extremely promising for nanoelectronics, because they undergo EM better than the majority of the conductors used in microelectronics (Al, Cu, Ag, etc.). However, existing problems with manipulating CNTs and the lack of accurate results from reliability tests still obstruct their use in commercial products today. Some papers claim CNTs can support current densities larger than 10^8 A/cm^2 without any problems, even at temperatures of 250 °C [443]; other papers report CNT failures after seconds of functioning at room temperature and similar current densities [444]. However, CNTs are among the most promising materials for the next generation of nanosystems.

The *nanowires* are possible elements for use in nanoelectronics. In January 2008, a group from Berkeley University reported the synthesis of silicon nanowires with electrical and thermal properties different form the bulk silicon. Their reliability is still under discussion [445]. The mechanical properties of metallic nanowires (polycrystalline copper) have been studied by a group from Georgia Institute of Technology, with this nanomaterial proving to be better than bulk copper, as verified by nanoindentation [446].

Last, but not least, *QD* is another material appropriate for use in nanoelectronics. In 2002, researchers from the Technical University in Berlin reported lifetimes greater than 3000 hours for laser diodes based on QD from InGaAs/GaAs, at output power of about 1.5 W and temperatures up to +50 °C [447]. But reliability studies on QDs refer only to the systems containing these nanostructures (e.g. [448]), with approaches specific to system reliability. Further studies of the reliability of QDs are needed, as they are basic elements in many systems.

5.8.2.2 Nanodevices

The design and fabrication of nanodevices is the subject of the discipline of *nanoelectronics*. Nanoelectronic architectures are created using special design rules which take into account the reliability of the future product. These are discrete and integrated devices with much smaller dimensions than those in microelectronics, and with their own specific problems, different from those at the micro-level [449]. An example is given by the *quantum structures*, which are in fact semiconductor devices with electrons confined in all three dimensions. Another example is represented by the *electromechanical systems at nano-level* (sensors + actuators + ICs), the so-called nano-MEMS or NEMS, which are similar to MEMS but smaller, and include devices for microfluidics, biocompatible devices and devices for biomedical applications [450]. In 2000, the first VLSI NEMS device was demonstrated by researchers from IBM.

The nanodevices are newcomers as electronic components and 'the reliability of nanodevices is still far from perfected', as noted in an excellent review on nano-reliability [451]. The degradation and FMs in microtechnologies (stiction, friction, wear) have different physical meanings at atomic and molecular scales. Certainly, the general reliability theory remains unchanged, but when it is used at nano-level, some adjustments must be made. The use of destructive testing continues [452], but nondestructive testing receives new valences at nano-level.

In NEMS metrology, the problem of reconstructing three-dimensional images at nano-level must be solved. At nano-level, the modelling of materials and structures must be conducted on new grounds. For nanodevices intended for biomedical applications, the reliability requirements are extremely severe. To pass from MEMS to NEMS implies the solution of complex issues about the degradation phenomena. For MEMS these issues are known in principle, but for NEMS new phenomena arise, due to:

- Modification of the physical and chemical properties for nanostructured materials (modelling of these material properties from the nanoscale to the final macroscopic form must be carried out).
- Transitory faults arising due to the diminishing of the noise tolerance at much lower levels of working current and voltage.
- Defects produced by ageing in using molecular techniques to create nanodevices.
- Manufacturing faults, which at nano-level become much more significant.

In FA at nano-level, the main problem is the lack of practical analytical tools. The major limitations and future prospects determined by industry roadmaps are discussed in [453], where the state-of-the-art microelectronic FA processes, instrumentation and principles are given. Specifically highlighted is the need for a fault-isolation methodology for FA of fully integrated nanodevices.

However, in recent years, in parallel with the effort to develop new tools for reliability analysis, various contributions to the reliability of nanodevices have been reported.

Bae *et al.* [454] provided basic physical modelling for MOSFET devices based on the nano-level degradation that takes place at defect sites in the MOSFET gate oxide. The authors investigated the distribution of hot-electron activation energies, and derived a logistic mixture distribution using physical principles on the nanoscale.

In [455], new FMs associated with breakdown in high-gate stacks on the case of Si oxides, or with oxynitrides of thickness ranging from some tens of nanometres down to about 1 nm, are detailed. In addition to DB-induced epitaxy commonly found in breakdowns in poly-Si/SiON and poly-Si/Si N MOSFETs, grain-boundary and field-assisted breakdowns near the poly-Si edge are found. The authors develop a model based on breakdown-induced thermo-chemical reactions to describe the physical microstructural damage triggered by breakdown in the high-gate stack and the associated post-breakdown electrical performance.

Hot-carrier reliability of gate-all-around twin-Si nano-wire FET (GAA TSNWFET) is reported and discussed in [456] with respect to the size and shape of the nano-wire channel, gate length, thickness and kind of gate dielectric. Smaller nano-wire channel size, shorter gate length and thinner gate oxide down to 2 nm thickness all show poorer hot-carrier reliability. The worst V_D for a 10-year guarantee, 1.31 V, satisfies the requirements of the ITRS roadmap.

References

1. Băjenescu, T.-M. and Bâzu, M. (2010) *Component Reliability for Electronic Systems*, Artech House, Boston and London.
2. Crow, L.H. (2006) Useful metrics for managing failure mode corrective action. Proceedings of Annual Reliability and Maintainability Symposium RAMS'06, pp. 247–252.
3. Goodman, D.L. (2000) *Prognostic Techniques for Semiconductor Failure Modes*, Internal Report of Ridgetop Group Inc., Tucson, AZ.
4. Hakim, E.B. (1989) *Microelectronics and Reliability*, Test and Diagnostics, Vol. 1, Artech House, Norwood.
5. Lee, J.H., Huang, Y.S. and Su, D.H. (2004) Wafer level failure analysis process flow, *Microelectronic Failure Analysis. Desk Reference*, 5th edn, ASM International, pp. 39–41.

6. Walton, A.J. and Smith, S. (2008) A review of test structures for characterising microelectronic and MEMS technology. *Adv. Sci. Technol.*, **54**, 356–365.
7. Schröpfer, G. *et al.* (2004) Designing manufacturable MEMS in CMOS compatible processes – methodology and case studies. SPIE's Photonics Europe, Conference 5455 – MEMS, MOEMS, and Micromachining, Strasbourg, France, April 26–30, 2004.
8. Dolphin Integration http://www.dolphin.fr/flip/teststructure/rycs.html. (Accessed 2010).
9. O'Sullivan, P. and Mathewson, A. (1993) Implications of a localized defect model for wafer level reliability measurements of thin dielectrics. *Microelectron. Reliab.*, **33** (11–12), 1679–1685.
10. Lee, J.-T., Won, J.-K. and Lee, E.-S. (2009) A study on the characteristics of a wafer-polishing process according to machining conditions. *Int. J. Precision Eng. Manuf.*, **10** (1), 23–28.
11. Wagner, D. (2004) Point defects in crystalline and amorphous silicon. *J. Optoelectron. Adv. Mater.*, **6** (1), 345–347.
12. Lazanu, S. and Lazanu, I. (2003) Role of oxygen and carbon impurities in the radiation resistance of silicon detectors. *J. Optoelectron. Adv. Mater.*, **5** (3), 647–652.
13. Henry, M.O. *et al.* (2001) The evolution of point defects in semiconductors studied using the decay of implanted radioactive isotopes. *Nucl. Instrum. Methods Phys. Res. B*, **18** (1–4), 256–259.
14. Vanvechten, J.A. and Wager, J.F. (1989) Point Defects in Semiconductors: Microscopic Identification, Metastable Properties, Defect Migration and Diffusion. Final Technical Report 31, Oregon State University, Corvallis. Department of Electrical and Computer Engineering, August 1986–March 1989.
15. O'Mara, W.C. *et al.* (1990) *Handbook of Semiconductor Silicon Technology*, William Andrew Publishing/Noyes.
16. Ravi, K.V. (1981) *Imperfections and Impurities in Semiconductor Silicon*, John Wiley & Sons, Inc.
17. Dubois, G., Volksen, W. and Miller, R.D. (2007) Spin-on dielectric materials, in *Dielectric Films for Advanced Microelectronics* (eds M. Baklanov, K. Maex and M. Green), John Wiley & Sons, Inc., New York, pp. 33–83.
18. Edelstein, D. *et al.* (2004) Reliability, yield, and performance of a 90 nm SOI/Cu/SiCOH technology. Proceedings IEEE International Interconnect Technology Conference, Burlingame, 7th June, 7–9, 2004, pp. 214–216.
19. Liu, J. *et al.* (2002) Porosity effect on the dielectric constant and thermomechanical properties of organosilicate films. *Appl. Phys. Lett.*, **81**, 4180–4182.
20. Hongxia, L., Yue, H. and Zhu, J. (2002) A study of hot-carrier-induced breakdown in partially depleted SIMOX MOSFETs. *J. Electron.*, **19** (1), 50–56.
21. Anon. (2000) Semiconductor Device Reliability Failure Models. International Sematech Technology Transfer No. 00053955A-XFR, http://www.sematech.org/docubase/document/3955axfr.pdf. (Accessed 2010).
22. Acovic, A., Rosa, G. and Sun, Y. (1992) A review of hot-carrier degradation mechanisms in MOSFETs. *Microelectron. Reliab.*, **36** (7), 845–869.
23. Shiyanovskii, Y. *et al.* (2010), Process reliability based trojans through NBTI and HCI effects, Proceedings of NASA/ESA Conference on Adaptive Hardware and Systems (AHS), 215–222.
24. Aresu, S. *et al.* (2004) Evidence for source side injection hot carrier effects on lateral DMOS transistors. *Microelectron. Reliab.*, **44** (9–11), 621–624.
25. Su, H. *et al.* (2007) Experimental study on the role of hot carrier induced damage on high frequency noise in deep submicron NMOSFETs. Proceedings of IEEE Radio Frequency Integrated Circuits (RFIC) Symposium, pp. 163–166.
26. Bez, R. *et al.* (2003) Introduction to flash memory. *Proc. IEEE*, **91** (4), 489–502.
27. Pagey, M.P. (2003) Hot-carrier reliability simulation in aggressively scaled MOS transistors, PhD thesis. Graduate School of Vanderbilt University, Nashville, TN.
28. Driussi, F. *et al.* (2000) Substrate enhanced degradation of CMOS Devices. Electron Devices Meeting, IEDM Technical Digest International, pp. 323–326.
29. Moens, P. *et al.* (2009) Hot carrier effects in trench-based integrated power transistors. Proceedings of IEEE International Physics Symposium, pp. 416–420.
30. Wittmann, R. (2007) Miniaturization problems in CMOS technology: investigation of doping profiles and reliability. PhD thesis. Technische Universität Wien, http://www.iue.tuwien.ac.at/phd/wittmann/diss.html. (Accessed 2010).
31. Rubaldo, R. *et al.*, Hot Carrier Reliability Improvement of PMOS I/O's Transistor in Advanced CMOS Technology, http://www.essderc2002.deis.unibo.it/data/pdf/Rubaldo.pdf. (Accessed 2010).

32. Wang, H.H.-C. *et al.* (2000) Hot carrier reliability improvement by utilizing phosphorus transient enhanced diffusion for input/output devices of deep submicron CMOS technology. *IEEE Electron. Device Lett.*, **21** (12), 598–600.
33. Kizilyalli, I.C. *et al.* (1998) Improvement of hot carrier reliability with deuterium anneals for manufacturing multilevel metal/dielectric MOS systems. *IEEE Electron. Device Lett.*, **19** (11), 444–446.
34. Ha, M.W. *et al.* (2007) Hot-carrier-stress-induced degradation of 1 kV AlGaN/GaN HEMTs by employing SiO2 passivation. Proceedings of the 19th International Symposium on Power Semiconductor Devices and IC's, ISPSD '07, pp. 129–132.
35. Miura, N. *et al.* (2002) TCAD driven drain engineering for hot carrier reduction of 3.3V I/O P-MOSFET. Proceedings of International Conference on Simulation of Semiconductor Processes and Devices, SISPAD, pp. 47–50.
36. Zhang, L. and Mitani, Y. (2006) A new insight into the breakdown mechanism in ultrathin gate oxides by conductive atomic force microscopy. IEEE 44th Annual International Reliability Physics Symposium, San Jose, pp. 585–589.
37. http://sirad.pd.infn.it/people/candelori/Tesi-Dottorato-PDF/TesiDottorato-Capitolo06.pdf. (Accessed 2010).
38. Li, X., Tung, C.H., Pey, K.L. and Lo, V.L. (2008) The chemistry of gate dielectric breakdown. Proceedings of IEEE International Electron Device Meeting, IEDM, pp. 1–4.
39. Pey, K.L. *et al.* (2004) Structural analysis of breakdown in ultrathin gate dielectrics using transmission electron microscopy. Proceedings of the 11th International Symposium on the Physics and Failure Analysis of Integrated Circuits, IFPA, pp. 11–16.
40. Tung, C.H. *et al.* (2005) Fundamental narrow MOSFET gate dielectric breakdown behaviors and their impacts on device performance. *IEEE Trans. Electron Devices*, **52** (4), 473–483.
41. Vogel, E.M. and Suehle, J.S. (2000) Degradation and breakdown of ultra-thin silicon dioxide by electron and hole injection. 31st IEEE Semiconductor Interface Specialists Conference, San Diego, December 7–9, 2000.
42. Pompl, T. *et al.* (2005) Change of acceleration behavior of time-dependent dielectric breakdown by the BEOL process: indications for hydrogen induced transition in dominant degradation mechanism. Proceedings of 43rd Annual IEEE Reliability Physics Symposium, San Jose, April 17–21, 2005, pp. 388–397.
43. Ohgata, K. *et al.* (2005) Universality of power-law voltage dependence for TDDB lifetime in thin gate oxide PMOSFETs. IEEE 43rd Annual International Reliability Physics Symposium, San Jose, pp. 372–376.
44. Changsoo, H. *et al.* (2005) Fast reliability evaluation of backend dielectrics using lifetime prediction techniques at wafer level. IEEE 43rd Annual International Reliability Physics Symposium, San Jose, pp. 588–589.
45. http://www.iirw.org/09/FEOL_tutorial_IRW2009_final_AK.pdf. (Accessed 2010).
46. Li, X. *et al.* (2009) Impact of gate dielectric breakdown induced microstructural defects on transistor reliability. *ECS Trans., Chall. ULSI Gate Dielectr.*, **22** (1), 11–25.
47. Zhang, H. and Solanki, R. (2001) Atomic layer deposition of high dielectric constant nanolaminates. *J. Elechrochem. Soc.*, **148** (4), F63–F66.
48. Tung, C.H. *et al.* (2006) Nanometal-oxide-semiconductor field-effect-transistor contact and gate silicide instability during gate dielectric breakdown. *Appl. Phys. Lett.*, **89** (2), 3 pp. doi:10.1063/1.2388242.
49. Keane, J. *et al.* (2008) An array-based test circuit for fully automated gate dielectric breakdown characterization. Proceedings of IEEE Custom Integrated Circuits Conference, pp. 121–124.
50. Lee, Y.-M. *et al.* (2003) Structural dependence of breakdown characteristics and electrical degradation in ultrathin RPECVD oxide/nitride gate dielectrics under constant voltage stress. *Solid-State Electron.*, **47**, 71–76.
51. Achanta, R.S., Plawsky, J.L. and Gill, W.N. (2008) A time dependent dielectric breakdown (TDDB) model for field accelerated low-k breakdown due to copper ions. Proceedings of the COMSOL Conference, Boston.
52. Balasubramanian, S. and Raghavan, S. (2008) Wet etching of heat treated atomic layer chemical vapor deposited Zirconium Oxide in HF based solutions. *Jpn. J. Appl. Phys.*, **47**, 4502–4504.
53. Sasse, G.T., Kuper, F.G. and Schmitz, J. (2008) MOSFET degradation under RF stress. *IEEE Trans. Electron Devices*, **55** (11), 3167–3174.
54. Meyer, M.A. *et al.* (2002) In Situ SEM observation of electromigration phenomena in fully embedded copper interconnect structures. *Microelectron. Eng.*, **64** (1–4), 375–382.
55. Schneider, G. *et al.* (2003) In situ X-ray microscopy studies of electromigration in copper interconnects. Proceedings of AIP 2003 International Conference on Characterization and Metrology for ULSI Technology, Vol. 683, pp. 480–484.
56. Kim, D.-Y. (2003) Study on reliability of VLSI interconnection structures. PhD thesis. Stanford University.
57. Ding. M. *et al.* (2005) A study of electromigration failure in Pb-Free solder joints. Proceedings of IEEE 43rd Annual International ReliabilityPhysics Symposium, San Jose, pp. 518–523.

58. Wilson, C.J. *et al.* (2006) Direct measurement of electromigration induced stress in interconnect structures. Proceedings of IEEE 44th Annual International Reliability Physics Symposium, San Jose, pp. 123–127.
59. Arnaud, L. *et al.* (2006) Analysis of electromigration voiding phenomena in Cu interconnects. Proceedings of IEEE 44th Annual International Reliability Physics Symposium, San Jose, pp. 675–676.
60. Zschech, E. *et al.* (2005) Electromigration-induced copper interconnect degradation and failure: the role of microstructure. Proceedings of the 12th International Symposium on the Physical and Failure Analysis of Integrated Circuits, IPFA, pp. 85–91.
61. Thompson, C.V. (2007) Void dynamics in Cu interconnects. Workshop on Electromigration Reliability, Vienna, September 2007.
62. Sukharev, V. *et al.* (2009) Microstructure effect on em-induced degradations in dual inlaid copper interconnects. *IEEE Trans. Device Mater. Reliab.*, **9** (1), 86–97.
63. Zschech, E. *et al.* (2009) Geometry and microstructure effect on EM-induced copper interconnect degradation. *IEEE Trans. Device Mater. Reliab.*, **9** (1), 20–30.
64. Lakatos, A. *et al.* (2010) Investigations of failure mechanisms at Ta and TaO diffusion barriers by secondary neutral mass spectrometry. *Vacuum*, **84**, 130–133.
65. Stoffler, D. *et al.* (2009) Scanning probe measurements and electromigration of metallic nanostructures under ultra-high vacuum conditions. Proceedings of Trends in NanoTechnology TNT, Barcelona, Spain.
66. von Hagen, J. (2000) New SWEAT method for fast, accurate and stable electromigration testing on wafer level. IEEE Integrated Reliability Workshop IRW Final Report, pp. 85–89.
67. Fiks, V.B. (1959) Directional atom-nucleus collisions in single crystals – a method of measuring the lifetimes of short-lived nuclei and investigating crystals, *Sov. Phys. Solid State*, **1**, 14.
68. Huntington, H.B. and Grone, A.R. (1961) Current induced marker motion in gold wires, *J. Phys. Chem. Solids*, **20**, 75.
69. Black, J.R. (1969) Electromigration failure modes in aluminium metallization for semiconductor devices. *Proc. IEEE*, **57**, 1587–1594.
70. Todorov, T.N. *et al.* (2001) Current-induced embrittlement of atomic wires. *Phys. Rev. Lett.*, **86**, 3606–3609.
71. Blech, I.A. (1976) Electromigration in thin aluminum films on titanium nitride, *J. Appl. Phys.*, **47**, 1203.
72. Federspiel, X. *et al.* (2005) Determination of the acceleration factor between wafer level and package level electromigration test. Proceedings of IEEE 43rd Annual International Reliability Physics Symposium, San Jose, pp. 658–659.
73. Tu, K.N. (2003) Recent advances on electromigration in very-large-scale-integration of interconnects. *J. Appl. Phys.*, **94** (9), 5451–5473.
74. Michael, N.L. *et al.* (2002) Mechanism of electromigration failure in submicron Cu interconnects. *J. Electron. Mater.*, **31** (10), 1004–1008.
75. Michael, N.L. *et al.* (2003) Electromigration failure in ultra-fine copper interconnects. *J. Electron. Mater.*, **32**, 988–993
76. Li, B. *et al.* (2005) Impact of via-line contact on Cu interconnect electromigration performance. Proceedings of IEEE 43rd Annual International Reliability Physics Symposium, San Jose, pp. 24–30.
77. Park, Y.-J. *et al.* (2005) Observation and restoration of negative electromigration activation energy behaviour due to thermo-mechanical effects. Proceedings of IEEE 43rd Annual International Reliability Physics Symposium, San Jose, pp. 18–23.
78. Cheng, Y.L. *et al.* (2007) Cu interconnect width effect, mechanism and resolution on down-stream stress electromigration. Proceedings of IEEE 45th Annual International Reliability Physics Symposium, Phoenix, pp. 128–133.
79. Ceric, H. *et al.* (2009) A comprehensive TCAD approach for assessing electromigration reliability of modern interconnects. *IEEE Trans. Device Mater. Reliab.*, **9** (1), 9–19.
80. Lee, K.-D. *et al.* (2006) Via processing effects on electromigration in 65 nm technology. IEEE 44th Annual International Reliability Physics Symposium, San Jose, pp. 103–106.
81. Aifantis, K.E. and Hackney, S.A. (2009) Morphological stability analysis of polycrystalline interconnects under the influence of electromigration. *Rev. Adv. Mater. Sci.*, **19**, 98–102.
82. Arijit, R. and Tan, C.M. (2008) Very high current density package level electromigration test for copper interconnects. *J. Appl. Phys.*, **103**, 7 pp. doi:10.1063/1.2917065.
83. Doyen. L. *et al.* (2008) Use of bidirectional current stress for in depth analysis of electromigration mechanism. Proceedings of IEEE 46th Annual International Reliability Physics Symposium, Phoenix, pp. 681–682.
84. Lin, M.H. *et al.* (2005) Copper interconnect electromigration behaviors in various structures and lifetime improvement by cap/dielectric interface treatment. *Microelectron. Reliab.*, **45** (7–8), 1061–1078.

85. Sutton, A.P. and Todorov, T.N. (2004) A Maxwell relation for current-induced forces. *Mol. Phys.*, **102**, 919–925.
86. Radhakrishnan, M.K. (2004) Device reliability and failure mechanisms related to gate dielectrics and interconnects. Proceedings of 17th International Conference on VLSI Design, pp. 805–808.
87. Kontos, D.K. *et al.* (2005) Interaction between electrostatic discharge and electromigration on copper interconnects for advanced CMOS technologies. Proceedings of IEEE 43rd Annual International Reliability Physics Symposium, San Jose, pp. 91–97.
88. White, M. *et al.* (2006) Product Reliability Trends, Derating Considerations and Failure Mechanisms with Scaled CMOS. IIRW Final Report, http://trs-new.jpl.nasa.gov/dspace/bitstream/2014/41122/3/06-3455.pdf. (Accessed 2010).
89. Wenbin, Z. *et al.* (2009) W-Plug via electromigration in CMOS process. *J. Semiconduct.*, **30** (5), 056001–056004.
90. Yokogawa, S. *et al.* (2008) Analysis of Al doping effects on resistivity and electromigration of copper interconnects. *IEEE Trans. Device Mater. Reliab.*, **8** (1), 216–221.
91. Lauerhaas, J.M. (2009) Suppression of Galvanic corrosion in advanced BEOL integration. *Solid State Technol.*, 12–15, www.fsi-intl.com/images/stories/pdf/1271.pdf
92. Wang, J.-P. *et al.* (2008) Effects of surface cleaning on stressvoiding and electromigration of Cu-damascene interconnection. *IEEE Trans. Device Mater. Reliab.*, **8** (1), 210–215.
93. Lin, M.H. *et al.* (2004) The improvement of copper interconnect electromigration resistance by cap/dielectric interface treatment and geometrical design. Proceedings of 42nd IEEE Annual International Reliability Physics Symposium, pp. 229–233.
94. Kazuyoski, U. *et al.* (2002) A high reliability copper dual-damascene interconnection with direct-contact via structure. AIP Conference Proceedings of the Sixth International Workshop on Stress-Induced Phenomena in Metallization, Vol. 612, April 2002, pp. 49–60.
95. Lemell, C. *et al.* (2007) On the Nano-Hillock formation induced by slow highly charged ions on insulator surfaces. *Solid State Electron.*, **51**, 1–7.
96. Ericson, F., Kristensen, N., Schweitz, J.-A. and Smith, U., (1991) A transmission electron microscopy study of Hillocks in thin aluminum films. *Surf. Interface Anal.*, **9** (1), 58–63.
97. Tan, S., Reed, M.L., Han, H. and Boudreau, R. (1996) Mechanisms of Etch Hillock formation. *J. Microelectromech. Syst.*, **5** (1), 66–72.
98. Sangchul, K. *et al.* (2007) Copper Hillock induced copper diffusion and corrosion behaviour in a dual damascene process. *Electrochem. Solid-State Lett.*, **10** (6), H193–H195.
99. Thomas, R.W. and Calabrese, D.W. (1983) Phenomenological observations on electromigration. IEEE Proceedings of IRPS, pp. 1–9.
100. Alers, G.B. *et al.* (2005) Stress migration and the mechanical properties of copper. Proceedings of 43rd Annual Reliability Physics Symposium, April 17–21, 2005, pp. 36–40.
101. Chang, C.H. *et al.* (1998) Grain orientation mapping of passivated aluminium interconnect lines by X-Ray micro-diffraction. AIP Conference Proceedings 449, pp. 424–426.
102. Puttlitz, A.F. *et al.* (1989) Semiconductor interlevel shorts caused by Hillock formation in Al-Cu metallization. *IEEE Trans. Comp., Hybrids, Manuf. Technol.*, **12** (4), 619–626.
103. Nazarpour, S. *et al.* (2009) Stress distribution and hillock formation in Au/Pd thin films as a function of aging treatment in capacitor applications. *Appl. Surf. Sci.*, **255** (22), 8995–8999.
104. Burton, B. (1998) A theoretical upper limit to coble creep strain resulting from concurrent grain growth. *J. Mater. Sci.*, **28** (18), 4900–4903.
105. Wei, T.B. *et al.* (2008) Hillocks and hexagonal pits in a thick film grown by HVPE. *Microelectron. J.*, **39** (12), 1556–1559.
106. Chaudhari, P., *et al.* (1977) Method of inhibiting Hillock formation in films and film thereby and multilayer structure therewith. US Patent No. 4,012,756, issued Mar. 15, 1977.
107. Mc Millan, L. and Shipley, R. (1976) Aluminum treatment to prevent hillocking. US Patent No. 3,986,897, issued Oct. 19, 1976.
108. Sato, K. *et al.* (1971) Hillock-free aluminum thin films for electronic devices. *Metalurg. Trans.*, **2**, 691–697.
109. Singh, A. (1985) A simple technique for eliminating Hillocks in integrated circuit metallization. *J. Vac. Sci. Technol. B*, **3** (3), 923–924.
110. Cadien, K.C. and Losee, D.L. (1984) A method for eliminating Hillocks in integrated-circuit metallizations. *Vac. Sci. Technol. B*, **2** (1), 82–83.

111. Wang, C., Irie, M. and Ito, T. (2000) Growth and characterization of Hillock-free high quality homoepitaxial diamond films. *Diamond Related Mater.*, **9** (9–10), 1650–1654.
112. Iwamura, E., Ohnishi, T. and Yoshikawa, K. (1995) A study of Hillock formation on Al-Ta alloy films for interconnections of TFT-LCDs. *Thin Solid Films*, **270** (1–2), 450–455.
113. Eungsoo, K., Kang, S.-H. and Lim, S.-K. (1994) Stress behavior of CVD-PSG films depending on deposition methods and hillock suppression. Symposium B2 from the 1994 MRS Fall Meeting.
114. Ueno, T. *et al.* (2008) Display unit – Hillocks is prevented when aluminium wiring is used in order to reduce line resistance in a display unit. Patent No. 20080272685, June 11, 2008.
115. Lee, Y.K. *et al.* Synthesis of Al-Y alloy films for ULSI metallization, in *Springer Proceedings in Physics, Polycrystalline Semiconductors II*, Vol. 54 (eds J.H. Werner and H.P. Strunk), Springer, http://www.polyse.de/pdf/1990/synthesis.pdf. (Accessed 2010).
116. Tokuda, N. *et al.* Hillock-Free heavily boron-doped homoepitaxial diamond films on misoriented (001) substrates. *Jpn. J. Appl. Phys.*, **46**, 1469–1470.
117. Kennedy, M.S. *et al.* (2007) The aging of metallic thin films: delamination, strain relaxation, and diffusion. *JOM*., **59** (9), 50–53, DOI: 10.1007/s11837-007-0117-1
118. Hideo, K. and Takefumi, I. (2001) Analysis of thin film delamination considering adhesive force of interface by boundary element method. *Nihon Kikai Gakkai Nenji Taikai Koen Ronbunshu*, **1**, 37–38.
119. Liu, P. *et al.* (2007) Finite element analysis of interface delamination and buckling in thin film systems by wedge indentation. *Eng. Fracture Mech.*, **7** (7), 1118–1125.
120. Kandula, S.S.V. *et al.* (2008) Dynamic delamination of patterned thin films. *Appl. Phys. Lett.*, **93**, 261902-1–261902-3.
121. Zhong, C. *et al.* (2001) A mechanical assessment of flexible optoelectronic devices. *Thin Solid Films*, **394** (1–2), 202–206.
122. Haixia, M. *et al.* (2008) Fracture, delamination, and buckling of elastic thin films on compliant substrates. Proceedings of the 11th Intersociety Conference on Thermal and Thermomechanical Phenomena in Electronic Systems, 2008, ITHERM, pp. 762–769.
123. US Fed News Service (2007) Dutch inventors develop thin film transistor device. *HighBeam Res.*, http://www.highbeam.com/doc/1P3-1373059821.html. (Accessed 2010).
124. Hara, T. and Balakumar, S. (2003) Innovation of the semiconductor technology. The latest topics of the CMP related technology. CMP of the copper plated wiring layer. *Electron. Parts Mater.*, **42**, 92–97.
125. Karabacak, T. and Lu, T.-M. (2004) Enhanced step coverage of thin films on patterned substrates by oblique angle physical vapor deposition. US Patent # 7244670, 22 May 2004.
126. Chiu, W.-C. and Tsu, S. (2003) Reduction of metal corrosion in semiconductor devices. US Patent 6515366, February 4, 2003.
127. Turner, T.E. (1989) Method of improving the corrosion resistance of aluminum contacts on semiconductors. US Patent 4818727, April 4, 1989.
128. Bâzu, M. (1994) A synergetic approach on the reliability assurance for semiconductor components. PhD thesis. Politechnica University of Bucharest, Romania.
129. Man, K.F. (1999) MEMS reliability for space applications by elimination of potential failure modes through testing and analysis. SPIE Proceedings Series, Santa Clara, CA, Vol. 3880, September 21, 1999, pp. 120–129.
130. Thomson, P. (1997) Chip-scale packaging. *IEEE Spectrum*, **34**, 36–43.
131. Czernohorsky, J. *et al.* (2007) Evaluation of the impact of solder die attach versus epoxy die attach in a state of the art power package. Proceedings of 13th International Workshop on Thermal Investigations of ICs and Systems Therminic, Budapest, Hungary.
132. Anon. (2007) Die Lifting, Siliconfareast.com, http://www.siliconfareast.com/die-lifting.htm. (Accessed 2010).
133. Winster, T., Borkowski, C. and Hobby, A. Wafer backside coating of die attach adhesives new method simplifies process, saves money, http://www.henkel.com/us/content_data/113841_WBCNewMethod.pdf. (Accessed 2010).
134. Holliday, R. Au-based Die Attach Materials, http://www.goldinnovationsblog.com/au-based-die-attach-materials. (Accessed 2010).
135. Huneke, T. *et al.* (2007) Film vs. Paste: die attach options for stacked die applications. *Global SMT Pack.*, 16–19.
136. Song, S.N., Tan, H.H. and Ong, P.L. (2005) Die attach film application in multi die stack package. Proceedings of 2005 Electronics Packaging Technology Conference, pp. 848–852.

137. Rosle, M.F. *et al.* (2009) Effect of manufacturing stresses to die attach film performance in quad flatpack no-lead stacked die packages. *Am. J. Eng. Appl. Sci.*, **2** (1), 17–24.
138. Polyak, Y. and Bastl, Z. (2009) XPS study of early stages of Al/Au interface formation. *Surf. Interface Anal.*, **41**, 830–833.
139. Minges, L.M. (1989) *Electronic Materials Handbook*, Packaging, Vol. 1, ASM International, p. 975.
140. Anon. (2008) Pushing the barriers of wafer level device integration to higher assembly speed. *EPP Europe*, October, 85–88.
141. Guo, Y., Lin, J.-K. and De Silva, A. (2002) Reliability evaluations of chip interconnect in lead-free solder systems. Proceedings of 52nd Electronic Components and Technology Conference, pp. 1275–1280.
142. Runibsztajn, S. *et al.* (2005) Development of novel filler technology for no-flow and wafer level underfill materials. *J. Electron. Pack.*, **127** (2), 77–86.
143. Blanchet, J.E. *et al.* (2008) Packaging reliability super chips. US Patent No. 7,348,792 B2, Mar. 25, 2008.
144. Quinones, H. and Babiarz, A. Electronic materials and packaging. EMAP 2000 International Symposium, pp. 398–405.
145. Johnson, S.C. (2009) Tessera working WLP, wafer-level optics. Semiconductor International, April 1, 2009.
146. Gao, G. *et al.* (2008) Low-cost Compliant Wafer-level Packaging Technology, Tessera Internal Report, 2008, http://www.tessera.com/technologies. (Accessed 2010).
147. Pathangey, B. *et al.* (2007) Application of TOFSIMS for contamination issues in the assembly world. *IEEE Trans. Device Mater. Reliab.*, **7** (1), 11–18.
148. Hoontrakul, P. *et al.* (2003) Understanding the strength of epoxy–polyimide interfaces for flip-chip packages. *IEEE Trans. Device Mater. Rel.*, **3** (4), 159–166.
149. Suhir, E., Lee, Y.C. and Wong, C.P. (2007) *Micro- and Opto-Electronic Materials and Structure: Physics, Mechanics, Design, Reliability, Packaging*, Springer, New York.
150. Bamal, M. *et al.* (2006) Performance comparison of interconnect technology and architecture options for deep submicron technology nodes. Proceedings of International Interconnect Technology Conference, pp. 202–204.
151. Heng, E. (2010) Nano-wafer-level packaging to revolutionise semiconductors. *Innov. – Mag. Res. Technol.*, **9** (1), www.innovationmagazine.com/innovation/volumes/v3n4/free/features2.shtml.
152. Stevens, C. Relay Failures Induced by the Growth of Tin Whiskers. A Case Study. http://nepp.nasa.gov/WHISKER/reference/tech_papers/stevens2001-relay-failures-induced-by-tin-whiskers.pdf. (Accessed 2010).
153. Savi, J. (2001) Reliability of electronic assemblies. 38th Annual Spring Reliability Symposium: Reliability and Safety, May, 2001.
154. LeBret, J.B. and Norton, M.G. (2003) Tin whiskers – a recurring industrial problem examined with electron microscopy. *Microsc. Microanal.*, **9** (Suppl. 2), 806–807.
155. Seelig, K. and Suraski, D. (2002) A study of lead contamination in lead-free electronics assembly and its impact on reliability. Proceedings of the IPC/JEDEC International Conference on Pb-free Electronic Assemblies, San Jose, CA, May 1–2, 2002, pp. 92–94.
156. Anon. (2002) Tin Whisker Risks, Appendix A, revision 1, 06/19/2002, Risks of Conductive Whiskers in High-reliability Electronics and Associated Hardware from Pure Tin Coatings, http://www.calce.umd.edu/tin-whiskers/TINWHISKERRISKS.pdf. (Accessed 2010).
157. Seelig, K. and Suraski, D. (2002) NASA Goddard Space Flight Tin Whisker Homepage. http://nepp.nasa.gov/whisker/2002. (Accessed 2010).
158. Qi, H., Ganesan, S. and Pecht, M. (2008) No-fault-found and intermittent failures in electronic products. *Microelectron. Reliab.*, **48** (5), 663–674.
159. http://www.speedlinetech.com/docs/Tin-Whiskers.pdf. (Accessed 2010).
160. McCoog, J.R. (1997) Commercial component integration plan for military equipment programs: reliability predictions and part procurement. Proceedings of the Annual Reliability and Maintainability Symposium, Philadelphia, January 13–16, 1997, pp. 100–110.
161. Chandrasekaran, A. (2003) Effect of encapsulant on high-temperature reliability of the gold wirebond – aluminum bondpad interface. Master of Science thesis. University of Maryland.
162. Lee, S.M. (2006) Filler-induced failure mechanism in plastic-encapsulated microelectronic packages. *Metals Mater. Int.*, **12** (6), 513–516.
163. Anon. (2006) Failure Mechanisms. Toshiba Internal Report, http://www.semicon.toshiba.co.jp/eng/product/reliability/device/failure/.(Accessed 2010).
164. Hillman, C., Castillo, B. and Pecht, M. (2003) Diffusion and absorption of corrosive gases in electronic encapsulants. *Microelectron. Reliab.*, **43** (4), 635–643.

165. Xie, J. and Pecht, M. (2001) Palladium-plated Packages; Creep Corrosion and its Impact on Reliability. Advanced Packaging, February 2001.
166. Ardebili, H. *et al.* (2002) A comparison of the theory of moisture diffusion in plastic encapsulated microelectronics with moisture sensor chip and weight-gain measurements. *IEEE Trans. Comp. Pack. Technol.*, **25**, 132–139.
167. Cai, X. *et al.* (2002) A study of moisture diffusion in plastic packaging. *J. Electron. Mater.*, **31**, 449–455.
168. Hu, S.J. and Cheang, F.T. (1990) Failure analysis of leakage current in plastic encapsulated packages. *J. Electron. Mater.*, **19** (11), 1319.
169. Kelly, G. (1999) *The Simulation of Thermomechanically Induced Stress in Plastic Encapsulated IC Packages*, Kluwer, Dordrecht.
170. Lea, C. and Tilbrook, D. (1993) Moisture induced failure in plastic surface mount packages. *Soldering Surf. Mount Technol.*, **1** (3), 30–34.
171. Fan, H. *et al.* (2005) An energy-based failure criterion for delamination initiation in electronic packaging. *J. Adhesion Sci. Technol.*, **19** (15), 1375–1386.
172. Kitano, K. *et al.* (1988) Analysis of package cracking during reflow soldering process. Proceedings of 26th International Reliability Physics Symposium, pp 90–95.
173. Yalamanchili, P. (1995) Evaluation of electronic packaging reliability using acoustic microscopy, PhD thesis, CALCE EPSC Graduate Student Thesis.
174. Lee, S.W.R. and Lau, D.C.Y. (2006) Fracture analysis on popcorning of plastic packages during solder reflow. Proceedings of the 16th European Conference of Fracture, Alexandroupolis, Greece, July 3–7, 2006.
175. Gannamani, R. and Pecht, M. (1996) An experimental study of popcorning in plastic encapsulated microcircuits. *IEEE Trans. Comp., Pack. Manuf. Technol. Part A*, **19**, 194–201.
176. Berta, R. (2007) Manufacturer's Corner; Dage X-ray: Popcorning. http://www.empf.org/empfasis/2007/Apr07/manf_corner-407.html. (Accessed 2010).
177. Gannamani, R. (1995) Assembly of PEMs without Popcorning. MS thesis. CALCE EPSC Graduate Student Thesis.
178. Shang, H. *et al.* (2008) Thermal performance of lead-free microelectronic BGA package with defects. Proceedings of 17th World Conference on Nondestructive Testing, Shanghai, China, October 25–28, 2008.
179. Arulvanan, P., Zhaowei, Z. and Xunqing, S. (2006) Effects of process conditions on reliability, microstructure evolution and failure modes of SnAgCu solder joints. *Microelectron. Reliab.*, **46** (2–4), 432–439.
180. Ong, K.C. *et al.* (2004) Dynamic materials testing and modeling of solder interconnects. Proceedings of 54th Electronic Components and Technology Conference 2004, Vol. 1, pp. 1075–1079.
181. Venkataraman, S. *et al.* (2005) Impact of proton irradiation on the RF performance of 0.12 pm, CMOS technology. Proceedings of 43rd Annual IEEE Reliability Physics Symposium, San Jose, CA, April 17–21, 2005, pp. 356–359.
182. Hanks, C.L. and Hamman, D.J. (1971) Radiation Effects Design Handbook, Section 1. Semiconductor Diodes. NASA Report. NASA-CR-1785.
183. Wang, X. *et al.* (2009) PF/EP/Nano-SiO2 composite paint for resistor. *Trans. Tianjin Univ.*, **15** (4), 283–287.
184. Cain, J.F., Hart, L. and McLean, J.R. (2003) Failures in thin metal film resistors – a case history. *Qual. Reliab. Eng. Int.*, **8** (2), 99–104.
185. Paulson, W.M. (1972) *Reliability of Thin Film Nichrome Resistors Used on Radiation Hardened Integrated Circuits*, Motorola.
186. Hahtela, O.M. *et al.* (2009) Atomic-layer-deposited alumina coating on thin-film cryoresistors. *IEEE Trans. Instrum. Meas.*, **58** (4), 1183–1187.
187. Gerke, D.R. and Ator, D. (2006) Embedded resistors and capacitors in organic and inorganic substrates. IEEE Aerospace Conference, 2006, Big Sky, MT.
188. Lee, T. *et al.* (2004) Characterization and reliability of TaN thin film resistors. Proceedings of IEEE 42nd Annual International Reliability Physics Symposium, Phoenix, pp. 502–508.
189. Brynsvold, R.R. and Manning, K. (2007) Constant-current stressing of SiCr-based thin-film resistors: initial 'Wearout' investigation. *IEEE Trans. Device Mater. Reliab.*, **7** (2), 259–269.
190. Robinett, W. *et al.* (2007) Demultiplexers for nanoelectronics constructed from nonlinear tunnelling resistors. *IEEE Trans. Nanotechnol.*, **6** (3), 280–290.
191. Shrivastava, A. *et al.* (2008) Thick film resistor failures. Proceedings of 34th International Symposium for Testing and Failure Analysis ISTFA.
192. Murthy, K.S.R.C. and Kumar, A.V. (1990) Failure mechanisms in bismuth ruthenate resistor systems. *J. Mater. Sci. Electron.*, **25** (1), 61–71.

193. Podda, S., Cassanelli, G., Fantini, F. and Vanzi, M. (2004) Failure analysis of RuO2 thick film chip resistors. *Microelectron. Reliab.*, **44** (9–11), 1763–1767.
194. Anon. (2008) The Capacitor Plague. Komp.com, http://209.85.129.132/search?q=cache:http://kkomp.com/archives/2186. (Accessed 2010).
195. Smith, P. (2003) Motherboard Capacitor Problem Blows Up. Silicon Chip, 176, http://www.siliconchip.com.au/cms/A_30328/article.html. (Accessed 2010).
196. Freeman, Y. *et al.* (2007) Reliability and critical applications of Tantalum capacitors. *Passive Comp. Ind.*, **9** (1), 22.
197. Vasina, P., Zednicek, T., Sikula, J. and Pavelka, J. (2002) Failure modes of Ta capacitors made by different technologies. *Microelectron. Reliab.*, **42** (6), 849–854.
198. Takada, D., Ishijima, M. and Nurajama, Y. (2006) Development of 'NeoCapacitor', a high-voltage functional polymer tantalum capacitor. *NEC Tech. J.*, **1** (5), 63–67.
199. Tonicello, F. (2005) Tantalum capacitors. *Eurocomp*, **8**, 2–7.
200. Prymak, J. (2003) Derating differences Ta/Ta-Polymer/Al-Polymer. CARTS 2003: 23rd Capacitor and Resistor Technology Symposium, San Antonio, March 31 – April 3, 2003,.
201. Post, H.A. *et al.* (2005) Failure mechanisms and qualification testing of passive components. *Microelectron. Reliab.*, **45** (9–11), 1626–1632.
202. Freeman, Y. *et al.* (2007) Reliability and critical applications of tantalum capacitors. Proceedings CARTS Europe 2007 Symposium, Barcelona, Spain, pp. 193–204.
203. Bonomo, W. *et al.* Failure Modes in Capacitors, http://www2.electronicproducts.com/Failure_modes_in_capacitors-article-farr_vishay_dec2007-html.aspx. (Accessed 2010).
204. Tarr, M. Failure Mechanisms in Ceramic Capacitor, http://www.ami.ac.uk/courses/topics/0179_fmcc/index.html. (Accessed 2010).
205. Carbone, R.A. and Roche, D.J. (2000) SMT ceramic capacitor failure mechanisms, isolation tools, techniques and analysis methods. Proceedings of the 26th International Symposium for Testing and Failure Analysis ISTFA 2000, Bellevue, pp. 347–353.
206. Wunderle, B. *et al.* (2007) Non-destructive failure analysis and modelling of encapsulated miniature SMD ceramic chip capacitors using thermal and mechanical loading. Proceedings of Therminic 2007, Budapest, Hungary, http://hal.archives-ouvertes.fr/docs/00/20/25/43/PDF/therm07104.pdf. (Accessed 2010).
207. Yang, S.J. *et al.* (2003) Reliability estimation and failure analysis of multilayer ceramic chip capacitors. *Int. J. Mod. Phys. B*, **17** (08–09), 1318–1323.
208. Blattau, N. and Hillman, C. (2006) Design guidelines for Ceramic Capacitors Attached with SAC Solder, DfR Solutions, College Park, http://www.dfrsolutions.com/uploads/publications/2006_DesignCeramCapSAC.pdf. (Accessed 2010).
209. Wang, Z. Sr. (2006) A study of low leakage failure mechanism of X7R multiple layer ceramic capacitor. Proceedings of 32nd International Symposium for Testing and Failure Analysis, Austin, November 2006, pp. 142–146.
210. Blattau, N., Barker, D. and Hillman, C. (2003) Design guidelines for preventing flex cracking failures in ceramic capacitors. Proceedings of the 23rd Capacitor and Resistor Technology Symposium, San Antonio, March 31 – April 3, 2003, pp. 156–162.
211. Donahoe, D.N. and Hillman, C.D. (2003) Failures in base metal electrode (BME) capacitors. Proceedings of the 23rd Capacitor and Resistor Technology Symposium, CARTS 2003, pp. 129–138.
212. Wenger, C. *et al.* (2009) Influence of the electrode material on HfO_2 metal-insulator-metal capacitors. *J. Vac. Sci. Technol., B*, **27** (1), 286–289.
213. Ng, C.H. *et al.* MIM capacitor integration for Mixed-Signal/RF applications. *IEEE Trans. Electron. Dev.*, **52** (7), 1399–1409.
214. Sheng, L. *et al.* (2008) Visualization and damage due to nano-dendrite defects in metal-insulator-metal capacitors. Proceedings of the IEEE 46th Annual International Reliability Physics Symposium, Phoenix, pp. 649–650.
215. Bing, M. *et al.* (2008) The role of carbon contamination in voltage linearity and leakage current in high-k metal-insulator-metal capacitors. *J. Appl. Phys.*, **104** (5), 054510–054510-8.
216. Martinez, V. *et al.* (2008) New insight into tantalum pentoxide metal-insulator-metal (MIM) capacitors: leakage current modeling, self-heating, reliability assessment and industrial applications. Proceedings of the IEEE 46th Annual International Reliability Physics Symposium, Phoenix, pp. 225–229.
217. Richard, M., Dean, T. and Delage, S. (2008) RF, DC, and reliability characteristics of Ta_2O_5 MIM capacitors. Proceedings of the 38th European Microwave Conference, EuMC 2008, Amsterdam, pp. 127–130.

218. Hung, C.C. *et al.* (2007) New understanding of metal-insulator-metal (MIM) capacitor degradation behaviour. Proceedings of the IEEE 45th Annual International Reliability Physics Symposium, Phoenix, pp. 630–631.
219. Martinez, V. *et al.* (2008) Impact of oxygen vacancies profile and fringe effect on leakage current instability of tantalum pentoxide metal-insulator-metal (MIM) capacitors. IEEE International Integrated Reliability Workshop Final Report 2008, IRW, pp. 21–24.
220. Ekanayake, S.R., Ford, M. and Cortie, M. (2004) Metal-insulator-metal (MIM) nanocapacitors and effects of material properties on their operation. *Mater. Forum*, **27**, 15–20.
221. Menou, N. (2004) FeRA M technology: reliability and failure mechanisms of elementary and integrated ferroelectric capacitors. PhD thesis. Université de Toulon.
222. Brennecka, G.L. et al. (2008) Multilayer thin and ultrathin film capacitors fabricated by chemical solution deposition. *J. Mater. Res.*, **23** (1), 176–181.
223. Anon. Transient Voltage Suppression Devices, Harris Suppression Products, DB450.
224. Yaacob, M.M. (2005) The behaviour and performance of metal oxide varistor under the application of multiple lightning impulses. PhD thesis. Universiti Teknology Malaysia.
225. Brown, K. (2004) Metal Oxide Varistor Degradation.
226. Ioannou, S.G. (2004) Comparative study of metal oxide varistors (MOVs) for failure mode identification. PhD thesis. University of South Florida.
227. Amicucci, G.L. and Mazzeti, C. (2004) Probabilistic method for reliability assessment of metal oxide varistors under lightning stress. *COMPEL Int. J. Comput. Math. Electr. Electron. Eng.*, **23** (1), 263–278.
228. Li, J. *et al.* (2003) The degradation of epoxy resin-coated ZnO varistors at elevated temperatures and ambient humidity conditions. *Active Passive Electron. Compon.*, **26** (4), 235–243.
229. Tada, T. (2008) The simulation of failure modes on ZnO varistors. *Trans. Inst. Electr. Eng. Jpn. B*, **128** (6), 860–870.
230. Jaroszewski, M. and Pospieszna, J. (2004) An assessment of ageing of oxide varistors exposed to pulse hazards using dielectric spectroscopy. Proceedings of the 2004 IEEE International Conference on Solid Dielectrics, Vol. 2, pp. 727–730.
231. Ramirez, M.A. *et al.* (2005) The failure analyses on ZnO varistors used in high tension devices. *J. Mater. Sci.*, **48**, 5591–5596.
232. Wang, X. and Xu, L.-J. (2007) Finite element model analysis of thermal failure in connector. *J. Zhejiang Univ. Sci. A*, **8** (3), 397–402.
233. Gao, J.C. and Zhang, J.G. (2002) Measurement of electrical charges carried by dust particles. Proceedings of the IEEE Holm Conference on Electrical Contacts, Orlando, pp. 191–196.
234. Zhang, J.G., Gao, J.C. and Feng, C. (2005) The "Selective" deposition of particles on electric contact and their effects on contact failure. Proceedings of the Fifty-First IEEE Holm Conference on Electrical Contacts, pp. 127–134.
235. Zhou, Y.-L. *et al.* (2007) Failure analysis of a kind of low power connector. *J. Zhejiang Univ. Sci. A*, **8** (3), 384–392.
236. Mroczkowski, R.S. (1993) Connector Design/Materials and Connector Reliability, AMP Incorporated Technical Paper.
237. Bahaj, A.S., James, P.A.B. and McBride, J.W. (2001) 17th European Solar Energy Conference and Exhibition, p. 630.
238. Peel, M. (2007) Gold flash Contacts – A Time for Technological Reflection, Internal Report, Contech Research Inc., November 24.
239. Lam, Y.-Z. *et al.* (2008) Displacement measurements at a connector contact interface employing a novel thick film sensor. *IEEE Trans. Compon. Pack. Technol.*, **31** (3), 566–573.
240. Swingler, J. and McBride, J.W. (2002) Fretting corrosion and the reliability of multicontact connector terminals. *IEEE Trans. Compon. Pack. Technol.*, **25** (4), 670–676.
241. Williamson, J.B.P., Greenwood, J.A. and Harris, J. (1956) The influence of dust particles on the contact of solids. *Proc. R. Soc. Lond. A Math. Phys. Sci.*, **237** (1211), 560–573.
242. Wang, D. and Xu, L.-J. (2007) Modeling of contact surface morphology and dust particles by using finite element method. *J. Zhejiang Univ. Sci. A*, **8** (3), 403–407.
243. Mueller, T. and Burns, N. (2009) Analysis and subsequent testing of cracked brass connector housing. *J. Fail. Anal. Prev.*, **9** (5), 466–469.
244. Stein, D., Hernandez, V. and Nailos, M.A. (2003) XRF correlation of board reseats due zo intermittent failures from the use of thin gold plating finish on the contact fingers. Proceedings of ISTFA, Santa Clara, pp. 125–130.

245. Xie, J.M. *et al.* (2004) Why gold flash can be detrimental to long-term reliability. *J. Electron. Pack.*, **126** (1), 37–41.
246. Ozarin, N. (2008) What's wrong with bent pin analysis, and what to do about it. Proceedings of the 2008 Annual Reliability and Maintainability Symposium, pp. 386–392.
247. Devin, A. (2002) Fibre optic connector failures, Connector Specifier, http://www.coastalcon.com/Failures.pdf. (Accessed 2010).
248. LeFevre, B.G. *et al.* (1993) Failure analysis of connector-terminated optical fibres: two case studies. *J. Lightwave Technol.*, **11** (4), 537–541.
249. Swingler, J., McBride, J.W. and Maul, C. (2000) Degradation of road tested automotive connectors. *IEEE Trans. Compon. Pack. Technol.*, **23** (1), 157–164.
250. Feng, C., Zhang, J.G., Luo, G. and Halkola, V. (2005) Inspection of the contaminants at failed connector contacts. Proceedings of the Fifty-First IEEE Holm Conference on Electrical Contacts, pp. 115–120.
251. Zhang, J.-G. (2007) Particle contamination, the disruption of electronic connectors in the signal transmission system. *J. Zhejiang Univ. Sci. A*, **8** (3), 361–369.
252. Rudner, V.I. (2006) Systematic Analysis of Induction Coil Failures. *Heat Treating Progress*, March/April, 21–26.
253. Silvus, S. (2007) Wire Ends Yield Failure Clues. Test & Measurement World, www.tmworld.com/article/323501-Wire_ends_yield_failure_clues.php.
254. Voldman, S.H. (2009) *ESD Failure Mechanisms and Models*, Chapter 7.9, John Wiley & Sons, Ltd, Chichester.
255. Lakshminarayanan, V. (2001) Failure Analysis Techniques for Semiconductors and Other Devices, p. 40, www.rfdesign.com. (Accessed 2010).
256. Anon. Varactor Diodes. Skyworks Solutions Application Note.
257. Obreja, V.V.N. (2007) The voltage dependence of reverse current of semiconductor PN junctions and its distribution over the device area. IEEE International Semiconductor Conference CAS, pp. 485–488.
258. Surface Mounted Glass Diodes. Handling Precautions, Rohm Semiconductors Specifications.
259. Varactor Diodes for VCO and Tunner, Toshiba Corporation, January 2008, http://www.toshiba.com/taec/components/docs/ProdBrief/Varactor_Diodes.pdf. (Accessed 2010).
260. Li Chuang, X. (2009) Electronic Technology Co., Ltd, http://www.aviation-lcelectronics.com/readnews.asp?id=19. (Accessed 2010).
261. Obreja, V.V.N. *et al.* (2005) Reverse leakage current instability of power fast switching diodes operating at high junction temperature. IEEE 36th Power Electronics Specialists Conference PESC '05, pp. 537–540.
262. Zener, C. (1934) A theory of electrical breakdown voltage of solid dielectrics. *Proc. R. Soc. London, Ser. A*, **145** (855), 523–529.
263. McKay, K.G. (1954) Avalanche breakdown in silicon. *Phys. Rev.*, **94** (4), 877–884.
264. Zhuang, Y. and Du, L. (2002) 1/f noise as a reliability indicator for subsurface zener diodes. *Microelectron. Reliab.*, **42** (3), 355–360.
265. Satoh, H. (2007) Transient temperature response of avalanche diodes against lightning surges. *Electron. Commun. Jpn. (Part II: Electron.)*, **72** (120), 33–39.
266. Huang, A.Q., Temple, V., Liu, Y. and Li, Y. (2003) Analysis of the turn-off failure mechanism of silicon power diode. *Solid-State Electron.*, **47** (4), 727–739.
267. Corvasce, C. (2006) Mobility and impact ionization in silicon at high temperature. PhD thesis. ETH Zurich.
268. Mahajan, S.V. (2006) Electro-thermal simulation studies of single-event burnout in power diodes. PhD thesis. Vanderbilt University, http://sameer-mahajan.com/research.aspx. (Accessed 2010).
269. Obreja, V.V.N. (2008) Transient surge voltage suppressors and their performance in circuit over-voltage protection. IEEE International Semiconductor Conference CAS 2008, Sinaia, Romania,October 13–15, pp. 321–324.
270. Schafft, H.A. and French, J.C. (1962) Second breakdown in transistors. *IRE Trans. Electron. Dev.*, **ED-9**, 129–136.
271. Snyder, M.E. (1988) Determination of the unstable states of the solid state plasma in semiconductor devices. PhD dissertation. Texas Tech University.
272. Menhart, S. (1988) Development of a second breakdown model for bipolar transistors. PhD thesis. Texas Tech University.
273. Domengès, B. *et al.* (2004) Comprehensive failure analysis of leakage faults in bipolar transistors. *Semicond. Sci. Technol.*, **19** (2), 191–197.

274. Goroll, M. *et al.* (2008) New aspects for lifetime prediction of bipolar transistors in automotive power wafer technologies by using a power law fitting procedure. *Microelectron. Reliab.*, **48** (8–9), 1509–1512.
275. Description of a SCR Controller, Control Concepts, http://www.ccipower.com/support/description.php Heatsink. (Accessed 4 March 2010).
276. Obreja, V. *et al.* (2005) The junction edge leakage current and the blocking I-V characteristics of commercial glass passivated thyristor devices. International Semiconductor Conference CAS, pp. 447–450.
277. Benmansour, A., Azzopardi, S., Martin, J.C. and Woirgard, E. (2006) Failure mechanism of trench IGBT under short-circuit after turn-off. *Microelectron. Reliab.*, **46**, 1700–1705.
278. Johnston, A.H. and Plaag, R.E. (1987) Models for total dose degradation of linear integrated circuits. *IEEE Trans. Nucl. Sci.*, **34**, 1474.
279. Fleetwood, D.M. *et al.* (1996) Radiation effects at low electric fields in thermal, SIMOX and bipolar base oxides. *IEEE Trans. Nucl. Sci.*, **43**, 2537.
280. Rax, B.G., Lee, C.I. and Johnston, A.H. (1998) Proton damage effects in linear integrated circuits. *IEEE Trans. Nucl. Sci.*, **45**, 2632.
281. Belaïd, M.A. *et al.* (2007) Reliability study of power RF LDMOS device under thermal stress. *Microelectron. J.*, **38** (2), 164–170.
282. Sinha, P. (2009) Contribution to failure mechanism driven qualification of electronic power devices and design guidelines for high temperature automative applications. PhD thesis. Institut Télécom-Télécom Bretagne.
283. Singh, P. (2004) Power MOSFET failure mechanisms. Proceedings of 26th Annual International Telecommunications Energy Conference, INTELEC, pp. 499–502.
284. De Souza, M.M. (2007) Design for reliability: the RF power LDMOSFET. *IEEE Trans. Dev. Mater. Reliab.*, **7** (1), 162–174.
285. Pecht, M. (2008) *Prognostics and Health Management of Electronics*, Wiley-Interscience, New York.
286. Tonti, W.R. (2008) MOS technology drivers. *IEEE Trans. Dev. Mater. Reliab.*, **8** (2), 406–415.
287. Koga, R. (1996) Single-event effect ground test issues. *IEEE Trans. Nucl. Sci.*, **43** (2, Part 1), 661–670.
288. Koga, R. *et al.* (1999) Single event burnout sensitivity of embedded field effect transistors. *IEEE Trans. Nucl. Sci.*, **46** (6), 1395–1402.
289. Walker, D.G. *et al.* (2000) Thermal characterization of single event burnout failure in semiconductor power devices. 16th Annual IEEE Semiconductor Thermal Measurement and Management Symposium, March 21–23, pp. 213–219.
290. Titus, J.L. and Wheatley, C.F. (2003) SEE characterization of vertical DMOSFETs: an updated test protocol. *IEEE Trans. Nucl. Sci.*, **50** (6, Part 1), 2341–2351.
291. Huard, V., Denais, M. and Parthasarathy, C. (2006) NBTI degradation: from physical mechanisms to modeling. *Microelectron. Reliab.*, **46** (1), 1–23.
292. Alam, M.A. and Mahapatra, S. (2005) A comprehensive model of PMOS NBTI degradation. *Microelectron. Reliab.*, **45** (1), 71–81.
293. Wang, Y. (2008) Effect of interface states and positive charges on NBTI in silicon-oxynitride pMOSFETS. *IEEE Trans. Dev. Mater. Reliab.*, **8** (1), 14–21.
294. Li, M.-F. *et al.* (2008) Understand NBTI mechanism by developing novel measurement technique. *IEEE Trans. Dev. Mater. Reliab.*, **8** (1), 62–71.
295. Kumar, M.J. *et al.* (2007) Impact of strain or Ge content on the threshold voltage of nanoscale strained-Si/SiGe bulk MOSFETs. *IEEE Trans. Dev. Mater. Reliab.*, **7** (1), 181–187.
296. Cho, S.-M. *et al.* (2008) High-pressure deuterium annealing effect on nanoscale strained CMOS devices. *IEEE Trans. Dev. Mater. Reliab.*, **8** (1), 153–159.
297. Li, X., Qin, J. and Bernstein, J.B. (2008) Compact modeling of MOSFET wearout mechanisms for circuit-reliability simulation. *IEEE Trans. Dev. Mater. Reliab.*, **8** (1), 98–121.
298. US Military Handbook DOD-HDBK-263.
299. Amerasekera, A. and Duvvury, C. (2002) *ESD in Silicon Integrated Circuits*, John Wiley and Sons, Ltd, Chichester.
300. Green, T. (1988) A review of EOS/ESD field failures in military equipment. Proceedings of the 10th EOS/ESD Symposium, pp. 7–14.
301. Carlson, J. (2009) Recommended ESD-CDM Target Levels, JEDEC Publication JEP 157, October 2009.
302. Bonfert, D. and Gieser, H. (1999) Transient induced latch-up triggered by very fast pulses. *Microelectron. Reliab.*, **39** (6–7), 875–878.
303. Becker, H.N., Miyahira, T.F. and Johnston, A.H. (2002) Latent damage in CMOS devices from single event latchup. *IEEE Trans. Nucl. Sci.*, **49** (6), 3009–3015.

304. Lin, I.-C. *et al.* (2004) Latchup test-induced failure within ESD protection diodes in a high-voltage CMOS IC product. Proceedings of 2004 Electrical Overstress/Electrostatic Discharge Symposium, September 19–23 2004.
305. Cheung, K.P. (2007) Advanced plasma and advanced gate dielectric – a charging damage prospective. *IEEE Trans. Dev. Mater. Reliab.*, **7** (1), 412–418.
306. Hansen, J.G. (2004) Design of CMOS cell libraries for minimal leakage currents. Master's thesis. Technical University of Denmark.
307. Yuan, X. *et al.* (2008) Gate-induced-drain-leakage current in 45-nm CMOS technology. *IEEE Trans. Dev. Mater. Reliab.*, **8** (3), 501–508.
308. Pompl, T. *et al.* (2006) Practical aspects of reliability analysis for IC designs. DAC 2006, San Francisco, July 24–28, 2006, pp. 193–198.
309. Segura, J. *et al.* (2002) Parametric failures in CMOS ICs – a defect-based analysis. International Test Conference 2002 (ITC'02), Baltimore, October 7–10.
310. Narasimham, B. *et al.* (2008) Quantifying the reduction in collected charge and soft errors in the presence of guard rings. *IEEE Trans. Dev. Mater. Reliab.*, **8** (1), 203–209.
311. Amusan, O.A. *et al.* (2009) Mitigation techniques for single-event-induced charge sharing in a 90-nm bulk CMOS process. *IEEE Trans. Dev. Mater. Reliab.*, **9** (2), 311–317.
312. Lu, Z. *et al.* (2004) Analysis of Temporal and Spatial Temperature Gradients for IC Reliability. Technical Report CS-2004-08, University of Virginia Department of Computer Science.
313. Suehle, J., Vogel, E. and Green, M. Device Design and Characterization Program, http://www.eeel.nist.gov/omp/Portfolio%202005%20Files%20for%20web/8_Device_Design_and_Characterization.pdf. (Accessed 2010).
314. Gusev, E.P. *et al.* (2006) Advanced high-j dielectric stacks with PolySi and metal gates: recent progress and current challenges. *IBM J. Res. Dev.*, **50** (4/5), 387–410.
315. Sim, J.H. *et al.* (2004) Hot carrier reliability of HfSiON NMOSFETs with poly and TiN metal gate. 62nd Device Research Digest, Vol. 1, pp. 99–100.
316. Okada, K. *et al.* (2006) Mechanism of gradual increase of gate current in high-K gate dielectrics and its application to reliability assessment. IEEE 44th Annual International Reliability Physics Symposium, San Jose, pp.189–194.
317. Park, H. *et al.* (2005) Effect of high pressure deuterium annealing on electrical and reliability characteristics of MOSFETs with high-k gate dielectric. IEEE 43th Annual International Reliability Physics Symposium, San Jose, pp. 646–647.
318. Magnone, P. *et al.* (2009) $1/f$ Noise in drain and gate current of MOSFETs with high-k gate stacks. *IEEE Trans. Dev. Mater. Reliab.*, **9** (2), 180–189.
319. Sato, M. *et al.* (2008) Fabrication process controlled pre-existing and charge-discharge effect of hole traps in nbti of high-K/metal gate pMOSFET. IEEE 46th Annual International Reliability Physics Symposium, Phoenix, pp. 655–656.
320. Semenov, O. *et al.* (2004) Evaluation of STI degradation using temperature dependence of leakage current in parasitic STI MOSFET. *Microelectron. Reliab.*, **44**, 1751–1755.
321. Shih, J.R. *et al.* (2004) Pattern density effect of trench isolation-induced mechanical stress on device reliability in sub-0.1 μm technology. IEEE 42th Annual International Reliability Physics Symposium, Phoenix.
322. Balasubramanian, A. *et al.* (2008) Implications of dopant-fluctuation-induced Vt variations on the radiation hardness of deep submicrometer CMOS SRAMs. *IEEE Trans. Dev. Mater. Reliab.*, **8** (1), 135–144.
323. Ho, P.S. *et al.* (1995) Thermal stress and relaxation behaviour of Al(Cu) submicron interconnects. Proceedings of the 4th International Conference on Solid-state and Integrated-Circuit Technology, Beijing, China, October 24–28, 1995, pp. 408–412.
324. Băjenescu, T.I. (1997) Status and trends of microprocessor design. Proceedings of the 1997 International Semiconductor Conference, 20th Edition, Sinaia, Romania, October 7–11, 1997.
325. Cristoloveanu, S. and Li, S.S. (1995) *Electrical Characterization of Silicon-on-insulator Materials and Devices*, Kluwer Academic Publishers, Boston, MA.
326. O'Leary, W. (1998) IBM Advances Chip Technology With Breakthrough For Making Faster, More Efficient Semiconductors, IBM Press releases.
327. Sim, J.H. *et al.* (2003) Dual work function metal gates using full nickel silicidation of doped poly-Si. *IEEE Electron. Dev. Lett.*, **24** (10), 631–633.
328. Liao, W.-S. *et al.* (2008) PMOS hole mobility enhancement through sige conductive channel and highly compressive ILD-SiNx Stressing layer. *IEEE Electron. Dev. Lett.*, **29** (1), 86–88.

329. Zhao, W. *et al.* (2004) Partially depleted SOI MOSFETs under uniaxial tensile strain. *IEEE Trans. Electron. Dev.*, **51** (3), 317–323.
330. Yeh, W.-K. *et al.* (2009) The impact of strain technology on FUSI gate SOI CMOSFET. *IEEE Trans. Dev. Mat. Reliab.*, **9** (1), 74–79.
331. Semenov, O. *et al.* (2006) Impact of self-heating effect on long-term reliability and performance degradation in CMOS circuits. *IEEE Trans. Dev. Mater. Reliab.*, **6** (1), 17–27.
332. Yang, C.-A. (2006) *Lifetime Test for Optical Transmitters in the ATLAS Liquid Argon Calorimeter Readout System*, SMU Physics Preprint SMU-HEP-05-13, Southern Metodist University.
333. Noh, Y.Y. (2005) Organic thin film phototransistors: materials and mechanism. Proceedings of SPIE, Vol. 5724, Bellingham, pp. 172–182.
334. Lee, M., Hillman, C. and Kim, D. (2005) Industry News: How to Predict Failure Mechanisms in LED and Laser Diodes. Military & Aerospace Electronics, pp. 1–5 www.militaryaerospace.com/index/display/article-display/230223/articles/military-aerospace-electronics/volume-16/issue-6/departments/electro-optics-watch/industry-news-how-to-predict-failure-mechanisms-in-led-and-laser-diodes.html.
335. Levada, S. *et al.* (2006) High brightness InGaN LEDs degradation at high injection current bias. Proceedings of the 44th IEEE Annual International Reliability Physics Symposium, San José, pp. 615–616.
336. Moreno, I. *et al.* (2010) Light-emitting diode spherical packages: an equation for the light transmission efficiency. *Appl. Opt.*, **49** (1), 12–20.
337. Deshayes, Y. *et al.* (2008) Selective activation of failure mechanisms in packaged double-heterostructure light emitting diodes using controlled neutron energy irradiation. *Microelectron. Reliab.*, **48** (8–9), 1354–1360.
338. Kumar, M.J. *et al.* (2010) Guest editorial; special issue on light-emitting diodes. *IEEE Trans. Electron. Dev.*, **57** (1), 7–11.
339. Tsukazaki, A. *et al.* (2005) Blue light-emitting diode based on ZnO. *Jpn. J. Appl. Phys.*, **44** (21), L643–L645.
340. Comizzoli, R.B. *et al.* (2001) Failure mechanism of avalanche photodiodes in the presence of water vapor. *J. Lightwave Technol.*, **19**, 252.
341. Laird, J.S. *et al.* (2005) Non-linear charge collection mechanisms in high-speed communication avalanche photodiodes. *Nucl. Instrum. Methods Phys. Res. A*, **541** (1–2), 228–235.
342. Djaja, S. *et al.* (2003) Implementation and testing of fault-tolerant photodiode-based active pixel sensor (APS). Proceedings of the 18th IEEE International Symposium on Defect and Fault Tolerance in VLSI Systems, p. 53.
343. Mawatari, H. *et al.* (1998) Reliability of planar waveguide photodiodes for optical subscriber systems. *J. Lightwave Technol.*, **16** (12), 2428.
344. Ishimura, E. *et al.* (2007) Degradation mode analysis on highly reliable guardring-free planar InAlAs avalanche photodiodes. *J. Lightwave Technol.*, **25** (12), 3686–3693.
345. Joo, H. *et al.* (2006) Experimental observation of the post-annealing effect on the dark current of InGaAs waveguide photodiodes. *Solid-State Electron.*, **50** (9–10), 1546–1550.
346. Joo, H.S. *et al.* (2005) Reliability of InGaAs waveguide photodiodes for 40-Gb/s optical receivers. *IEEE Trans. Dev. Mater. Reliab.*, **5** (2), 262–267.
347. Yun, I. *et al.* (1995) Reliability assessment of multiple quantum well avalanche photodiodes. IEEE International Reliability Physics Symposium, Las Vegas, April 4–6, 1995, pp. 200–204.
348. Park, S. *et al.* (2006) Temperature, current, and voltage dependencies of junction failure in PIN photodiodes. *ETRI J.*, **28** (5), 555–560.
349. Flint, S. (1980) Failure Rates for Fibre Optic Assemblies. Final Technical Report. IIT Research Institute, Chicago, IL.
350. Gilard, O. *et al.* (2007) Bipolar phototransistors reliability assessment for space applications. Proceedings of IEEE Radiation Effects Data Workshop, pp. 85–91.
351. Bajenesco, T.I. Aging problem of optocouplers. Proceedings of the 7th Mediterranean Electrotechnical Conference, MELECON, Vol. 2, pp. 571–574.
352. Johnston, A.H. and Miyahira, T.F. (2005) Hardness assurance methods for radiation degradation of optocouplers. *IEEE Trans. Nucl. Sci.*, **52** (6), 2649–2656.
353. Stawarz-Graczyk, B. *et al.* (2008) The noise macromodel of an optocoupler including 1/f-noise source. *Bull. Pol. Acad. Sci. Techn. Sci.*, **56** (1), 59–64.
354. Jevtić, M.M. (2007) Low frequency noise as a tool to study optocouplers with phototransistors. *Microelectron. Reliab.*, **44** (7), 1123–1129.

355. Elliott, S., Gordon, J. and Plourde, P. (2003) Indium Tin Oxide (ITO) film removal technique for failure analysis on packaged optoelectronic devices. Proceedings of the 29th International Symposium for Testing and Failure Analysis, Santa Clara, November 2–6, 2003, pp. 431–435.
356. Hwang, N., Naidu, P.S.R. and Trigg, A. (2003) Failure analysis of plastic packaged optocoupler light emitting diodes. Proceedings of 5th Electronics Packaging Technology Conference, EPTC, pp. 346–349.
357. Miyahira, T.F. and Johnston, A.H. (2002) Trends in optocoupler radiation degradation. *IEEE Trans. Nucl. Sci.*, **49** (6), 2868–2973.
358. Johnston, A.H. *et al.* (1998) Breakdown of gate oxides during irradiation with heavy ions. *IEEE Trans. Nucl. Sci.*, **45** (6), 2500.
359. Kitagawa, K. *et al.* (1991) Mechanism of misalignment failure of liquid-crystal display devices. *IEEE Trans. Reliab.*, **40** (3), 296–301.
360. Blattau, N. and Hillman, S. (2004) Failure MECHANISM in Electronic Products at High Altitudes, DfR Solutions, http://www.dfrsolutions.com/uploads/publications/2004_HighAltitude_Hillman-Blattau.pdf. (Accessed 2010).
361. Kwon, S. *et al.* (2007) Investigation of the failure of a liquid crystal display panel under mechanical shock. *Proc. Inst. Mech. Eng. Part C*, **221** (11), 1475–1482.
362. Crawford, G.P. (2005) *Flexible Flat Panel Display*, John Wiley & Sons, Ltd.
363. de Goede, J. *et al.* (2004) Failure of brittle functional layers in flexible electronic devices. *Mater. Res. Soc. Symp. Proc.*, **854**, 190–195.
364. Osada, T. *et al.* (2010) Development of liquid crystal display panel integrated with drivers using amorphous in–Ga–Zn-oxide thin film transistors. *Jpn. J. Appl. Phys.*, **49**, 03CC02.
365. Tsuji, S. *et al.* (2004) Application of focused ion beam techniques and transmission electron microscopy to thin-film transistor failure analysis. *J. Electron. Microsc.*, **53** (5), 465–470.
366. Choi, S.-H. *et al.* (2010) Reliability in short-channel p-type polycrystalline silicon thin-film transistor under high gate and drain bias stress. *Jpn. J. Appl. Phys.*, **49**, 03CA04-1–03CA04-6.
367. Lee, M. and Pecht, M.G. (2004) Thermal assessment of glass-metal composition plasma display panels using design of experiments. *IEEE Trans. Compon. Pack. Technol.*, **27** (1), 210–216.
368. Jeong, J.S. *et al.* (2004) Electrical overstress failure mechanism investigation about high-power scan driver IC for plasma display panel. *IEIC Techn. Rep.*, **104** (153), 95–98.
369. Quintana, M.A., King, D.L., Mcmahon, T.J. and Osterwald, C.R. (2002) Commonly observed degradation in field-aged photovoltaic modules. Conference Record of the 29th IEEE Photovoltaic Specialists Conference No. 29, pp. 1436–1439.
370. Abdelal, G.F. and Atef, A. (2008) Thermal fatigue analysis of solar panel structure for micro-satellite applications. *Int. J. Mech. Mater. Des.*, **4** (1), 53–62.
371. Green, M.A. *et al.* (2004) Crystalline silicon on glass (CSG) thin-film solar cell modules. *Solar Energy*, **77**, 857–863.
372. Grätzel, M. (2003) Dye-sensitized solar cells. *J. Photochem. Photobiol. C Photochem. Rev.*, **4**, 145–153.
373. Greijer, A.H., Lindgren, J. and Hagfeldt, A. (2003) Degradation mechanisms in a dye-sensitized solar cell studied by UV-VIS and IR spectroscopy. *Solar Energy*, **5** (2), 169–180.
374. Zhang, X.-T. *et al.* (2007) Investigation of the stability of solid-state dye-sensitized solar cells. *Res. Chem. Intermediat.*, **33** (1–2), 5–11.
375. Myers, B., Bernardi, M. and Grossman, J.C. (2010) Three-dimensional photovoltaics. *Appl. Phys. Lett.*, **96** (7), 1902.
376. Krebs, F.C. and Norrman, K. (2007) Analysis of the failure mechanism for a stable organic photovoltaic during 10000~hours of testing. *Prog. Photovoltaics Res. Appl.*, **15** (8), 697–712.
377. Khlil, R. *et al.* (2004) Soft breakdown of MOS tunnel diodes with a spatially non-uniform oxide thickness. *Microelectron. Reliab.*, **44** (3), 543–546.
378. Vogt, A. *et al.* (1997) Characteristics of degradation mechanisms in resonant tunnelling diodes. *Microelectron. Reliab.*, **37** (10), 1691–1694.
379. Jeppsson, B. and Marklund, I. (1967) Failure mechanisms in gun diodes. *Electron. Lett.*, **3** (5), 213–214.
380. Mojzes, I. *et al.* (1989) Comparative reliability study of n+-n and n+-n-n+ gunn diodes. *Microelectron. Reliab.*, **29** (2), 131–132.
381. Lutz, J. and Feller, M. (2007) Some aspects on power cycling induced failure mechanisms. Annual Report 2007. Center for Microtechnologies, Chemnitz University of Technology, p. 77.
382. Staecker, P. (1973) Ka-band impatt diode reliability. International Electron Devices Meeting 1973, Vol. 19, Washington, DC, December 4–5, 1973, pp. 493–496.

383. Peck, D.S. and Zierdt, C.H. Jr. (1974) The reliability of semiconductor devices in the bell system. *Proc. IEEE*, **62** (2), 185–211.
384. Lutz, J. and Domeij, M. (2003) Dynamic avalanche and reliability of high voltage diodes. *Microelectron. Reliab.*, **43** (4), 529–536.
385. Henning, J., Ward, A. and Kierstead, P. (2008) The new standard for high power semiconductors. *Power Electron. Eur.*, (8), 18–19.
386. Singh, R., Hefner, A.R. and McNutt, T.R. (2003) Reliability concerns in contemporary SiC power devices. Proceedings of the International Semiconductor Device Research Symposium, pp. 368–369.
387. Udal, A. and Velmre, F. (1997) SiC-diodes forward surge current failure mechanisms: experiment and simulation. *Microelectron. Reliab.*, **37** (10–11), 1671–1674.
388. Fischer, G. (2007) Derating of Schottky Diodes, White paper, DfR Solutions.
389. Soelkner, G. *et al.* (2007) Reliability of SiC power devices against cosmic radiation induced failures. *Mater. Sci. Forum*, **556–557**, 851–856.
390. Hanks, C.L. and Hamman, D.J. (1971) Radiation Effects Design Handbook, Section 1. Semiconductor Diodes, NASA-CR-1785.
391. Albadri, A.M., Schrimpf, R.D., Galloway, K.F. and Walker, D.G. (2006) Single event burnout in power diodes: mechanisms and models. *Microelectron. Reliab.*, **46** (2–4), 317–325.
392. Albadri, A.M. *et al.* (2005) Coupled electro-thermal simulation of single event burnout in power diodes. *IEEE Trans. Nucl. Sci.*, **52** (6), 2194–2199.
393. Ohyanagi, T. *et al.* (2009) EBIC analysis of breakdown failure point in 4H-SiC PiN diodes. *Mater. Sci. Forum*, **615–617**, 707–710.
394. Livingstone, H. (2003) *A Survey of Heterojunction Bipolar Transistor (HBT) Device Reliability*, JPL Publication 96-25, California Institute of Technology, Pasadena, CA.
395. Chakraborty, P.S. *et al.* (2009) Mixed-mode stress degradation mechanisms in PNP SiGe HBTs. Proceedings of IEEE International Reliability Physics Symposium, pp. 83–88.
396. Welser, R.E. and DeLuca, P.M. (2001) Exploring physical mechanisms for sudden beta degradation in GaAs-based HBTs. Proceedings of GaAs Reliability Workshop, pp. 135–157.
397. Feinberg, A. *et al.* (2005) Parametric degradation in transistors. Proceedings of Annual Reliability and Maintainability Symposium RAMS'05, pp. 266–270.
398. Panko, D. *et al.* (2006) *Time-to-fail Extraction Model for the 'Mixed-Mode' Reliability of High-performance SiGe Bipolar Transistors*, IRPS, San Diego, CA, pp. 512–515.
399. Grens, C.M. *et al.* (2005) Reliability issues associated with operating voltage constraints in advanced SiGe HBTs. Proceedings of IEEE International Reliability Physics Symposium, San Jose, pp. 409–414.
400. Gao, G.-B. *et al.* (1989) Thermal design studies of high-power heterojunction bipolar transistors. *IEEE Trans. Electron Dev.*, **36** (5), 854–863.
401. Guogong, W. *et al.* (2007) Power performance characteristics of SiGe power HBTs at extreme temperatures. Proceedings of 45th Annual IEEE International Reliability Physics Symposium, April 15–19, 2007, pp. 584–585.
402. Chau, H.-F. *et al.* (2006) Reliability study of InGaP/GaAs HBT for 28V operation. 28th IEEE Compound Semiconductor IC (CSIC) Symposium, San Antonio, November 12–15, 2006.
403. Trew, R.J., Liu, Y., Kuang, W.W. and Bilbro, G.L. (2006) The physics of reliability for high voltage AlGaN/GaN HFETapos's. Proceedings of IEEE Compound Semiconductor Integrated Circuit Symposium, CSIC, pp. 103–106.
404. Mazzanti, A. (2002) Physical investigation of trap-related effects in power HFETs and their reliability implications. *IEEE Trans. Dev. Mater. Reliab.*, **2** (3), 65–71.
405. Matsushita, K. *et al.* (2007) Reliability study of AlGaN/GaN HEMTs device. Proceedings of CS MANTECH Conference, Austin, May 14–17, 2007, pp. 87–89.
406. Vitusevich, S.A. *et al.* (2008) AlGaN/GaN high electron mobility transistor structures: self-heating effect and performance degradation. *IEEE Trans. Dev. Mater. Reliab.*, **8** (3), 543–548.
407. Bhasin, K.B. and Connolly, D.J. (1986) Advances in gallium arsenide monolithic microwave integrated-circuit technology for space communications systems. *IEEE Trans. Microw. Theory Tech.*, **34** (10), 994–1001.
408. Muraro, J.L. (1998) Conditions optimales de fonctionnement pour la fiabilité des transistors à effet de champ micro-ondes de puissance, PhD thesis, Université Paul Sabatier – Toulouse III.
409. Brierley, C.J. and Pedder, D.J. (1993) Surface mounted IC packages – their attachment and reliability on PWBs. *Circuit World*, **10** (2), 28–31.

410. Long, N.J. and Eccles, A.J. (2006) Thin film characterization using MiniSIMS and ToF-MiniSIMS. Mikkeli International Industrial Coating Seminar, March 16–18, 2006.
411. Dixon, J.B., Rooney, D.T. and Castello, N.T. Metal Migration in a Hybrid Device: Interaction on Package Materials, Process Residuals, and Ambient Gases, http://www.ors-labs.com/pdf/Metal%20Migration%20Failures%20in%20a%20Hybrid%20Device.pdf. (Accessed 2010).
412. Liao, T.W. (2004) An investigation of a hybrid CBR method for failure mechanisms identification. *Eng. Appl. Artif. Intell.*, **17** (1), 123–134.
413. Moore, G.E. (1965) Cramming more components onto integrated circuits. *Electron. Mag.*, **38**, 114–117.
414. Strukov, D.B. and Likharev, K.K. (2005) FPGA: a reconfigurable architecture for hybrid digital circuits with two-terminal nanodevices. *Nanotechnology*, **16**, 888–900.
415. Snider, G.S. and Williams, R.S. (2007) Nano/CMOS architectures using a field programmable nanowire interconnect. *Nanotechnology*, **18**, 035204–035210.
416. Strukov, D.B., Snider, G.S., Stewart, D.R. and Williams, R.S. (2008) The missing memristor found. *Nature*, **453**, 80–83.
417. Xia, Q. *et al.* (2009) Memristor-CMOS hybrid integrated circuits for reconfigurable logic. *Nano Lett.*, **9** (10), 3640–3645.
418. Zhang, W., Shang, L. and Jha, N.K. (2007) NanoMap: an integrated design optimization flow for a hybrid nanotube/CMOS dynamically reconfigurable architecture. Proceedings of 44th ACM/IEEE Design Automation Conference, DAC '07, pp. 300–305.
419. Eccofey, S. (2007) Ultra-thin nanograin polysilicon devices for hybrid CMOS-nano integrated circuits. PhD thesis, EPFL, Switzerland.
420. White, N.M. and Turner, J.D. (1997) Thick-film sensors: past, present and future. *Meas. Sci. Technol.*, **8** (1), 1–20.
421. Feinberg, A.A. and Widom, A. (1996) Connecting parametric aging to catastrophic failure through thermodynamics. *IEEE Trans. Reliab.*, **45** (1), 28–33.
422. Manca, J. *et al.* (2007) In situ failure detection in thick film multilayer systems. *Q. Reliab. Eng. Int.*, **11** (4), 307–311.
423. Kiilunen, J., Kuusiluoma, S. and Heino, P. (2007) Reliability of adhesive attachments on thick film hybrid substrate. Proceedings of 9th Electronics Packaging Technology Conference, EPTC, pp. 764–769.
424. Bâzu, M. *et al.* (2007) Quantitative accelerated life testing of MEMS accelerometers. *Sensors*, **7**, 2846–2859.
425. Bâzu, M. (2004) Concurrent engineering – a tool for improving MEMS research and manufacturing. 24th International Conference on Microelectronics (MIEL), Nis, Serbia and Montenegro, 2004, May 16–19, pp. 41–48.
426. Bâzu, M. (2003) Degradation phenomena in polymers used in microtechnologies. Symposium Micro/Nano Interactions and Systems Based on Natural and Synthetic Polymers, Petru Poni Institute, Iassy, Romania.
427. van Spangen, W.M. (2003) MEMS reliability from a failure mechanisms perspective. *Microelectron. Reliab.*, **43** (7), 1049–1060.
428. Walraven, J.A. (2003) Failure mechanisms in MEMS. International Test Conference 2003 (ITC'03), Charlotte, pp. 828–830.
429. Schmitt, P. *et al.* (2003) MEMS behavioral simulation: a potential use for physics of failure (PoF) modeling. *Microelectron. Reliab.*, **43** (7), 1061–1083.
430. Bhushan, B. (ed.) (2006) *Nanotechnology*, Springer, New York.
431. Patton, S.T. and Zabinski, J.S. (2006) Failure mechanisms of DC and capacitive RF MEMS switches. Proceedings SPIE, Mechanisms, Structures, and Models, Vol. 6111.
432. Tazzoli, A. *et al.* (2005) Reliability issues in RF-MEMS switches submitted to cycling and ESD test. IEEE 43rd Annual International Reliability Physics Symposium, San Jose, pp. 410–415.
433. Lin, L. (2000) MEMS post-packaging by localized heating and bonding. *IEEE Trans. Adv. Pack.*, **23** (4), 608–616.
434. Malshe, A.P. *et al.* (2001) Packaging and integration of MEMS and related microsystems for system-on-a-package (SOP). Proceedings of SPIE Symposium on Smart Structures and Devices, Vol. 4235, pp. 198–208.
435. Li, Q. *et al.* (2008) Failure analysis of a thin-film nitride MEMS package. *Microelectron. Reliab.*, **48** (8–9), 1557–1561.
436. De Pasquale, G., Somà, A. and Ballestra, A. (2009) Mechanical fatigue analysis of gold microbeams for RF-MEMS applications by pull-in voltage monitoring. *Analog Integr. Circuits Signal Process.*, **61** (3), 215–222.
437. Hsu, T.R. (2006) Reliability in MEMS packaging. 44th International Reliability Physics Symposium, San Jose, March 26–30, 2006.

438. Ping, C.W., Izumi, S. and Sakai, S. (2004) Strength and reliability analysis of MEMS micromirror. Proceedings of the Annual Meeting of JSME/MMD, pp. 273–274.
439. Broue, A. *et al.* (2009) Methodology to analyze failure mechanisms of ohmic contacts on mems switches. Proceedings of IEEE International Reliability Physics Symposium, IRPS, pp. 869–873.
440. Tanner, D.M. *et al.* (2005) Accelerated aging failures in MEMS device. IEEE 43th Annual International Reliability Physics Symposium, San Jose, pp. 317–324.
441. Meneghesso, G. and Tazzoli, A. (2007) Reliability of RF-MEMS for high frequency applications. *IEEE Trans. Dev. Mater. Reliab.*, **7** (3), 429–437.
442. Nakaoka, T. *et al.* (2004) Manipulation of electronic states in single quantum dots by micro machined air-bridge. *Appl. Phys. Lett.*, **84**, 1392–1394.
443. Wei, B. *et al.* (2001) Reliability and current carrying capacity of carbon nanotubes. *Appl. Phys. Lett.*, **79** (8), 1172–1174.
444. Collins, P. *et al.* (2001) Temperature measurements of shock compressed liquid deuterium up to 230 GPa. *Phys. Rev. Lett.*, **86**, 3128–3131.
445. Hochbaum, A. *et al.* (2008) Enhanced thermoelectric performance of rough silicon nanowires. *Nature*, **451**, 163–167.
446. Bansal, S. (2005) Nanoindentation of single crystal and polycrystalline copper nanowires. Electronic Components and Technology Conference, May 31 – June 3, 2005, pp. 71–76.
447. Sellin, R. *et al.* (2002) High-reliability MOCVD – grown quantum dot laser. *Electron. Lett.*, **38** (16), 883–884.
448. Liu, M. and Lent, C. (2007) Reliability and defect tolerance in metallic quantum-dot cellular automata. *J. Electron. Test. Theory Appl.*, **23** (2–3), 211–218.
449. Bhaduri, D. (2004) Tools and techniques for evaluating reliability trade-offs for nano-architectures. Thesis submitted to the Faculty of Virginia Polytechnic Institute and State University, Blacksburg, Virginia.
450. Michel, B. (2005) Nanoreliability – combined simulation and experiment. European Conference on Mechanical and Multiphysics Simulation and Experiments in Microelectronics and Microsystems, Invited Plenary Lecture, Berlin, April 18–20 2005.
451. Jeng, S.-L. (2007) A review of reliability research on nanotechnology. *IEEE Trans. Reliab.*, **56** (3), 401–410.
452. Reiner, J.C., Gasser, P. and Sennhauser, U. (2002) Novel FIB-based sample preparation technique for TEM analysis of ultra-thin gate oxide breakdown. *Microelectron. Reliab.*, **42** (9–10), 1753–1757.
453. Vallett, D.P. (2002) Failure analysis requirements for nanoelectronics. *IEEE Trans. Nanotechnol.*, **1**, 117–121.
454. Bae, S.J. *et al.* (2007) Statistical models for hot electron degradation in nano-scaled MOSFET devices. *IEEE Trans. Reliab.*, **56**, 392–400.
455. Lombardo, S. *et al.* (2005) Dielectric breakdown mechanisms in gate oxides. *J. Appl. Phys.*, **98**, 121301–121301-36.
456. Yeoh, Y.Y. *et al.* (2009) Investigation on hot carrier reliability of gate-all-around twin Si nanowire field effect transistor. Proceedings of IEEE International Reliability Physics Symposium, pp. 400–404.

6

Case Studies

In this chapter, 12 case studies on the failure analysis (FA) of electronic components and systems are detailed. These case studies offer the reader practical examples of the use of FA techniques (previously presented in Chapter 4) in identifying failure mechanisms (FMs) by starting from the failure modes (FMos) – as detailed in Chapter 5.

The authors of this book intended to increase the 'objectivity' of this chapter by using case studies already reported in literature. Obviously, the basic information about the case studies, as contained in various papers and reports (which are mentioned in each case as references), was processed according to the needs of this book, and the figures are preserved in their original form. Special thanks are due to the Lab Microsystem Characterization & Reliability from CEA-Léti, led by Dr Didier Bloch, which furnished us unpublished results.

As the reader will see, the case studies cover a broad range of electronic components, starting with passive components ($PbZr_xTi_{1-x}O_3$ (PZT) capacitors) and continuing with active electronic components fabricated by bipolar technologies (insulated gate bipolar transistors, IGBTs), MOS (metal-oxide semiconductor) technologies (complementary metal-oxide semiconductor, CMOS, microcomputer and metal-oxide semiconductor field-effect transistor, MOSFET), photonic technologies (thin-film transistors, TFTs), non-silicon technologies (heterojunction field effect transistors, HFETs) and microsystem technologies (resonators, micro-cantilevers, magnetic and nonmagnetic switches). The last two case studies cover packages issues (for a modern solution, chip-scale packages, CSPs) and the mounting of electronic components on printed circuit boards (PCBs) to fabricate electronic systems.

6.1 Case Study No. 1: Capacitors

6.1.1 *Subject*

High-density capacitors made by PZT material with very high dielectric permittivity (>800), reaching up to 30 nF/mm^2 values. These capacitors are used in integrated passive and active device (IPAD) technology for fabricated mobile phones, which combines different types of passive and active elements on a monolithic silicon substrate.

6.1.2 *Goal*

An extended investigation of PZT capacitor reliability properties, including both lifetime determination and early failure-rate control. The study was performed by researchers from ST Microelectronics, Tours and ST Microelectronics, Crolles, France [1].

Failure Analysis: A Practical Guide for Manufacturers of Electronic Components and Systems, First Edition.
Marius I. Bâzu and Titu-Marius I. Băjenescu.
© 2011 John Wiley & Sons, Ltd. Published 2011 by John Wiley & Sons, Ltd.

6.1.3 Input Data

A possible FM for PZT capacitors is time-dependent dielectric breakdown (TDDB). Time-to-breakdown distributions are determined by the Weibull law, with the shape parameter β being a key factor in the statistical analysis of dielectric breakdown. According to the percolation theory [2], the β value is related to the critical defect density that triggers the breakdown. So, the β value is characteristic of a given FM. The percolation model is a widely accepted statistical model that describes how the accumulation of microscopic defects under stress conditions leads to dielectric breakdown.

6.1.4 Sample Preparation

- PZT capacitors with Zr : Ti ratio of 52 : 48 were prepared by spinon sol gel-processing. After each spin, films were fired at 350 °C for 5 minutes in air and then processed by rapid thermal annealing (RTA) at 700 °C. The final multi-layered PZT film thickness was around 250 nm. The capacitors' top-electrode contact area ranged from 0.033 to 1 mm^2, corresponding to integrated capacitor values from 1 to 30 nF.
- Constant-voltage stress (CVS) tests were performed, using a Keithley 236 source measure unit (SMU). The voltage stress was reached at the end of a rapid staircase voltage ramp and then kept constant until breakdown detection.
- Two types of test were performed on 1 nF PZT capacitors: (i) voltage acceleration from 30 to 36 V, at 85 °C and (ii) temperature acceleration from 125 to 220 °C, at low voltage (12 V).

6.1.5 Working Procedure and Results

- The time-to-breakdown Weibull distributions resulting from CVS tests were obtained for both types of test. For the temperature-acceleration test, all distributions exhibited the same β value, which means the degradation mechanism is not modified by the temperature acceleration. The β values of the Weibull distributions were very different when obtained from voltage acceleration ($\beta = 1.3$) rather than from temperature acceleration at low voltage ($\beta = 3.7$), revealing the existence of two different FMs depending on the applied electrical fields. Since the real-life operating voltage of these capacitors is relatively low, around 3 V, the reliability study was focused on the 'low-voltage' mechanism ($\beta = 3.7$).
- From the statistical analysis of time-to-breakdown data, two different TDDB mechanisms for PZT capacitors were identified, depending on the applied voltage levels. From the time evolution of leakage current recorded during the CVS tests, it seems that the 'high-voltage' breakdown can be attributed to a thermal runaway process, which occurs during the resistance degradation phenomena [3], whereas the 'low-voltage' breakdown follows the percolation theory [2], with an electrical-field-driven creation of defects, enhanced at the periphery of capacitor structures.
- As a conclusion to the above, when estimating the reliability of PZT capacitors, CVS tests performed at low voltages (10–14 V) and high temperatures (125–220 °C) must be performed.
- But first, a screening procedure to eliminate those items affected by infant mortality (10% of the whole population, as obtained by statistical analysis) is required: applying a bias pulse to all items. The level of this bias pulse was optimised by using three FA techniques to identify the FMs of items affected by infant mortality: (i) visual inspection, (ii) SEM (scanning electron microscope) analysis and (iii) electrical analysis.
- The visual inspection showed the apparition of black spots (Figure 6.1a), which are a direct effect of bias pulse on the weakest items, representing about 10% of capacitors, which correspond to the proportion of extrinsic population.
- The SEM view cross-section (Figure 6.1b) indicated that the back spot formation resulted in a kind of crater, impacting the whole capacitor stack. This meant that the apparition of black spots could not be detected by a simple leakage measurement, since no contact existed between the electrodes,

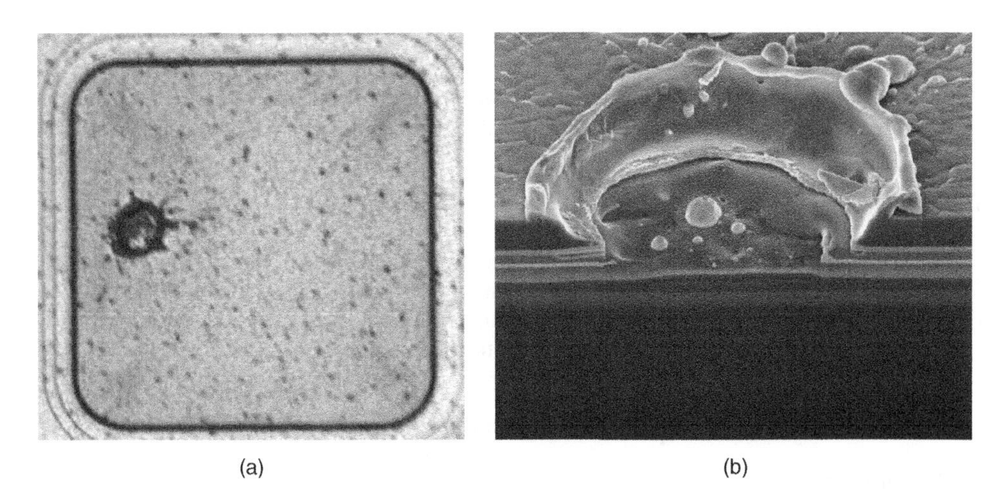

Figure 6.1 Optical view (a) and SEM cross-section (b) of black spots observed on extrinsic capacitors after pulse application. Reprinted from Bouyssou, E., *et al.*, Extended Reliability Study of High Density PZT Capacitors: Intrinsic Lifetime Determination and Wafer Level Screening Strategy, 45th Annual International Reliability Physics Symposium, Phoenix, 2007, pp. 433–438, Figure 8. © 2007 IEEE [1]

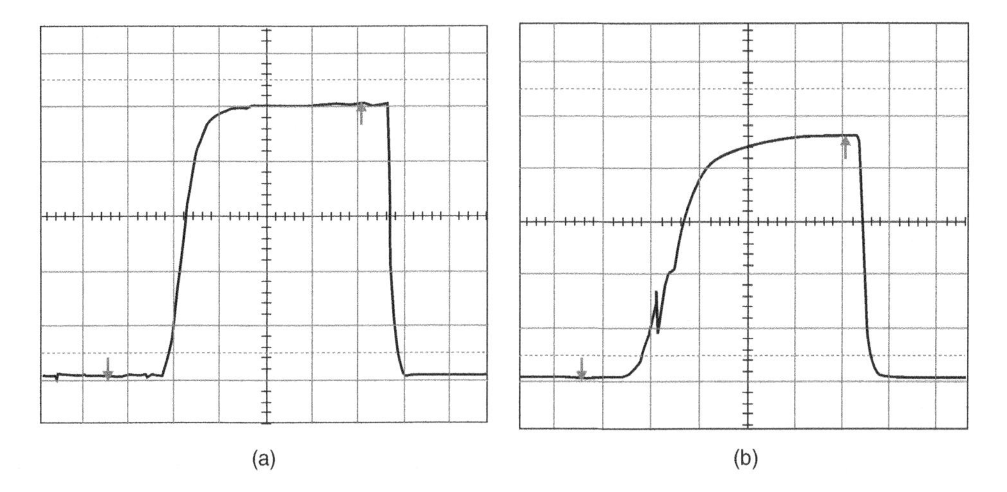

Figure 6.2 Charging curves recorded during bias pulse application, for intrinsic (a) and extrinsic (b) capacitors. Reprinted from Bouyssou, E. *et al.*, Extended Reliability Study of High Density PZT Capacitors: Intrinsic Lifetime Determination and Wafer Level Screening Strategy, 45th Annual International Reliability Physics Symposium, Phoenix, 2007, pp. 433–438, Figure 9. © 2007 IEEE [1]

and thus no short-cut was created in the capacitor structure. Moreover, the craters were quite small compared to the capacitor area, so the failures couldn't be detected by capacitance measurement.

- The charging curves recorded during the bias-pulse application were analysed and it was noticed that the items with black spots had small discharging peaks (Figure 6.2b). No such peaks were observed for items without black spots (Figure 6.2a).
- Consequently, an electrical method was developed to identify the weak items: the detection of small dV/dt discharging peaks recorded during the pulse application. This allowed the level of the bias pulse to be optimised for screening, in order to eliminate the weak items without affecting the main population.

- After using the optimised-screening procedure, it was possible to estimate the reliability level of the capacitors, by performing accelerated tests at high temperatures, keeping the voltage below 14 V. The activation energies were evaluated for extrapolation of lifetime data at lower temperatures and the voltage-extrapolation procedure was completed by considering the well-known E model.

6.1.6 Output Data

- The PZT capacitors' lifetime was found to exceed 10 years in voltage and temperature operating conditions, for capacitors ranging from 0.03 to 1 mm^2 area, with a failure rate of 0.01%.
- A screening procedure to eliminate infant mortality was developed (application of a bias pulse), including a method, based on FA, for optimising the pulse level.

6.2 Case Study No. 2: Bipolar Power Devices

6.2.1 Subject

A popular commercial standard-speed IGBT, optimised for minimum saturation voltage and low operating frequencies, mounted in a TO-220AB package.

6.2.2 Goal

To find out the root cause of the field failure of an IGBT fabricated by a small–medium enterprise (SME) and employed in the on-board electronics of a racing car. The failure happened after very few hours of intense operation. Note that this was the second FA performed in this case, the first one being unsatisfactory. The study was performed by researchers from the University of Cagliari and the University of Modena and Reggio Emilia [4].

6.2.3 Input Data

- It is difficult to handle such field failures, because the small or medium volume of the production makes even a few failures of statistical significance. There is however accurate reliability data about previous events.
- In such a situation the role of FA is increased, as the only solution for identifying the root cause of the failure, and in order to elaborate proper corrective actions in design and/or manufacture and to indicate sounder values for reliability parameters.
- The FMo is associated with violent thermal phenomena. After one day of repeated tests on the race circuit and regular switching off at night, the system failed due to short circuit of the IGBT as soon as it was switched on the next morning. An overall short circuit was noticed at electrical measurements.
- The first FA proposed some transient electrical overstress (EOS) or some unexpected electrostatic discharge (ESD) as possible FMs and corrective actions were directed towards identifying and removing possible transients on currents, voltages and electromagnetic fields during the operation of the system. This FA was considered unsatisfactory.

6.2.4 Working Procedure for FA and Results

- The plastic package was partially removed by mixed grinding and chemical etching. Extended thermal damage to the plastic itself all around the thick emitter wire was noticed, which is an indication for catastrophic overheating. It was obvious that the short circuit was a bulk failure, because there were no electrical connections from the top of the chip to the collector.

- X-ray energy-dispersive spectrometry (EDS) inspection of the exposed section of the wire revealed a central core nearly completely made of Si, surrounded by a region rich in Si and O. This situation indicates some strong vertical current flow, closer to an EOS event than to ESD. However, no indication about the type of mechanism (sudden or progressive) was obtained.
- The plastic package was completely removed and a cross-section of the damaged area was performed.
- The general view of the area, obtained by optical microscopy, indicated some Si extrusion into the Al metallisation, but also Al intrusion into Si (Figure 6.3).
- More detailed SEM analysis of the Al intrusions (Figure 6.4) showed the polyhedric (pyramidal) shape of the Al-filled cavities, which indicates a solid-state interdiffusion of Si into the overlaying

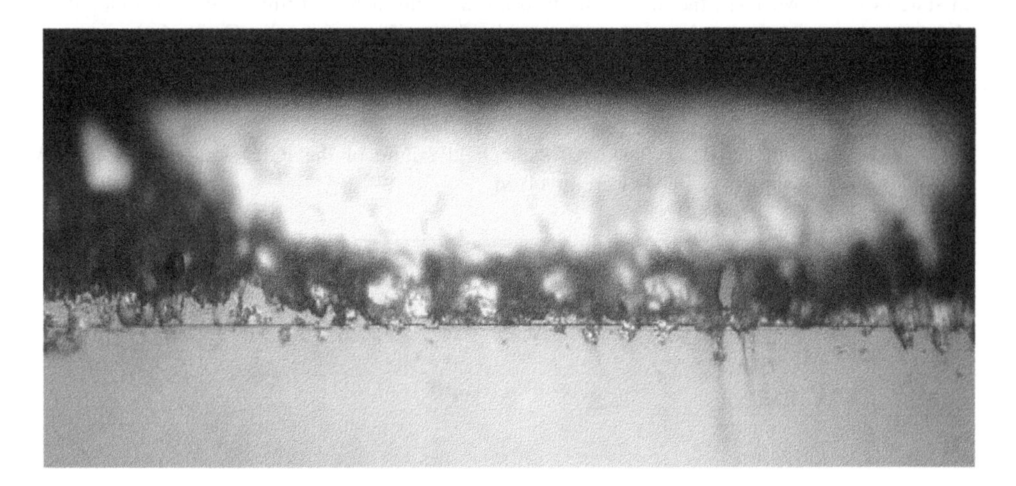

Figure 6.3 General view of the sectioned area under the emitter wire. Silicon extrusion into the Al metallisation and Al intrusion into the silicon crystal are displayed. Reprinted from Mura, G. and Cassanelli, G., Failure Analysis and Field Failures: A Real Shortcut to Reliability Improvements, Therminic 2006, Nice, Côte d'Azur, France, 27–29 September 2006 [4], Figure 3a. Reproduced by permission of Prof. Bernard Courtois (TIMA Editions)

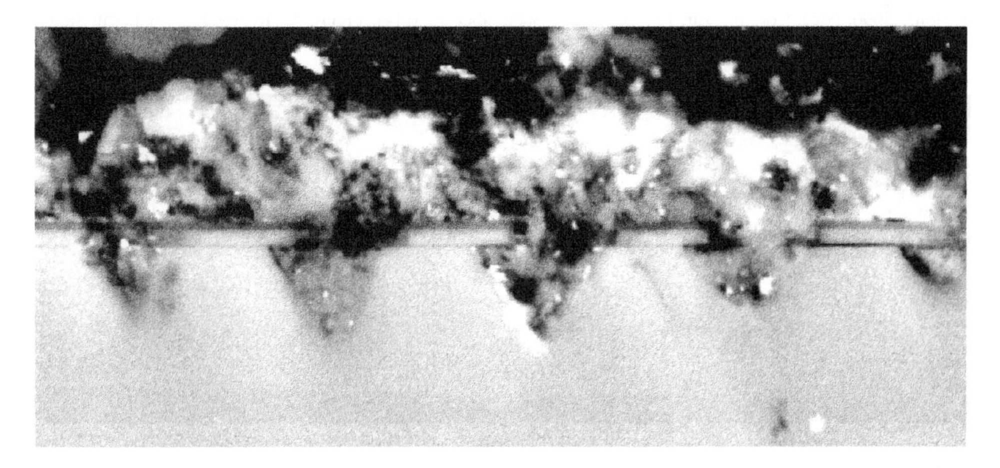

Figure 6.4 SEM detail of the Al intrusions. The polyhedric shape of the Al-filled cavities is a fingerprint of solid-state interdiffusion of Si into the overlaying metallic thin film. Reprinted from Mura, G. and Cassanelli, G., Failure Analysis and Field Failures: A Real Shortcut to Reliability Improvements, Therminic 2006, Nice, Côte d'Azur, France, 27–29 September 2006 [4], Figure 3b. Reproduced by permission of Prof. Bernard Courtois (TIMA Editions)

metallic thin film. SEM analysis of the Si extrusions (Figure 6.5) showed that Si had precipitated inside the original upper Al thin film.
- However, it seems two FMs were superposed, because interdiffusion (which was the first FM) is not compatible with the sudden, transient phenomenon that would have resulted in fusion and chaotic recrystallisation of the vertical structure.
- A second FM might be linked to an overcurrent that was allowed to flow, undetected, to expose the die to excessive temperature. In any case, a strong and sustained current must be indicted for the failure, independent of the sudden, and late, manifestation of the final catastrophic step. The origin of the overcurrent could be a latch-up FM, which is one of the most frequent problems of IGBTs. In spite of these two FMs, the device survived up to its ultimate and irreversible short circuit, as shown by the interaction at the wire contact and by the cracking of the die (Figure 6.6).

6.2.5 Output Data

- A complete and satisfactory explanation for the field failure of the IGBT was given using FA methodology: two superposed FMs were involved.

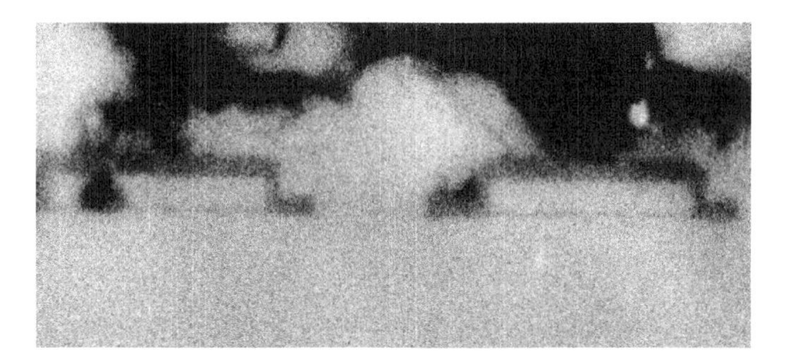

Figure 6.5 SEM details of the Si extrusions. The polysilicon gates (encapsulated in SiO_2) look intact. Reprinted from Mura, G. and Cassanelli, G., Failure Analysis and Field Failures: A Real Shortcut to Reliability Improvements, Therminic 2006, Nice, Côte d'Azur, France, 27–29 September 2006 [4], Figure 3c. Reproduced by permission of Prof. Bernard Courtois (TIMA Editions)

Figure 6.6 Section of the chip/thermal sinks structure. The creeks roughly crowd along the expected current/heat-flux lines from the upper emitter-wire contact to the lower collector surface. Reprinted from Mura, G. and Cassanelli, G., Failure Analysis and Field Failures: A Real Shortcut to Reliability Improvements, Therminic 2006, Nice, Côte d'Azur, France, 27–29 September 2006 [4], Figure 4. Reproduced by permission of Prof. Bernard Courtois (TIMA Editions)

6.3 Case Study No. 3: CMOS Devices

6.3.1 Subject

A single-chip 8-bit CMOS microcomputer.

6.3.2 Goal

To find out the root cause of the field failure of the CMOS microcomputer, employed on the control board of a boiler device for domestic application, failed in field operation after passing the final production tests. Note that this was the second FA performed in this case, the first one being unsatisfactory. The study was performed by researchers from the University of Cagliari and the University of Modena and Reggio Emilia [4].

6.3.3 Input Data

- It is difficult to handle such field failures, because the small or medium volume of the production makes even a few failures of statistical significance. There is however accurate reliability data about previous events.
- In such a situation the role of FA is increased, as the only solution for identifying the root cause of the failure, and in order to elaborate proper corrective actions in design and/or manufacture and to indicate sounder values for reliability parameters.
- The FMo is associated with violent thermal phenomena. After one day of repeated tests on the race circuit and regular switching off at night, the system failed due to short circuit of the IGBT as soon as it was switched on the next morning. An overall short circuit was noticed at electrical measurements.
- The first FA proposed some transient electrical overstress (EOS) or some unexpected electrostatic discharge (ESD) as possible FMs and corrective actions were directed towards identifying and removing possible transients on currents, voltages and electromagnetic fields during the operation of the system. This FA was considered unsatisfactory.

6.3.4 Working Procedure for FA and Results

- On inspecting the surface of one of the failed chips, only minor damage was noticed (Figure 6.7), not sufficient to explain the severity of the measured short circuit: the top metal layer was 'burned' in the small region of the most strongly failed input pin.
- The chip was chemically delayered at the 'burned' region (Figure 6.8), but no signs of melted elements were identified.
- The investigation of a second chip confirmed this phenomenon (Figure 6.9). The upper metal layer seemed to be intact (Figure 6.9a) and edges of the hole opened by the etch in nitride passivation are characteristic for a polycrystalline structure, as expected for the Al layer (Figure 6.9b). In the SEM picture of Figure 6.9c, the hole shows an intact lower-level metallisation.
- The surface was cleaned with HF and a small spot (arrowed) could be seen in the Si, surrounded by Si penetrating from the substrate into the metal layers, branching in four different directions (Figure 6.10). By corroborating this result with the pictures from Figure 6.9, the phenomenon could be explained: amorphous Si was present at the top metal level, which caused its removal under the etch of the passivation, while Al penetrated the Si bulk, and was removed during the HF etch. A 3D plot [6] of the SEM images (Figure 6.11) clearly showed the deep penetration of the damage (the interdiffusion of Al and Si).

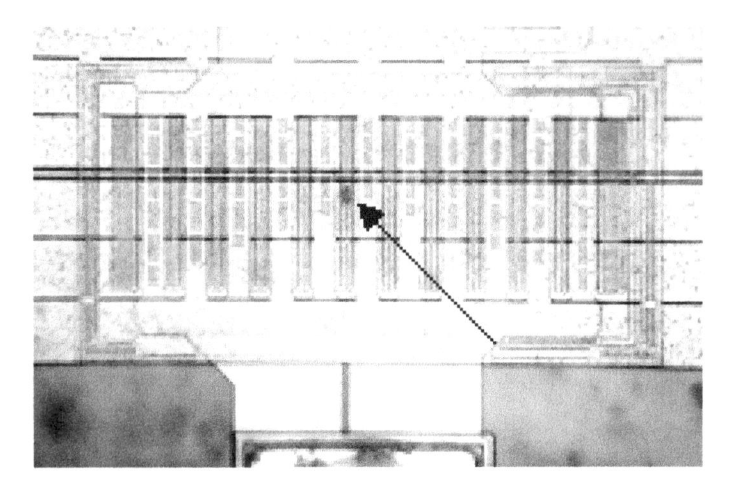

Figure 6.7 General view of the surface of the failed input pin of the first chip, upon removal of the plastic package. The arrow points to a small 'burnt' area. Reprinted from Mura, G. and Cassanelli, G., Failure Analysis and Field Failures: A Real Shortcut to Reliability Improvements, Therminic 2006, Nice, Côte d'Azur, France, 27–29 September 2006, [4] Figure 6a. Reproduced by permission of Prof. Bernard Courtois (TIMA Editions)

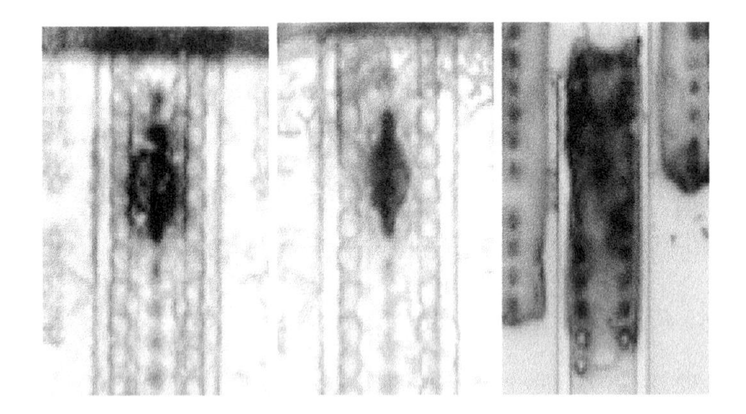

Figure 6.8 Optical micrograph of the damaged area after sequential removal of polyamide (a) and nitride passivation (b), and finally after HF (c). Reprinted from Mura, G. and Cassanelli, G., Failure Analysis and Field Failures: A Real Shortcut to Reliability Improvements, Therminic 2006, Nice, Côte d'Azur, France, 27–29 September 2006 [4], Figure 6b. Reproduced by permission of Prof. Bernard Courtois (TIMA Editions)

- Based on the above elements, the root cause could be described. There were two superposed FMs. An original ESD event caused a local short circuit, perhaps not sufficient to directly bring the circuit to fail, but enough to set up a small parasitic current, locally dense enough to fire the Al-Si interdiffusion FM (which had previously developed) and eventually leading to failure. Because the discharge occurred at some internal point of the circuit, but always under the wide upper metal layer, an I/O instability may not have caused the damage, but rather some capacitive coupling of the large parasitic capacitor, perhaps of the upper metal layer and the substrate, and

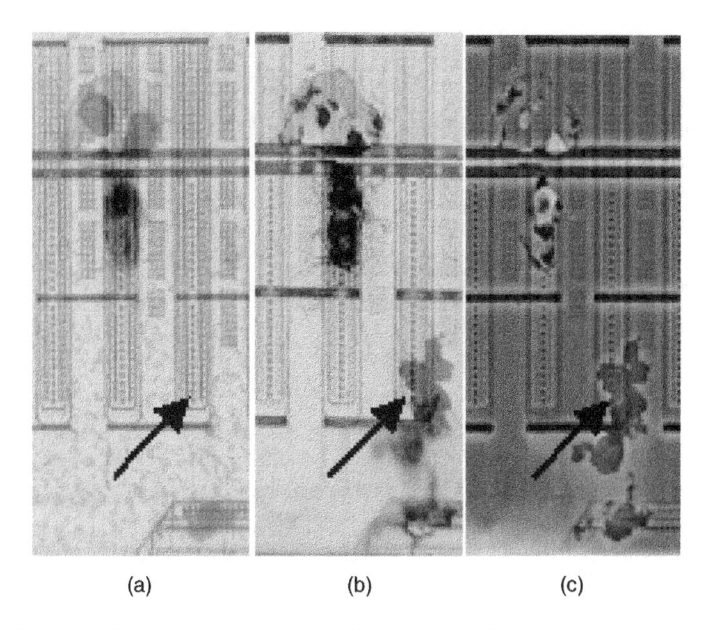

Figure 6.9 Optical view of the damaged area after selective polyamide (a) and nitride (b) removal. The SEM image corresponding to (b) is displayed in (c). The arrows point to the interdiffusion region on the top metal layer. Reprinted from Mura, G. and Cassanelli, G., Failure Analysis and Field Failures: A Real Shortcut to Reliability Improvements, Therminic 2006, Nice, Côte d'Azur, France, 27–29 September 2006 [4], Figure 7. Reproduced by permission of Prof. Bernard Courtois (TIMA Editions)

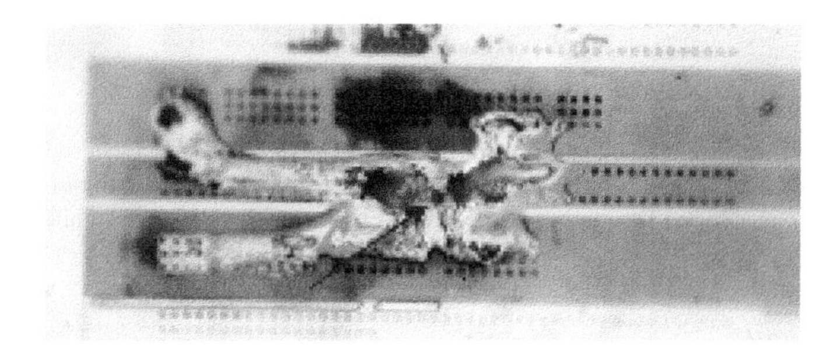

Figure 6.10 Optical view of the damaged area after complete HF etching. Reprinted from Mura, G. and Cassanelli, G., Failure Analysis and Field Failures: A Real Shortcut to Reliability Improvements, Therminic 2006, Nice, Côte d'Azur, France, 27–29 September 2006 [4], Figure 8a. Reproduced by permission of Prof. Bernard Courtois (TIMA Editions)

some huge electromagnetic disturbance near the device (possible a power switch very close to the board).

6.3.5 Output Data

- A complete and satisfactory explanation for the field failures of a CMOS microcomputer was given by the FA methodology: two superposed FMs were involved.

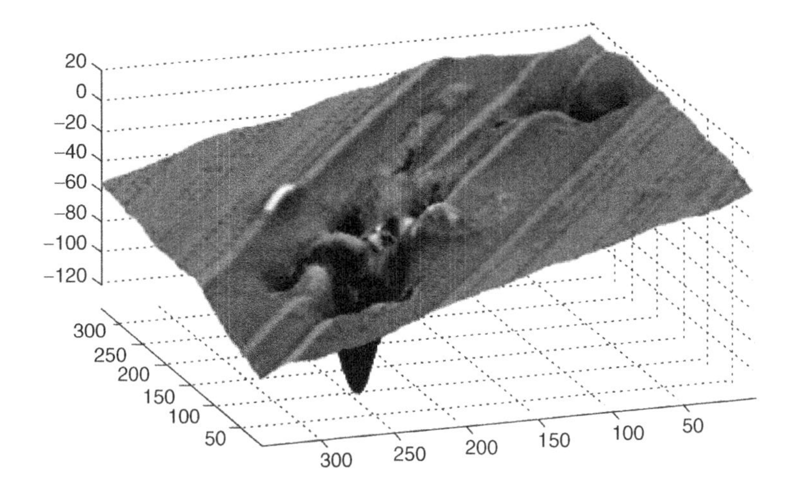

Figure 6.11 SEM 3D reconstruction of the same detail, showing the central source of the Si material. Reprinted from Mura, G. and Cassanelli, G., Failure Analysis and Field Failures: A Real Shortcut to Reliability Improvements, Therminic 2006, Nice, Côte d'Azur, France, 27–29 September 2006 [4], Figure 8b. Reproduced by permission of Prof. Bernard Courtois (TIMA Editions)

6.4 Case Study No. 4: MOS Field-Effect Transistors

6.4.1 Subject

An FDU6644-type *n*-channel MOSFET, encapsulated in TO-252.

6.4.2 Goal

To develop a reliable FA methodology to discover the FM responsible for a gate-open failure, characterised by high or over-ranged V_{th}, low BV_{DSS} and high I_{DSS}, and discontinuity of the gate contact towards the gate connection. The study was performed by researchers from Fairchild Semiconductor Philippines, Incorporated [7].

6.4.3 Input Data

- The analysis of gate-open failures is difficult, because: (i) the FMo is intermittent; (ii) the EOS may cloud the FM; and (iii) there are some uncertainties concerning the destructive chemical decapsulation process, which may introduce defects.
- A possible FM was an open or intermittently open contact in one or more of the interfaces, such as the gate lead post–wire bond, the die bond pad–wire bond, or within the gate wire or wedge itself.

6.4.4 Sample Preparation

- The first sample (S1) was represented by two items failed at the customer, during functional tests.
- Two types of reliability test were performed in the laboratory: temperature-cycling and high-temperature reverse-biasing (HTRB). Two other samples were obtained: (i) S2 – one item failed after 1000 temperature cycles and (ii) S3 – five items failed after 1000 hours' HTRB.

6.4.5 *Working Procedure for FA*

- Visual inspection of top-package markings, whole package and lead formation of the packages, using a 15× magnification optical microscope.
- X-ray inspection to check on package anomaly.
- Electrical verification with curve tracer, using a special curve-trace test sequence which enables differentiation between an item with a drain-source leakage failure from that with a gate-open failure [8].
- Inspection on delamination, using a scanning acoustic microscope (SAM) to identify voids and delaminations on such interfaces as compound–die, compound–DAP, compound–G-D-S lead posts. Delamination on gate lead post and on die gate pad could indicate gate-open.
- For items with gate-open validated by electrical verification, package-moulding and partial grinding were performed, which removed enough of the moulding compound to partially expose the internal gate wire, and removed part of the epoxy to expose the external gate lead. Through this gate-continuity verification procedure, potential damage to the device leads and internal structures could be eliminated. Electrical continuity was then performed between the gate lead and exposed gate wire to verify the gate lead post–wedge interface. The gate-bond pad and gate–wedge interface were verified by extending the continuity testing between the exposed gate wires and exposed source wires, exposed gate wires and source lead posts, and drain metal, which was the heat sink.
- Partial decapsulation on the gate lead post and bond pad meant the mould compound was chemically etched, exposing specific parts inside the package, until the gate-open was evidenced. Visual inspection, SEM analysis and electrical verification were performed to identify the FMs.

6.4.6 *Results*

- For S1, no defects were noticed at steps (a) and (b). At step (c), one item had an open-gate FMo (characterised by high V_{th}) and the other item an intermittent open-gate (characterised by low BV_{DSS} and high V_{th}). No defects were found and steps (d) and (e) were executed. No continuity was observed between the exposed gate wire and the external gate lead post on either item. After step (f), the items were visually inspected and then subjected to SEM inspection focusing on anomalies observed. A lifted gate wedge was noticed for S1 (Figure 6.12) and a broken gate wire, both at lead post and at die pad, for S2 (Figures 6.13 and 6.14). These were the failure roots of the observed FMo.
- For S2, the same steps were followed as above. For the single failed item of this sample, the electrical FMo was an intermittent open-gate, characterised by low BV_{DSS} and high V_{th}. The SEM inspection at step (f) allowed the FM to be identified: a broken wire at post (gate) (see Figure 6.15).
- For S3, the same procedure was used again. At step (c), the FMo identified for all five items was an intermittent open-gate, characterised by low BV_{DSS} and high V_{th}. At step (f), the FM was identified by SEM analysis: a lifted gate bond at post (Figure 6.16).

6.4.7 *Output Data*

- An FA methodology was developed to discover the FMs responsible for the open-circuit FMo of various types of electronic component.
- For an *n*-channel MOSFET transistor encapsulated in TO-252, the typical FMs producing open-gate FMo were identified as being produced by the wire-bonding process.

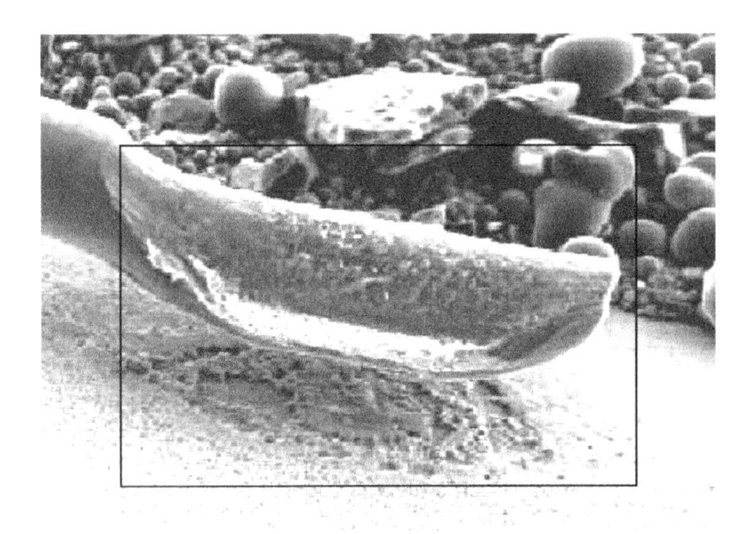

Figure 6.12 SEM photo showing lifted gate wedge. Reprinted from Remo, N.C. and Fernandez, J.C.M., A Reliable Failure Analysis Methodology in Analyzing the Elusive Gate-Open Failures, Proceedings of 12th IPFA 2005, Singapore, pp. 185–189, Figure 9 [7]. © 2005 IEEE

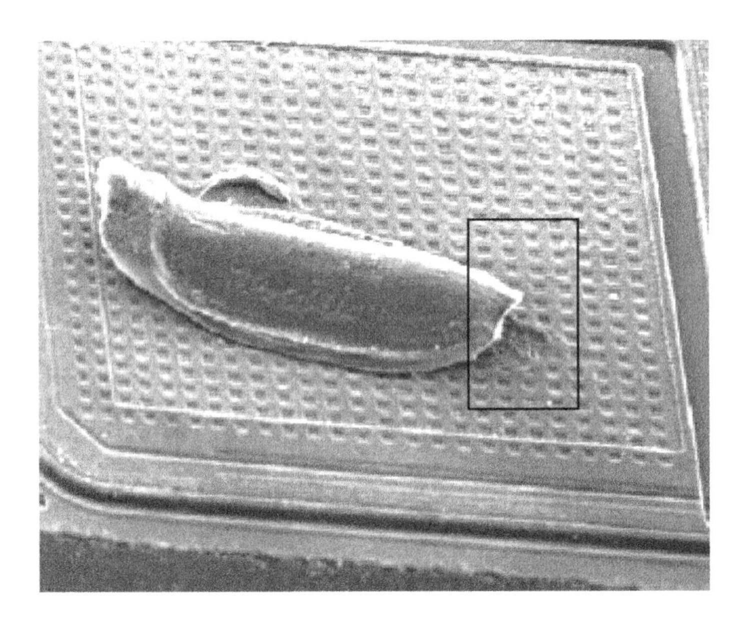

Figure 6.13 SEM photo of a wedge on a gate die pad. Reprinted from Remo, N.C. and Fernandez, J.C.M., A Reliable Failure Analysis Methodology in Analyzing the Elusive Gate-Open Failures, Proceedings of 12th IPFA 2005, Singapore, pp. 185–189, Figure 11 [7]. © 2005 IEEE

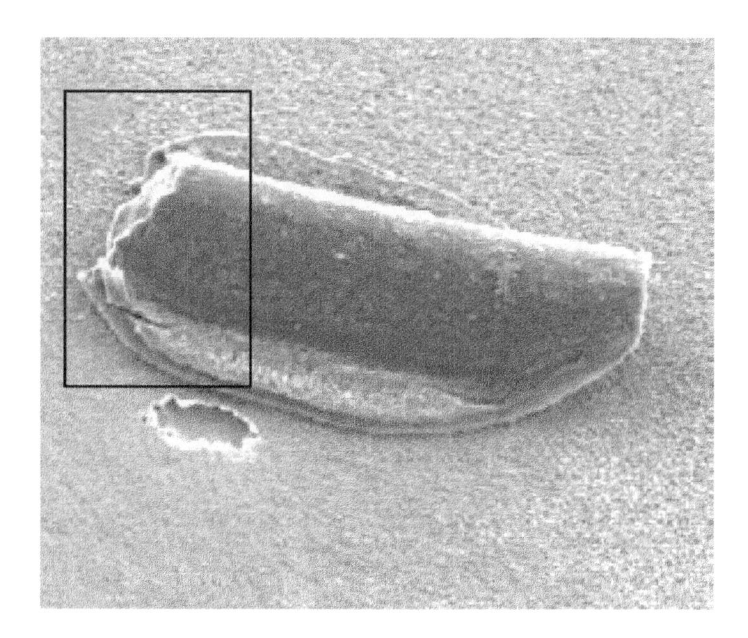

Figure 6.14 SEM photo of a gate wedge at post. Reprinted from Remo, N.C. and Fernandez, J.C.M., A Reliable Failure Analysis Methodology in Analyzing the Elusive Gate-Open Failures, Proceedings of 12th IPFA 2005, Singapore, pp. 185–189, Figure 12 [7]. © 2005 IEEE

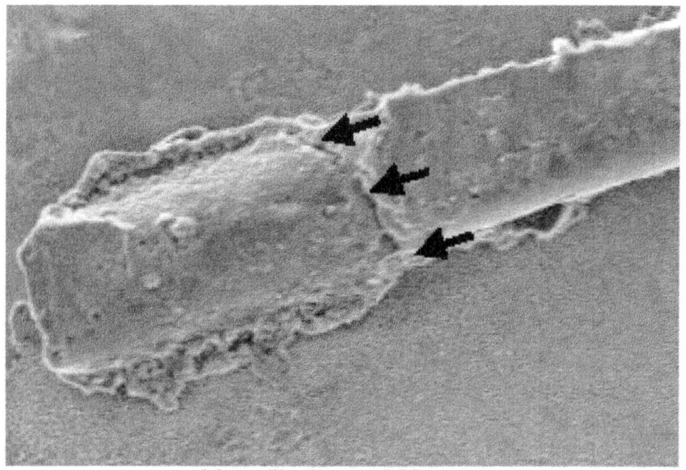

Figure 6.15 SEM photo of the broken wire at post. Reprinted from Remo, N.C. and Fernandez, J.C.M., A Reliable Failure Analysis Methodology in Analyzing the Elusive Gate-Open Failures, Proceedings of 12th IPFA 2005, Singapore, pp. 185–189, Figure 14 [7]. © 2005 IEEE

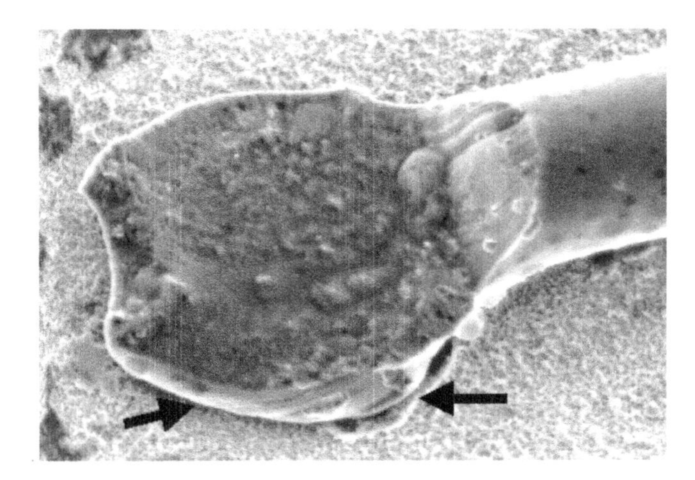

Figure 6.16 SEM photo showing an exposed gate bond after partial decapsulation, with a gap under the bond. Reprinted from Remo, N.C. and Fernandez, J.C.M., A Reliable Failure Analysis Methodology in Analyzing the Elusive Gate-Open Failures, Proceedings of 12th IPFA 2005, Singapore, pp. 185–189, Figure 16 [7]. © 2005 IEEE

6.5 Case Study No. 5: Thin-Film Transistors

6.5.1 Subject

Amorphous silicon (a-Si:H) TFTs.

6.5.2 Goal

To study the degradation mechanisms of TFTs, and to propose a model for degradation. The weak point of amorphous silicon TFTs is the electric-field-induced threshold voltage shift, which limits the lifetime of active matrix backplanes, particularly for low-temperature processes required for flexible backplanes [9]. The possibility of using TFTs in viable applications is linked to the achievable reliability. Corrective actions are needed to either slow or reverse the increase in threshold voltage of a-Si:H TFTs. The study was performed by researchers from the Flexible Display Center at Arizona State University, Tempe, AZ, USA [10].

6.5.3 Input Data

- If a DC voltage is applied to the gate of an a-Si:H TFT, the threshold voltage, V_{th}, increases by several volts over a few hours [11].
- The shift (which is linearly proportional to the electric field in the channel or equivalent to the charge in the channel) can be reduced by an order of magnitude if the process temperature is raised from 150 to 350 °C, which is typical of glass LCD processes [9].
- The V_{th} shift can be reduced by increasing the drain to source voltage during gate-voltage stress, with the largest shifts occurring for devices biased in the linear region. Increased drain voltages correspond to reduced field and charge in the channel near the drain.
- The V_{th} shift over time follows a power law and is typically proportional to $t^{0.3}$. The shift is duty-cycle dependent. Reducing the duty cycle for a pulsed stress improves the lifetime in more than inverse proportion to the reduced 'on' time.

- The two mechanisms responsible for the V_{th} degradation in a-Si:H TFT are: (i) charge injection in the silicon nitride (SiN_x) gate insulator (a reversible process, produced by carrier-tunnelling from extended states in the a-Si:H layer to the trap states in the nitride) and (ii) creation of defect states in the a-Si:H conducting channel (a much less reversible process). Mobile carriers are responsible for breaking the weak Si–H bonds resulting in the creation of charged defect states, named 'dangling bonds'. The density of created defect states is proportional to the number of mobile carriers, which affects the Fermi-level position, resulting in a V_{th} shift.
- Depending on stress conditions, one mechanism is dominant: under low to medium positive V_{GS} stress, defect creation is dominant, whereas for negative V_{GS}, charge-trapping is the dominant instability mechanism.

6.5.4 *Sample Preparation*

- An amorphous silicon TFT was fabricated by a low-temperature process (180 °C) on flexible substrates such as stainless steel and heat-stabilised PEN (polyethylene naphthalate).
- The gate was made of molybdenum, the gate dielectric was silicon nitride and the active layer was hydrogenated amorphous silicon deposited with plasma-enhanced chemical-vapour deposition. Before etching the contacts, a nitride passivation step was performed.
- Source/drain metal was sputtered on as an $n+$ amorphous silicon/aluminum bilayer.
- After fabrication, the devices were annealed for 3 hours, at 180 °C, in nitrogen atmosphere.

6.5.5 *Working Procedure for FA and Results*

- The hot electron degradation in a-Si:H TFTs was localised by an electrical method developed for MOSFET. When the source and drain are reversed, I_{DS} is degraded for all V_{DS}. So by reversing the source and drain during measurement, the location of the hot electron degradation can be identified, because carriers are forced away from the Si-SiO_2 interface in saturation and the trapped oxide charge at the drain has no effect.
- The V_{th} degradation was localised at the gate to a-Si:H channel interface region and further into the channel. Carrier flow was at the surface until the pinch-off point. Consequently, essentially normal post-stress saturation current magnitude was observed for all V_{DS} in the forward direction. Reversing the source and drain maximised current flow in the unaffected region and allowed the pinch-off point to sweep across the traps that created the V_{th} degradation. The source–channel barrier appeared to be unaffected by the ΔV_{th} mechanism since the drain current in the linear operating region following the linear-mode or saturation-mode stress was less affected in the reverse-measurement configuration.
- It was shown that reduced ΔV_{th} occurred at the drain end, where electric fields drove the carriers away from the surface after the channel pinch-off point. If the device was biased further into saturation, defect creation occurred over a smaller area, which was always at the source side of the pinch-off point.
- The threshold voltage of stressed a-Si:H substantially recovered after room-temperature rest without applied voltage. However, subsequent stress rapidly degraded the TFT to its pre-rest condition [12].

6.5.6 *Output Data*

- The hot electron degradation in a-Si:H TFTs was studied using electrical methods and a model was proposed for this phenomenon.

6.6 Case Study No. 6: Heterojunction Field-Effect Transistors

6.6.1 Subject

AlGaN/GaN HFETs.

6.6.2 Goals

(i) To identify the degradation and FMs of 36 mm AlGaN/GaN HFETs-on-Si under DC stress conditions on a large number of nominally identical devices that were chosen randomly across a production process. (ii) To propose and check corrective actions to improve reliability. The study was performed by researchers from Nitronex Corporation, Raleigh, NC, USA [13].

6.6.3 Input Data

GaN's capability for high-power RF applications has already been proved in a number of commercial and military markets, including cellular infrastructure, broadband wireless access, radar and communications applications. The only problem is the reliability of the technology and understanding the key FMs.

6.6.4 Sample Preparation

- Six wafers randomly sampled from a 150-wafer baseline distribution were extensively characterised and subjected to a complete battery of tests.
- A transistor die attached to high-thermal-conductivity CuW single-ended ceramic packages using AuSi eutectic process was used as a test vehicle. For packaging, a non-hermetic epoxy sealed lid was used.

6.6.5 Working Procedure and Results

- DC stress tests were performed at three junction temperatures (260, 285 and 310 °C). High-temperature operating-life (HTOL) data from a fourth T_j of 200 °C was overlaid and exhibited consistent behaviour with the three-temperature data. In all cases, devices were stressed at V_{ds} of 28 V and I_{ds} of 2.3 A, with the ambient temperature adjusted to achieve the desired T_j.
- Both electrical and physical characterisation were used for FA, which was performed on unstressed and stressed (1000 hours' HTOL) devices. The electrical characterisation consisted of electrical analysis of field-effect transistor (FET) transfer curves, revealing an increase in pinch-off voltage (V_p) with time of stress that seemed to be consistent with the I_{dss} degradation. The proposed explanation for this V_p shift was a permanent Schottky barrier height (SBH) increase.
- Physical analysis was performed using scanning tunnelling electron microscopy (STEM) on a series of cuts made by a focused ion beam (FIB). It was discovered that there was an interfacial layer between the gate and the semiconductor on unstressed samples, which seemed to be modified on the stressed samples (Figure 6.17), being the possible cause of the previously described SBH.
- In studying this possible FM, forward-diode characteristics for several single-finger 100 μm devices were obtained on-wafer. These devices were then stressed for an hour under conditions similar to the HTOL test (28 V with I_{ds} adjusted to provide T_j of 200 °C). The forward-diode

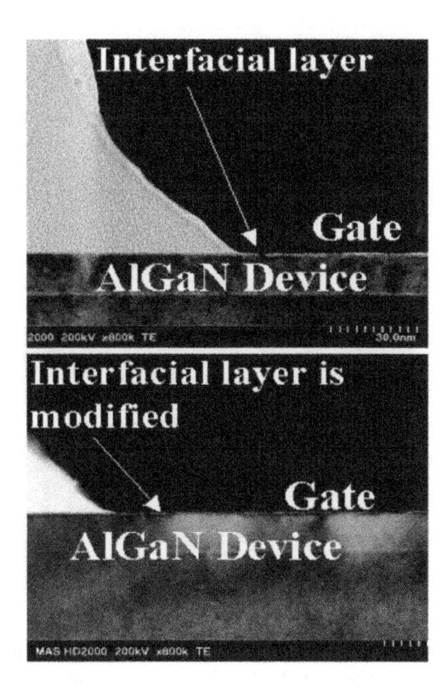

Figure 6.17 STEM pictures of an unstressed (a) and a stressed (b) sample showing change in the interfacial layer after stress. Reprinted from Singhal, S., Ga-on-Si Failure Mechanisms and Reliability Improvements, 44th Annual International Reliability Physics Symposium, IRPS 2006, San Jose, CA, pp. 95–98, Figure 4 [13]. © 2006 IEEE

curves were then re-measured. This procedure confirmed that the SBH did indeed increase after stress.

- A corrective action was proposed to reduce the effect of the interfacial layer: an annealing step was added to the gate-fabrication process, which created a more ideal semiconductor–metal interface under the gate before device operation and stress.
- The previously described physical analysis was used to confirm the improved interface, through STEM analyses of FIB cuts on three gates taken from different areas of the wafer before and after gate annealing. The images shown in Figure 6.18 reveal a similar modification to the interfacial layer as seen during HTOL stress.
- The previously described DC stress tests were performed on items fabricated by the new gate process: one wafer was processed with no anneal and one wafer received the gate anneal. Ten devices were selected from each wafer with different gate processes and were packaged and stressed. A dramatic reduction in I_{dss} change over time was noticed relative to the old gate process. Specifically, devices that had gate anneal degraded 50% less during the first 24 hours than the devices without the anneal.

6.6.6 *Output Data*

- The typical FM of 36 mm AlGaN/GaN HFETs-on-Si under DC stress conditions was identified: a thin interfacial layer under the gate is diminished with stress, causing an SBH shift.
- A corrective action was proposed and verified: a new gate process was inserted into the fabrication flow, resulting in a negligible SBH shift and a dramatic improvement in the large periphery HTOL results.

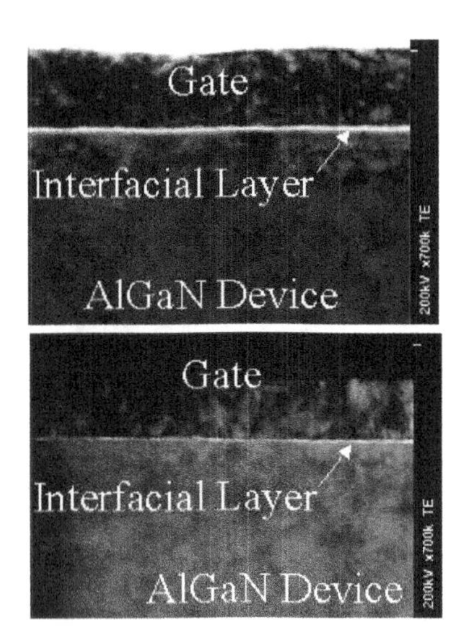

Figure 6.18 STEM pictures of a device before (a) and after (b) gate anneal showing change in the interfacial layer with anneal. Reprinted from Singhal, S., Ga-on-Si Failure Mechanisms and Reliability Improvements, 44th Annual International Reliability Physics Symposium, IRPS 2006, San Jose, CA, pp. 95–98, Figure 6 [13]. © 2006 IEEE

6.7 Case Study No. 7: MEMS Resonators

6.7.1 Subject

Silicon-based micro-electro mechanical system (MEMS) resonators.

6.7.2 Goal

To study the stability of resonant frequency for single-wafer, thin-film-encapsulated silicon MEMS resonators, for both long-term operation and temperature-cycling. MEMS resonators are candidates for the replacement of quartz resonators, due to their potential for reduced size, cost and power consumption, as well as integration with circuitry on the same wafer and with the IC chip, which can reduce parasitic losses from and the cost of higher-level packaging. The study was performed by researchers from Stanford University, Departments of Mechanical and Electrical Engineering, Stanford, CA, USA [14].

6.7.3 Input Data

Existing data concerning long-term stability of MEMS resonators is insufficient for many applications. Previous work has investigated possible fatigue in thin-film silicon. It seems MEMS resonators may show better performances in frequency stability only if operating in a vacuum isolated from outside atmosphere and minimising exposure to oxygen or humidity. Thus, safer packages are needed.

6.7.4 Sample Preparation

- A new wafer-scale package was developed for MEMS resonators and inertial sensors: a single-wafer polysilicon thin-film encapsulation process, which involved covering unreleased MEMS devices with a sacrificial oxide layer (etched with a vapour-phase HF etch process) and a 2 μm-thick epitaxial polysilicon encapsulation, released by a 25 μm-thick epitaxial polysilicon deposition and planarised via chemical-mechanical polishing (CMP).

- Two resonator designs were used, both specifically designed with a single mechanical support in the middle of the structure to minimise the possibility of induced stress from thermal expansion of different materials or residual stress of adjacent layers. The temperature coefficient of resonant frequency (resonant-frequency sensitivity to the environmental temperature change) was measured at room temperature as −27 ppm/°C for design A resonators and −33 ppm/°C for design B resonators. These values depend on the temperature coefficient of Young's modulus of silicon and resonator geometry.

6.7.5 *Working Procedure for FA and Results*

- Three types of test were performed and the results showed the good stability of the resonant frequency of the MEMS resonators.
- Long-term frequency drift was examined by monitoring six separate resonators: each was excited and measured approximately every 30 minutes for more than 1 year (about 10 000 hours) using an Agilent 4395A network analyser, in a temperature-controlled chamber which provided a test temperature within ±0.1 °C error range. The resonant frequency for both types of resonator remained in error ranges.
- The resonant-frequency stability after rapid environmental temperature changes was investigated. The MEMS resonators were single anchored, being immune to axial stress derived from differential expansion of layers in the encapsulated silicon die. Seven hundred cycles from −50 to +80 °C were performed and the resonant frequency of two A-type resonators was measured, at 30 °C (after 30 minutes to reach thermal equilibrium in the temperature chamber), in between each temperature cycle (Figure 6.19). The drift in resonant frequency remained in measurement error range (Figure 6.20). In addition, the resonant-frequency difference between measurements after high-temperature cycles and after low-temperature cycles stayed within the measurement error. The observed frequency shift was most likely related to slight temperature gradients within the apparatus at the time of the measurements.
- The resonant frequency was measured for A-type resonators, every 10 °C, while the temperature ramped up and down between −10 and +80 °C. At each step of 10 °C, the chamber was held for 30 minutes and allowed to reach thermal equilibrium before the resonant frequency was measured. No trend between temperature-increasing and -decreasing cases was found and the difference in

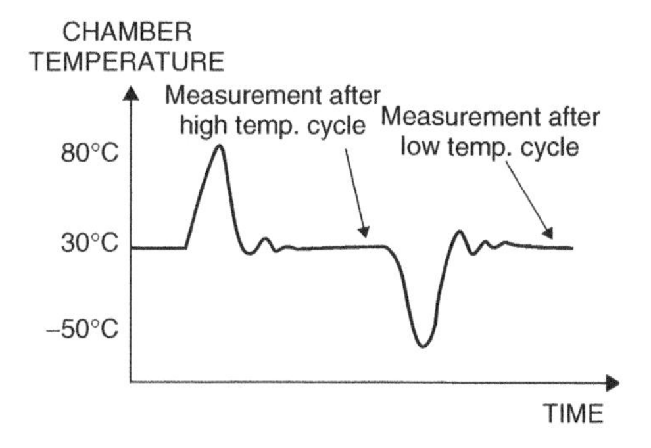

Figure 6.19 Resonant frequency of the MEMS resonators, measured at 30 °C after the temperature inside the chamber was cycled between −50 and +80 °C. Every measurement took place after holding for about 30 minutes to reach thermal equilibrium. Reprinted from Kim, B. *et al.*, Frequency Stability of Wafer-Scale Film Encapsulated Silicon Based MEMS Resonators, Sensors and Actuators A: Physical, Vol.136, Issue 1, 1 May 2007, pp. 125–131, Figure 7a [14]. Copyright 2007 with permission from Elsevier

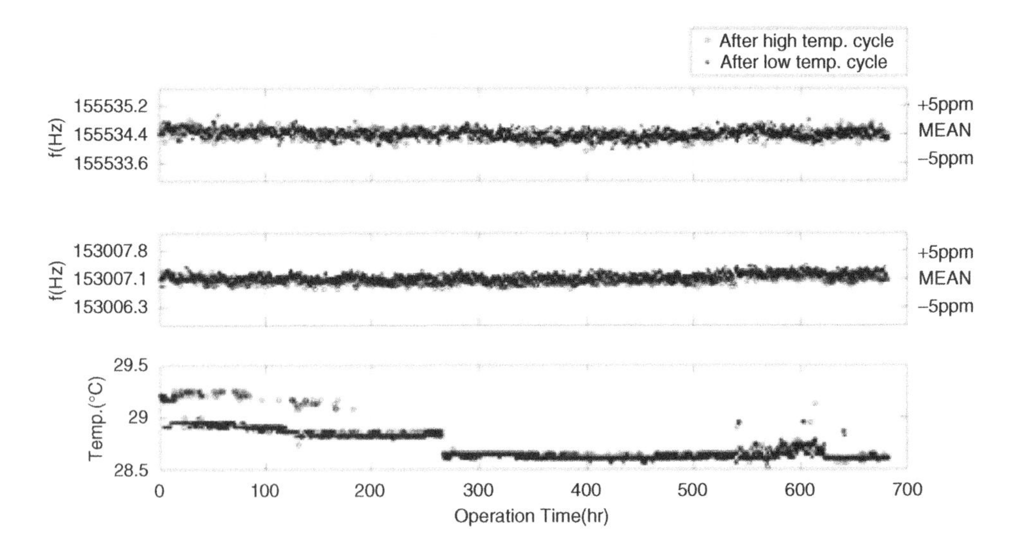

Figure 6.20 Plot of resonant frequency vs temperature for temperature-cycling test. Resonant frequencies are measured at 30 °C after high-temperature cycle (O) and after low-temperature cycle (X). This result suggests long-term stability of resonant frequency even after a large number of wide temperature cycles. Reprinted from Kim, B. *et al.*, Frequency Stability of Wafer-Scale Film Encapsulated Silicon Based MEMS Resonators, Sensors and Actuators A: Physical, Vol.136, Issue 1, 1 May 2007, pp. 125–131, Figure 7b [14]. Copyright 2007 with permission from Elsevier

resonant frequency again remained in measurement-error range. The results from these separate experiments indicated that there was no significant residual stress or differential thermal expansion that might affect resonant-frequency stability for rapid changes in environmental temperature.

6.7.6 Output Data

- The long-term ageing rate of MEMS resonators was found to be equivalent to commercial quartz-crystal resonators.
- Moreover, unlike quartz resonators, no initial ageing or stabilisation period was found and there was no measurable hysteresis.
- It seems this improved performance was due to the clean, high-temperature fabrication process used to encapsulate these resonators, which suggests that the drift and hysteresis problems observed in other MEMS resonators were initiated by their operating environment.

6.8 Case Study No. 8: MEMS Micro-Cantilevers

6.8.1 Subject

Micro-cantilever beams achieved by bulk micro-machining MEMS technology.

6.8.2 Goal

To develop a method for studying the mechanical behaviour of single-crystal silicon (SCS) micro-cantilever beams and to identify the typical FMs of bent beams. The study was performed by researchers from Feng Chia University and the National Chin-Yi Institute of Technology, Taichung, Taiwan, ROC [15].

6.8.3 *Input Data*

- Micro-scale silicon structures may not behave like bulk silicon structures and the mechanical properties of silicon at the micro scale are not well known.
- The development of both standardised test methods and material property databases has lagged behind that of design and simulation tools, limiting their utility. This can be proved by the discrepancy in reported experimental values, probably only due to differences in experimental technique and associated measurement error at the MEMS scale.
- A standard test method is needed to characterise the mechanical properties of micro-fabricated material produced by the same processes and at the same scales as the intended application.

6.8.4 *Sample Preparation and Working Procedure*

- Eight sets of micro-cantilever beams with the same width (100 μm), four lengths (400, 500, 600 and 700 μm) and two variants of thickness (50 and 60 μm) were fabricated by a standard MEMS process from single-crystal Si p-type (100) wafer.
- Flexural testing was performed on the beams, using the MTS Tytron 250 micro-force testing machine, with a special design of the fixture. The flexural test was conducted by first touching the probe on the free end of the cantilever beam and then moving the probe toward the beam at a constant rate of 0.1 μm/second. During the test, the flexural strength, failure strain and Young's modulus were computer-controlled. Five duplicate tests were conducted in order to obtain one valid result for stress, strain and Young's modulus.
- The FMs of bent beams were studied using SEM.
- The influence of surface roughness on strength was measured by a surface profiler.

6.8.5 *Results and Discussion*

- The stress–strain relationship approximately followed a linear behaviour, indicating that the brittle SCS beam fails in an elastic manner when subjected to bending.
- The local zigzag in the stress–strain curve noted when the length of the beam increased might be caused either by a stick–slip mechanism at the probe–beam interface, or just by numerical noise: measurement noise caused by the transduction of deformation and force into an electrical voltage close to system resolution.
- The values of Young's modulus (obtained from the slope of stress–strain curves) for eight sets of beam dimensions were in the range 159.2–188.6 GPa (consistent with previous data obtained by other experiments).
- The flexural strength and strain of beams with a constant beam width of 100 μm and thickness of 50 or 60 μm decreased as beam length increased from 400 to 700 μm. The explanation for this was that for brittle materials the probability of finding a flaw of a given size decreases with the volume (or area) of material under loading. If a flaw exists at the fixed end of longer beams, this seems to be more harmful to the beam, because it fails with larger displacement, leading to lower strength.
- The results of the SEM analyses allowed a better understanding of the failure process.
- In Figure 6.21 (bent fracture of the beam with 400 μm length and 50 μm thickness), the dominant FM is the cleavage initiated at, or near, the surface flaw along the {111} plane. It seems that the bending crack initiates at the fixed end of beam's top flat surface due to maximum tensile stress. High pitch striations can be seen in the central area of the broken surface. A zigzag pattern is found at the lower-right end of the fracture surface, which may have been produced by a compressive

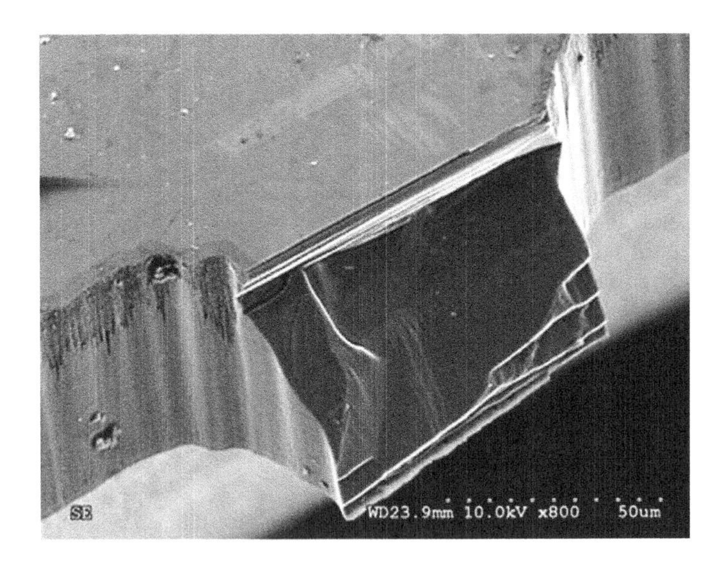

Figure 6.21 Bending fracture of the micro-cantilever beam with width 100 μm, thickness 50 μm and length 400 μm. Reprinted from Pan, C.H. and Liu, P.P., Dimension effect on mechanical behavior of silicon micro-cantilever beams, Measurement, Vol. 41, 2008, pp. 885–895, Figure 6a [15]. Copyright 2008, with permission from Elsevier

stress at the upper and lower parts of the broken surface. The undulated surface from the left end of the broken surface may be caused by crack initiation from a larger surface flaw.

- In Figure 6.22 (bent fracture of the beam with 700 μm length and 50 μm thickness), the FM seems to be different: the {111} fracture plane is considerably flat. Correlated with the previous result showing a lower strength for a longer beam, it seems that the crack propagated easily along the upper smooth {111} plane due to a combination effect of surface flaws and larger displacement during testing.
- In Figure 6.23 (bent fracture of the beam with 400 μm length and 60 μm thickness), a single fracture surface along the {111} plane can be seen. Comparison with the thinner beam of 50 μm (from Figure 6.21) shows that the upper surface of the thicker beam is flatter, without the track of crack propagation caused by surface flaws. This could mean that surface flaws cause higher stress concentrations in thinner beams, resulting in lower strength.
- In Figure 6.24 (bent fracture of the beam with 700 μm length and 60 μm thickness), the morphology indicates that a larger flat {111} fracture plane is initiated from the top surface, though not as flat as that in Figure 6.22. The crack propagation slows down at the lower portion of the broken surface subjected to compression, and changes as local peaks and valleys until final failure. It seems this FM is initiated by its having the lowest strength among beams with different lengths, though it is stronger than the thinner beam with the same length.
- In conclusion, it seems three factors influence the mode in a synergistic manner: (i) dimension of the beam, (ii) surface flaws and (iii) loading condition.
  - For thin beams, a large stress gradient occurs because stress distribution changes from tension to compression, leading to lower strength when coupled with large edge roughness caused by mask.
  - For thick beams, the situation becomes less serious and the beam strength has a higher value.
  - For short beams, the rough fracture surface caused by surface flaws indicates the difficulty of crack propagation and leads to a higher strength of beam.
  - For long beams, stress concentration at the fixed end of the beam caused by surface flaws is enhanced by the large displacement, leading to a lower strength of beam.

Figure 6.22 Bending fracture of the micro-cantilever beam with width 100 μm, thickness 50 μm and length 700 μm. Reprinted from Pan, C.H. and Liu, P.P., Dimension effect on mechanical behavior of silicon micro-cantilever beams, Measurement, Vol. 41, 2008, pp. 885–895, Figure 6b [15]. Copyright 2008, with permission from Elsevier

Figure 6.23 Bending fracture of the micro-cantilever beam with width 100 μm, thickness 60 μm and length 400 μm. Reprinted from Pan, C.H. and Liu, P.P., Dimension effect on mechanical behavior of silicon micro-cantilever beams, Measurement, Vol. 41, 2008, pp. 885–895, Figure 7a [15]. Copyright 2008, with permission from Elsevier

6.8.6 Output Data

- A methodology for characterising the mechanical properties of MEMS structures was developed. Mechanical loads were applied to the specimens via a testing microprobe, using a highly precise micro-force testing apparatus. This method allows various specimens to be tested in a

Figure 6.24 Bending fracture of the micro-cantilever beam with width 100 μm, thickness 60 μm and length 700 μm. Reprinted from Pan, C.H. and Liu, P.P., Dimension effect on mechanical behavior of silicon micro-cantilever beams, Measurement, Vol. 41, 2008, pp. 885–895, Figure 7b [15]. Copyright 2008, with permission from Elsevier

larger range of forces and displacements. The mechanical loading is applied to the beam by direct contact between the probe and the free end of the beam. The methodology is able to obtain mechanical properties and stress–strain curves of SCS micro-cantilever beams with various dimensions.

- The typical FMs of SCS micro-cantilever beams and the influence of the dimensions on the failure risks were identified based on SEM analyses.
- By using precise mechanical loading via microprobes, the advantage of the proposed methodology is that it can simultaneously apply a large force and displacement to MEMS devices, especially for thicker structures fabricated by bulk micro-machining and LIGA processes. This indicates that it can conduct testing on both force sensors and actuators such as micro-accelerometers and micro-mirrors, and the methodology might be extended to evaluate other MEMS devices (e.g. optical lenses and sensors). The testing method can also easily be extended to nano-scale specimens by adding a force-magnification lever mechanism.

6.9 Case Study No. 9: MEMS Switches

6.9.1 Subject

MEMS switches.

6.9.2 Goal

To identify the FM leading to contact-resistance increase of MEMS switches during static and cyclic loading. The study was performed by researchers from Centre de Microélectronique de Provence, Georges Charpak, Gardanne, from CEA-Léti, Minatec, Grenoble and from Schneider Electric Industries, Grenoble, France [16].

6.9.3 *Input Data*

- A MEMS switch is composed of a mobile part (bridge or beam) which closes and opens an air gap on a flat or spherical contact area of a few μm^2, with loads ranging between tens of μN up to fractions of mN [17].
- Previous reliability studies revealed two primary FMos: (i) contact-resistance increase and (ii) stiction [18, 19]. Only the first FMo was investigated in this study.
- Gold, which is preferred as a contact material due to its low resistivity and low sensitivity to oxidation, has a relatively low hardness and low softening temperature, which may be the origin of contact degradation by stiction [18, 19]. The solution was to alloy gold with platinum-group metals [18–20], leading to a sevenfold lifetime increase [19], but the same FMs still occur.
- There are some previous results on contact degradation (which is the goal of this investigation): (i) it seems that degradation area increases with current intensity [21] and under mechanical cycling at a force of 200 μN as a function of the number of cycles, being linked to an increase in contact resistance [22]; and (ii) the deformation in the contact area seems to be linked with partial flattening of asperities and pile-up formation occurs during actuation [23].

6.9.4 *Sample Preparation*

- The samples under investigation consisted of a thin metallic film layer sputtered on silicon substrate, which was similar to that found in MEMS-switch contact. The thin layer was composed of 1 μm of gold deposited over a 50 nm tungsten nitride diffusion barrier and a 20 nm tungsten adhesion layer.

6.9.5 *Working Procedure for FA and Results*

- High-resolution electron back-scattering diffraction (EBSD) analysis[1] in field-emission scanning electron microscopes (FESEMs) [24] coupled with mechanical-loading experiments were used to gain an understanding of how mechanically induced microstructural changes can be identified and linked to the degradation of contacts by contact-resistance increase.
- After static and cyclic loading (see below) under spherical indentation of a sputtered gold thin film, high-resolution EBSD was used to investigate microstructural changes in the contact area. The crystallographic texture of the film was determined using EBSD, and a strong {111} fibre texture normal to the gold surface was exhibited with an average 90 nm grain size. It should be noted that electron-penetration depth for EBSD analysis is around 5 nm beneath the surface of gold thin films.
- Complementary atomic-force microscopy (AFM) topography characterisation of the contact area was performed, as was localised nano-indentation measurement of hardness evolution linked to observed microstructural changes.
- Nano-indentation was used to measure the mechanical properties of the gold thin film with a Berkovich indenter using the continuous stiffness measurement (CSM) technique [25]. At a depth of 150 nm, the hardness of the gold thin film was 1.6 GPa and its elastic modulus was 80 GPa.
- Two types of experiment were performed, based on static loading and cyclic loading, respectively.
- Static loads ranging from 50 μN up to 130 mN were applied on the gold thin film at different locations, using the nano-indenter with a 50 μm-radius spherical diamond tip. The corresponding apparent average pressures calculated using Hertz elastic contact law gave an upper bound to

[1] EBSD allows a high spatial resolution as small as 10 nm to be reached.

the applied pressure ranging from 210 Mpa for 0.05 mN static loading up to 2920 Mpa for 130 mN.

- AFM measurements allowed the geometry of the residual indent mark created at the contact location to be determined. A transition from purely elastic deformation at low pressure to plastically dominated behaviour at high pressure [26] was noticed.
- The EBSD characterisation was used to investigate the indent marks already analysed by AFM (Figure 6.25). Comparison of the images demonstrates the one-to-one equivalency between topographic asperities of the grain shapes observed by EBSD. For example, the grain circled in the middle of the indent mark obtained under a 1.6 mN static load suffers 10° rotation in comparison to the {111} initial fibre texture. Only grains in the middle of the indent present a rotation. Therefore, observed deformation is heterogeneous and a $1 \times 1\ \mu m^2$ area is analysed for each indent to determine average grain rotation as a function of apparent average pressure.
- No grain-size modification was observed on any indent. The rotation angle of the grains increased with pressure above 500 MPa. This rotation reached 21° under an equivalent pressure of 3 GPa. The ratio (s : h) of pile-up height to indent depth and the grain rotation followed the same evolution with applied pressure and were representative of the level of plastic deformation (Figure 6.26). These rotations induced the {111} fibre-texture degradation. The central spot and external fringe on the {111} pole figure were larger. In extreme cases, the central spot transformed in a new circle. This type of crystallographic texture deformation is also observed on polycrystals under uniaxial compression [27].
- Cycle loads between 10 and 500 μN were applied. For a contact-bump curvature radius of 10 μm, the MEMS switch pressure was between 240 and 900 MPa. Endurance-cycling experiments were carried out at 470 MPa using a 20 μm curvature-radius conducting gold tip with a dedicated apparatus described elsewhere [28].
- Two types of cycling test were performed: (i) 500 000 cycles in the cold-switching mode contact (opening and closing with contact polarisation set at 0 V and no current allowed) and (ii) 26 000 cycles in the hot-switching mode (under a 5 V constant applied voltage and a maximum allowed current of 1 mA).

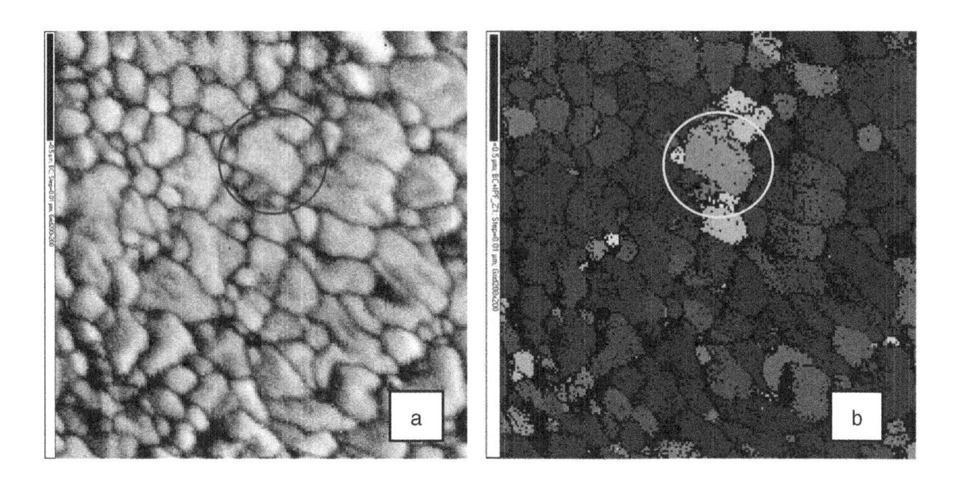

Figure 6.25 Images of the centre-of-indent mark under a 1.6 mN static load, obtained with $2 \times 2\ \mu m^2$; AFM (a) and corresponding EBSD (b). Reprinted from Mandrillon, V., Arrazat, B., Inal, K., Vincent, M. and Poulain, C., MEMS Switch Gold Ohmic Contact: An Investigation of Possible Failure Mechanisms Linked with Microstructural Evolutions Using EBSD, CEA-LETI Internal Report, June 2010, Figure 1 [16]. Reproduced by permission of Vincent Mandrillon (CEA-Léti, Minatec, Grenoble, France)

- After cold-switching, a grain rotation of 30° was noticed, which was higher than that under a 3 GPa static-loading apparent contact pressure. The {111} pole figure obtained had a strong crystallographic texture and was slightly different to static loads. Mechanical cycling induced a larger central spot and external fringe on the {111} pole figure. A nano-indentation measurement was performed in the degradation area. The measured hardness increased by 12% at 150 nm depth compared to

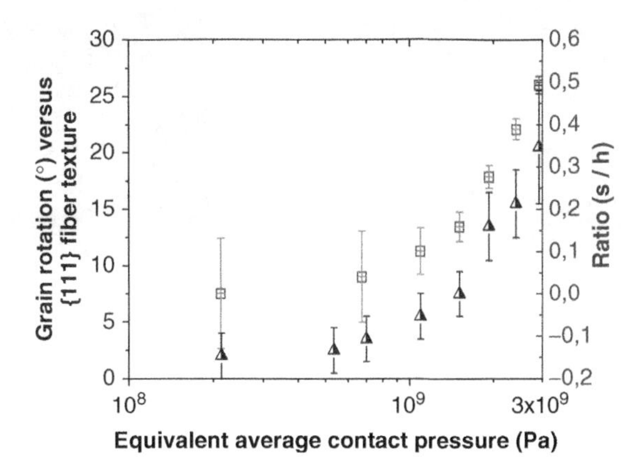

Figure 6.26 Ratio s : h between height of pile-up and indent depth, and grain rotation relative to {111} fibre texture as a function of equivalent average contact pressure (Pa) of spherical indentation (50 μm radius) in the gold thin film. Reprinted from Mandrillon, V., Arrazat, B., Inal, K., Vincent, M. and Poulain, C., MEMS Switch Gold Ohmic Contact: An Investigation of Possible Failure Mechanisms Linked with Microstructural Evolutions Using EBSD, CEA-LETI Internal Report, June 2010, Figure 2 [16]. Reproduced by permission of Vincent Mandrillon (CEA-Léti, Minatec, Grenoble, France)

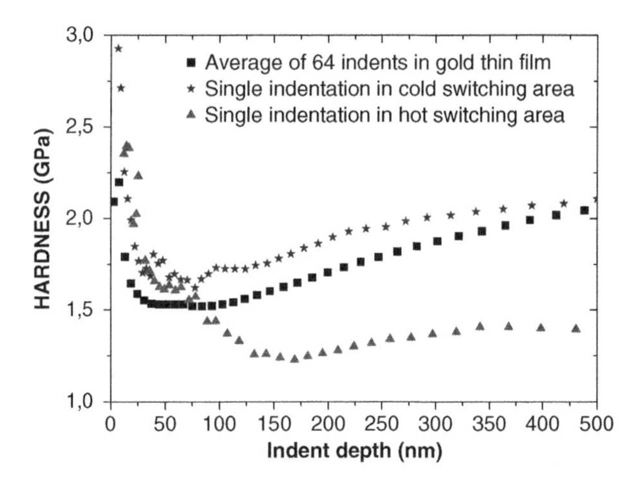

Figure 6.27 Gold thin-film hardness measurement as a function of depth before and after contact loading. Reprinted from Mandrillon, V., Arrazat, B., Inal, K., Vincent, M. and Poulain, C., MEMS Switch Gold Ohmic Contact: An Investigation of Possible Failure Mechanisms Linked with Microstructural Evolutions Using EBSD, CEA-LETI Internal Report, June 2010, Figure 4 [16]. Reproduced by permission of Vincent Mandrillon (CEA-Léti, Minatec, Grenoble, France)

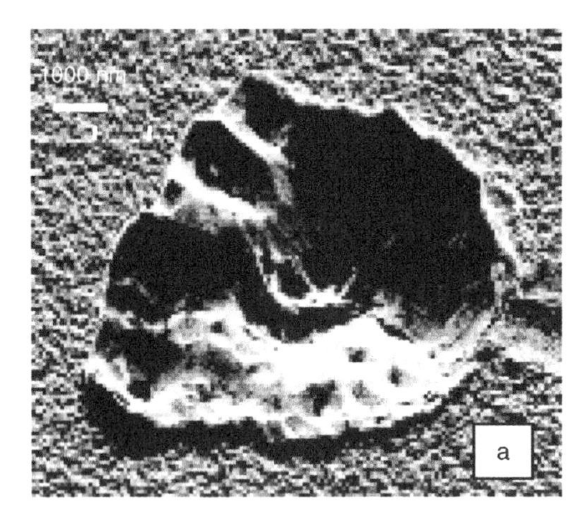

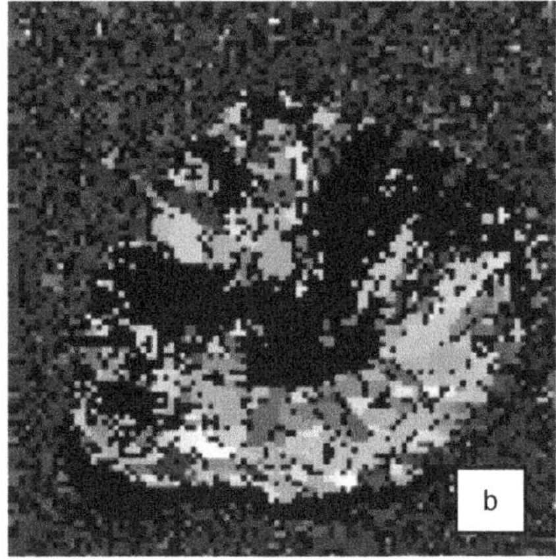

Figure 6.28 10 × 10 μm SEM image (a) and EBSD map (b) of the contact area of a micro-switch after 26 000 hot-switching cycles under 300 μN load. Dark areas in EBSD image correspond to orientation detection failure due to local surface tilt. Reprinted from Mandrillon, V., Arrazat, B., Inal, K., Vincent, M. and Poulain, C., MEMS Switch Gold Ohmic Contact: An Investigation of Possible Failure Mechanisms Linked with Microstructural Evolutions Using EBSD, CEA-LETI Internal Report, June 2010, Figure 3 [16]. Reproduced by permission of Vincent Mandrillon (CEA-Léti, Minatec, Grenoble, France)

the original gold thin film (Figure 6.27). No grain-size modification was observed. Therefore, the measured hardness increase was due to strain-hardening.

- Following hot-switching mode, a grain rotation of 25° was obtained, close to the value obtained after cold-switching. Moreover, the resulting crystallographic texture of the contact area was extremely different from that in previous tests (Figure 6.28). A strong degradation of the {111} fibre texture was identified, but no particular orientation was predominant. The grain size increased in contact area. The average grain size was around 250 nm, whereas the initial average grain size of gold thin film is around 90 nm. The thermal energy produced during hot-switching (arcing or melting) induced an abnormal recrystallisation. The grain size increased and the resulting crystallographic texture differed from previous solicitations and was the expression of a different mechanism of degradation.
- A nano-indentation measurement in the degradation area showed that, at a depth of 150 nm, the measured hardness decreased by 22% compared to the initial gold thin film. A comparison with the results obtained in a Cao study [41] into the decrease of gold hardness with grain size is presented in Figure 6.29.

6.9.6 Output Data

- Experimental results led to the identification of the basic degradation mechanisms of MEMS-type switches, under static and cyclic loads.
- The hardness measurements were coherent with microstructural evolution observed by EBSD.
- The EBSD-determined grain size confirmed the idea that thermal events occurring during hot-switching are strong enough to be significant for all grain volume and to induce grain growth, resulting in a localised decrease in hardness. This decrease in hardness together with grain growth corresponding to a decrease in contact roughness may be an explanation for contact failure by stiction.

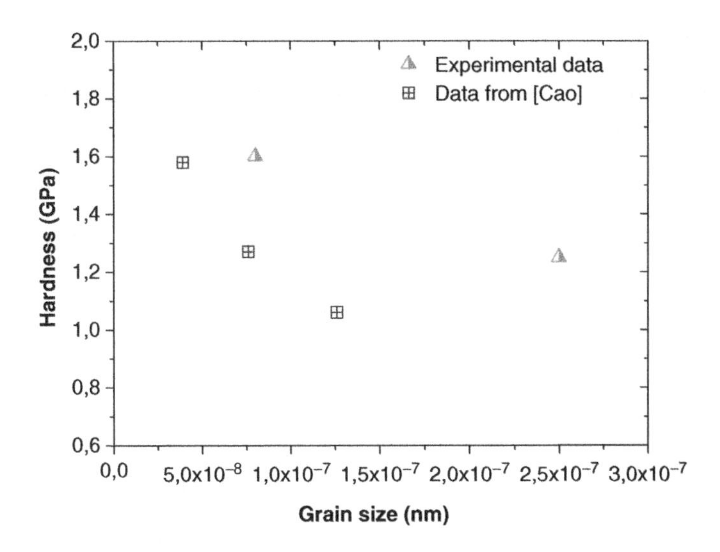

Figure 6.29 Measured hardness vs grain size. Reprinted from Mandrillon, V., Arrazat, B., Inal, K., Vincent, M. and Poulain, C., MEMS Switch Gold Ohmic Contact: An Investigation of Possible Failure Mechanisms Linked with Microstructural Evolutions Using EBSD, CEA-LETI Internal Report, June 2010, Figure 5 [16]. Reproduced by permission of Vincent Mandrillon (CEA-Léti, Minatec, Grenoble, France)

6.10 Case Study No. 10: Magnetic MEMS Switches

6.10.1 Subject

An ultra-miniature magnetic switch fabricated using MEMS technology.

6.10.2 Goal

To identify the FMs responsible for the stiction of the ultra-miniature switch, developed by Schneider Electric in collaboration with the CEA-LETI [29, 30], and to elaborate corrective actions for diminishing (or avoiding) this FM. The study was performed by researchers from CEA- CEA-Léti, Minatec, Grenoble, from Schneider Electric Industries, Grenoble, from Grenoble Electrical Engineering Lab, Grenoble INP, Saint Martin d'Hères and from Laboratoire de Génie Electrique de Paris, Universités Paris 6 et Paris 11, Gif-sur-Yvette, France [31].

6.10.3 Input Data

- The magnetic MEMS switch has to remain highly reliable after performing 10^8 cycles in hot-switching conditions, at 3 V/10 μA (resistive load).
- The main reliability criterion is the contact resistance, which has to remain lower than 10 kΩ over the whole lifetime.

6.10.4 Sample Preparation

- The MEMS switch under study is a magnetically actuated on/off switch composed of a 800 × 800 μm moving ferromagnetic plate supported by two torsion arms, which allow its rotation above the silicon substrate, as shown in Figure 6.30. The two bumps at the extremity of the plate form the mobile contacts that come into contact with an electrical line patterned on the substrate, representing the fixed contacts. The plate is actuated by an external moving magnet which applies a contact force of typically 200 μN on each contact pair.

- Two variants of the MEMS switch were fabricated. The first was made with 1.5 μm-thick gold contacts obtained by electro-deposition (mobile contact) and sputtering (fixed contact). After the first series of experiments, a second variant was fabricated, with a thin layer (100 nm) of Ruthenium (Ru) deposited by sputtering on top of the mobile and fixed gold contacts. This allowed a low initial contact resistance (typically 0.9 Ω for an applied contact force of 200 μN) and a hard contact surface.

6.10.5 *Working Procedure for FA and Results*

- For the first variant, the initial contact resistance was around 0.4 Ω for an applied contact force of 200 μN. No significant contact resistance increase was noticed after hot-switching tests at 3 V/10 μA (resistive load). However, 16% of the components exhibited an FMo caused by permanent stiction of the contacts.
- The surface of the electrical contacts after endurance tests was examined by SEM, with the results shown in Figure 6.31. Both mobile and fixed contact surfaces revealed important changes in the topography: material wear and melted areas. This behaviour might be explained by the low hardness and high ductility of gold [32], leading to important material deformation – even under low contact forces – and a large contact area. Adhesion forces between the two contacts are consequently high and overcome the opening force available in the switch. This phenomenon is typical to MEMS components, in which restoring forces are rather small and are close to adhesion forces.
- Consequently, a second fabrication variant, with Ru as an alternative contact material, was introduced and tested. Ruthenium has a high hardness and a low ductility but is significantly more resistive than gold. So in order to maintain a low contact resistance, a thin layer (100 nm) of Ru was deposited by

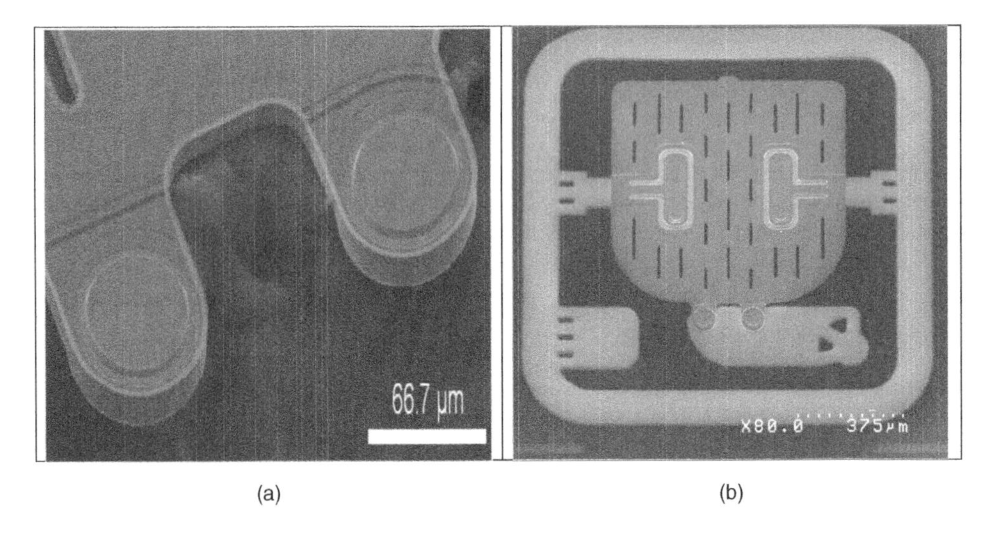

Figure 6.30 The magnetic MEMS switch: (a) details of the bifurcated contacts and (b) global view. Reprinted from Poulain, C., Vincent, M., Chiesi, L., Delamare, J., Houzé, F. and Sibuet, H., Improvement of Electrical Contacts Reliability in a Magnetic MEMS Switch, CEA-LETI Internal Report, 2010, Figure 1 [31]. Reproduced by permission of Cristophe Poulain (CEA-Léti, Minatec, Grenoble, France)

sputtering on top of the mobile and fixed gold contacts. This allowed a low initial contact resistance (typically 0.9 Ω for an applied contact force of 200 μN) and a hard contact surface.

- Components with Ru-coated contacts have been tested in the same conditions as described above. During the tests, none of the switches encountered permanent sticking of the contacts. Most of the components had, on average, a higher contact resistance at the end of the tests, but without reaching the failure criterion (>10 k Ω).
- The SEM analysis of the contact surfaces following the endurance tests is presented in Figure 6.32. No topographic modifications were identified on Ru contact surfaces, compared to Au–Au contacts. Hence, Ru thin-film coatings seem to be highly efficient in limiting material deformation and wear.

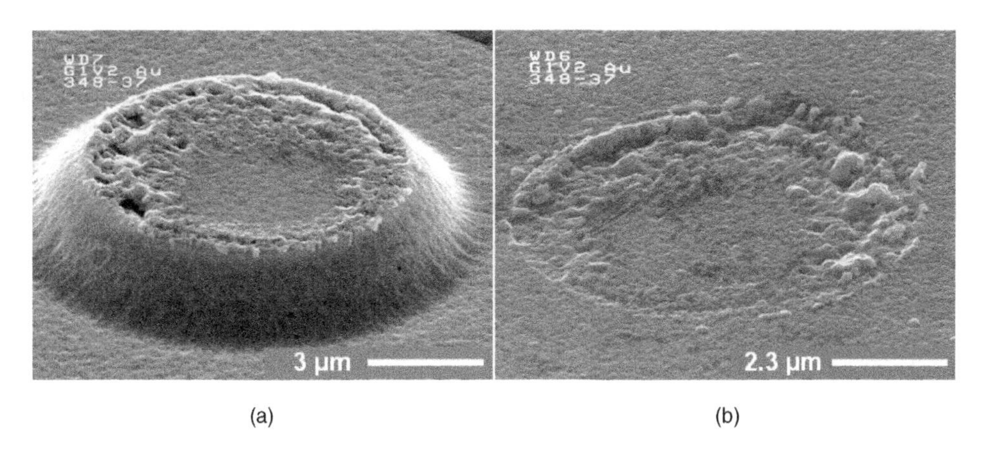

Figure 6.31 SEM observation of the gold electrical contacts after 108 cycles under hot-switching 3 V/10 μA: (a) mobile contact and (b) fixed contact. Reprinted from Poulain, C., Vincent, M., Chiesi, L., Delamare, J., Houzé, F. and Sibuet, H., Improvement of Electrical Contacts Reliability in a Magnetic MEMS Switch, CEA-LETI Internal Report, 2010, Figure 2 [31]. Reproduced by permission of Cristophe Poulain (CEA-Léti, Minatec, Grenoble, France)

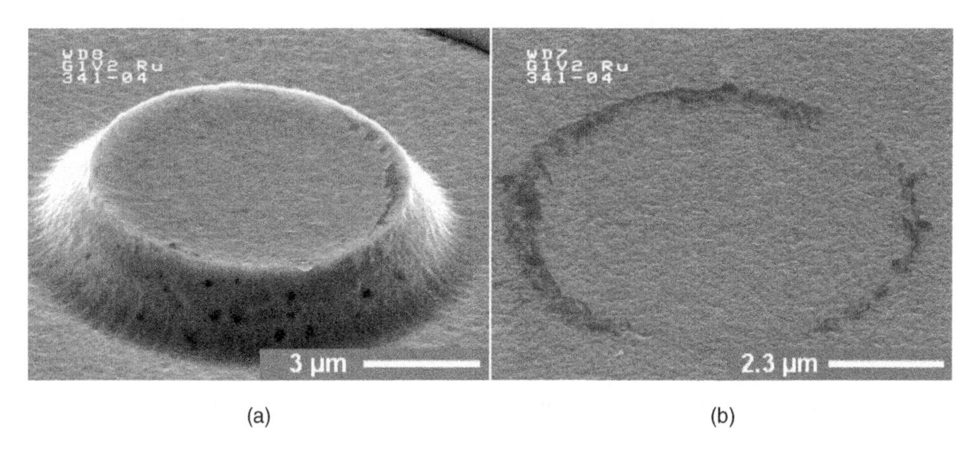

Figure 6.32 SEM observation of the ruthenium-coated electrical contacts after 108 cycles under hot-switching 3 V/10 μA: (a) mobile contact and (b) fixed contact. Reprinted from Poulain, C., Vincent, M., Chiesi, L., Delamare, J., Houzé, F. and Sibuet, H., Improvement of Electrical Contacts Reliability in a Magnetic MEMS Switch, CEA-LETI Internal Report, 2010, Figure 3 [31]. Reproduced by permission of Cristophe Poulain (CEA-Léti, Minatec, Grenoble, France)

This limits adhesion forces between the two contact surfaces to a rather low value; enough to avoid contact stiction phenomena.
- Black areas could be observed on the contact surfaces, mostly on the edges of the contact. These black traces are called 'frictional polymers' and are generated when catalytic contact materials, such as Ru, are repeatedly brought into contact in the presence of carbonaceous contamination [33]. These polymers are resistive and can lead to high contact resistance. However, in this case the failure criterion of 10 k Ω was never reached; contact resistance always remained under 800 Ω during the tests. Ru, despite being an active catalytic material, is well suited to solving the problem of contact stiction of MEMS switches.

6.10.6 Output Data

- The use of a multilayer contact material made of a thick Au coating covered by an additional thin Ru layer solves the main issue encountered during MEMS-switch operation: permanent contact stiction.

6.11 Case Study No. 11: Chip-Scale Packages

6.11.1 Subject

A lead-free chip-scale package (CSP).

6.11.2 Goal

To study the FMs for board-level lead-free CSPs and to increase the reliability of solder interconnections in electronics assemblies through an improved understanding of failure. The study was performed by researchers from Cal Poly State University, San Luis Obispo, CA, USA [34].

6.11.3 Input Data

- The solder provides the mechanical and electrical interconnection between the package and printed circuit board (PCB), with each failure of a solder interconnection leading to the failure of a whole PCB. Any accidental drop of the PCB greatly increases the failure risk. Therefore, many studies have been performed on the drop reliability of solder interconnections.
- In recent years, the European Union's Restriction of Hazardous Substances (RoHS) and similar legislation around the world has determined a shift toward lead-free solder alloys (e.g. tinsilver-copper-SnAgCu). Poor drop-test reliability of lead-free solders has been reported [35], because SnAgCu alloys have higher strength and modulus than SnPb. Two methods have been proposed to improve drop-test reliability: lower Ag content [36] and the addition of micro-alloying [37]. Both methods have resulted in many SAC alloys.
- However, the FMs of lead-free solder interconnections remain unclear.

6.11.4 Sample Preparation

- The test vehicle was designed according to the requirements of the Joint Electron Device Engineering Council (JEDEC) for drop tests: 15 CSPs were assembled, each with 228 daisy-chained 0.5 mm pitch solder joints using Sn-3.0 wt% Ag-0.5 wt% Cu (SAC305) lead-free solder.
- The electrical continuity of each daisy-chained CSP was verified before drop-testing.
- The drop tests were performed using three different peak accelerations of 900, 1500 and 2900 g, with 0.7, 0.5 and 0.3 ms pulse durations, respectively, until a majority of the CSPs experienced electrical failure.

6.11.5 Working Procedure for FA

- A dye-penetrant method was used to examine pre-existing cracks in solder joints, by submerging them into a thermal-curable dye so that the dye penetrated into the crack area. After the dye was cured, the components were pulled off and the cracks were examined using metallographic microscopes.
- The specimens were prepared by cutting sections out of the PCB with a diamond abrasive saw, then the CSP and board section were mounted in a clear epoxy, polished to expose the solder-joint cross-section, and fine-polished using 0.3 and 0.05 alumina slurries. The last steps were etching and sputter-coating with a very thin layer of gold. Cross-sectioned samples were etched with a 1% nitol etchant to evaluate the microstructure.
- SEM with EDS was used to investigate the crack location and intermetallic composition in solder joints.
- Micro-hardness tests were carried out on both the solder and the resin to compare the strengths of the materials. The tests used a Vickers micro-hardness indenter at 100 g of force.
- The FMs and solder-joint locations were mapped.

6.11.6 Results and Discussion

- The FA was performed for 60 components on four boards, which were dye-penetrated. Five FMs have been identified (presented from most to least frequent): (i) pad-cratering (more than 80% of the failed items); (ii) solder-cracking near the board side; (iii) solder-cracking near the component side; (iv) input/output (I/O) trace fracture; and (v) daisy-chain trace fracture. The results imply that the solder joint is more reliable than the PCB during a drop test.
- Pad-cratering was produced by the cracking of the thin resin-rich region beneath the copper pads and traces, as shown in Figure 6.33. It should be pointed out that pad-cratering does not necessarily lead to electrical failure of solder joints.
- It seems that the component location plays a significant role, with the components along the board centre tending to fail earliest and most frequently. For example, pad-cratering was present on nearly every CSP, mainly in the corners of the CPS, after relatively few drop impacts, and may have existed

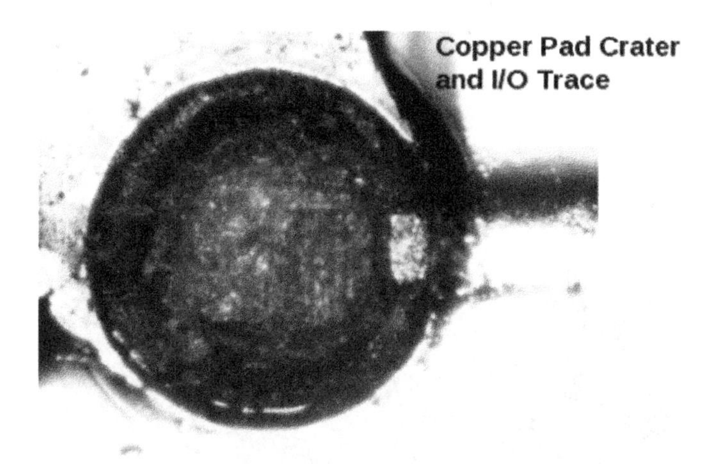

Figure 6.33 Copper pad-cratering with dyed board fibres, 100× magnification. Reprinted from Vickers, N. *et al.*, Board Level Failure Analysis of Chip Scale Package Drop Test Assemblies, Proceedings of the 41st International Symposium on Microelectronics, Providence, RI, Nov. 2008, Figure 3 [34]. Reproduced by permission of International Microelectronics and Packaging Society (IMAPS)

before electrical failure occurred. In a few cases electrical failures occurred through trace cracking without pad-cratering or solder fracture.

- The second most common FM is I/O trace fracture (Figure 6.34). All components that exhibit I/O trace failures also exhibit pad-cratering, which may lead to the conclusion that the two FMs are coupled. Similarly, all daisy-chain trace-fractured components exhibit pad-cratering, implying that daisy-chain trace fracture is caused by pad-cratering.
- Cross-sectioning SEM allows identification of three FMs:
  - *Pad-cratering* (Figure 6.35), which was observed near the corners of the component. This phenomenon is caused by resin cracks, which never penetrated into the weave region and occurred only in the resin-rich region above the fibre weave near copper pads and traces. Resin cracks exhibited brittle fracture characteristics.
  - *Daisy-chain trace-cracking* and *solder-cracking* (Figure 6.36), which are less common FMs.

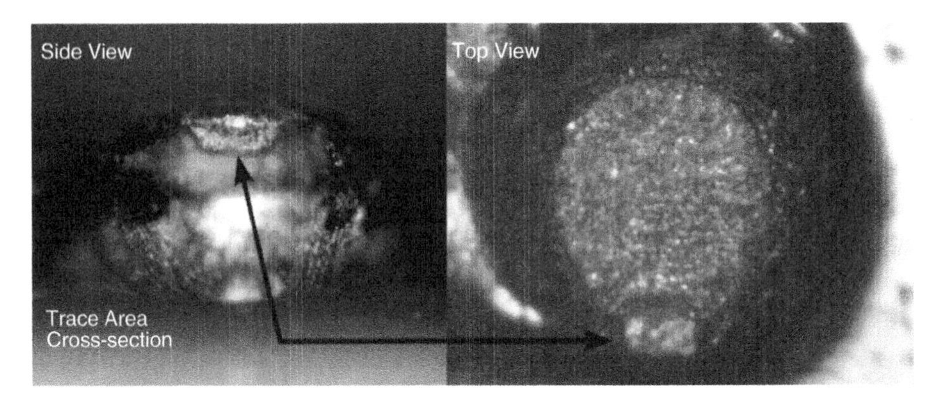

Figure 6.34 I/O trace cracked away from solder joint, 100× magnification. Reprinted from Vickers, N. *et al.*, Board Level Failure Analysis of Chip Scale Package Drop Test Assemblies, Proceedings of the 41st International Symposium on Microelectronics, Providence, RI, Nov. 2008, Figure 9 [34]. Reproduced by permission of International Microelectronics and Packaging Society (IMAPS)

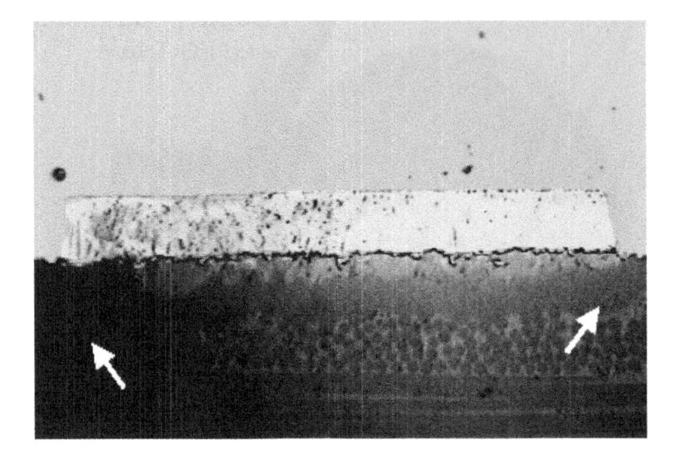

Figure 6.35 Resin cracks beneath copper pad, 500× magnification. Reprinted from Vickers, N. *et al.*, Board Level Failure Analysis of Chip Scale Package Drop Test Assemblies, Proceedings of the 41st International Symposium on Microelectronics, Providence, RI, Nov. 2008, Figure 12 [34]. Reproduced by permission of International Microelectronics and Packaging Society (IMAPS)

- The solder ball microstructure was investigated. As shown in Figures 6.37 and 6.38, small dendritic grain growth was noticed. The smaller grains strengthen the solder material. Consequently, the failure process is shifted to the weaker material, which is the resin (as found by comparing the relative strengths of the solder material and the resin, the average micro-hardness values being 72 and 33, respectively).

Figure 6.36 Solder crack near board-side IMC shown on the right side of the picture. On the left is evidence of a daisy-chain trace fracture on the board side at 100× magnification. Reprinted from Vickers, N. *et al.*, Board Level Failure Analysis of Chip Scale Package Drop Test Assemblies, Proceedings of the 41st International Symposium on Microelectronics, Providence, RI, Nov. 2008, Figure 13 [34]. Reproduced by permission of International Microelectronics and Packaging Society (IMAPS)

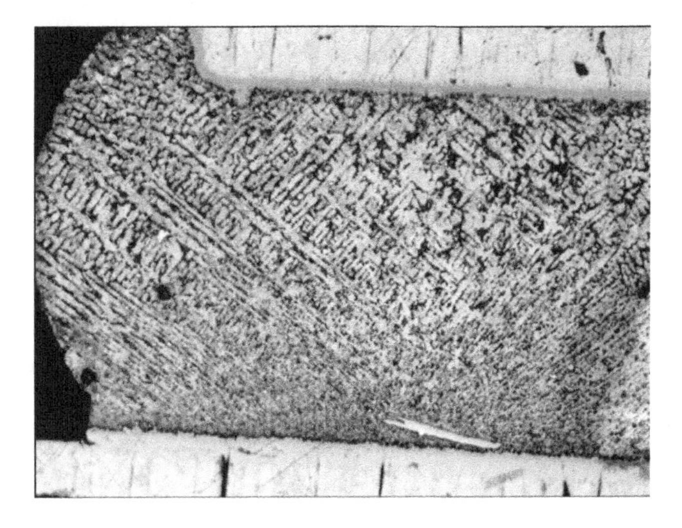

Figure 6.37 Image of solder-ball microstructure at 200× magnification. Reprinted from Vickers, N. *et al.*, Board Level Failure Analysis of Chip Scale Package Drop Test Assemblies, Proceedings of the 41st International Symposium on Microelectronics, Providence, RI, Nov. 2008, Figure 14a [34]. Reproduced by permission of International Microelectronics and Packaging Society (IMAPS)

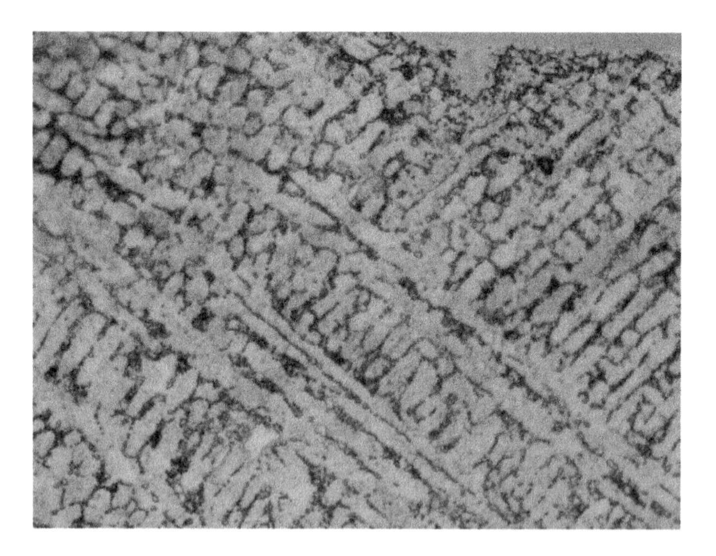

Figure 6.38 Image of solder-ball microstructure at 500× magnification. Reprinted from Vickers, N. *et al.*, Board Level Failure Analysis of Chip Scale Package Drop Test Assemblies, Proceedings of the 41st International Symposium on Microelectronics, Providence, RI, Nov. 2008, Figure 14b [34]. Reproduced by permission of International Microelectronics and Packaging Society (IMAPS)

6.11.7 Output Data

- The primary FM of the material is pad-cratering.
- The weakest link in the electronic assembly is the board material, which will fail first, and will initiate other FMs such as trace-cracking.
- Trace-cracking due to pad-cratering is a more common FM than solder-cracking, especially where traces are connected to outer-array solder joints and run outward from the CSP.
- The majority of failures occur at the corners of the packages, implying that the corners of the components experience the most strain and stress.

6.12 Case Study No. 12: Solder Joints

6.12.1 Subject

Solder joints of active electronic components.

6.12.2 Goal

To study the reliability of the eutectic Sn-Ag-Cu as a candidate for replacing the eutectic Sn-Pb-Ag solder joints, in an effort to promote lead-free solders[2] with a comparable or, preferably, better reliability. The study was performed by the Interuniversitary MicroElectronics Centre (IMEC), Leuven, Belgium [38].

6.12.3 Input Data

- Possible FMs: (i) thermally activated solder fatigue and (ii) formation of intermetallic layers at the under-bump metallisation (UBM), which causes a premature brittle interfacial fracture.

[2] In Europe, any use of lead in electronics has been forbidden since July 2006, except in some special applications.

6.12.4 Sample Preparation

- Four kinds of sample were prepared by combining two types of package (polymer stud grid array (PSGA) and optimised[3] PSGA packages), both soldered on a PCB using either eutectic Sn-Ag-Cu solder or eutectic Sn-Pb-Ag solder.
- Reliability tests: thermal cycling in the range −40 to +125 °C (1 hour cycle time, air to air, 15 minutes dwell time).

6.12.5 Working Procedure for FA and Results

- The results of reliability tests demonstrated the higher reliability of lead-free solder by comparing the number of thermal cycles for 50% failures: 1800 for Sn-Pb-Ag (4500 for optimised package) and 7000 for Sn-Ag-Cu (10 500 for optimised package).
- Cross-sectioning and SEM analyses (including backscattered electron imaging (BEI) mode) allowed the typical FMs to be identified.
- For Sn-Pb-Ag solder and non-optimised package, two FMs were identified (Figure 6.39), obtained at a corner joint, after 5900 cycles: (i) fatigue crack formation between the Pb and Sn grains in the bulk of the solder, along a rather straight horizontal path following the direction of the highest strain; and (ii) fracture at the interface with the UBM, where brittle Au-containing intermetallics are formed during the thermal cycling.
- Quantitative EDS microanalysis allowed the intermetallics to be identified (Figure 6.40): a thin continuous layer of Ni_3Sn4 phase covered by two Au-containing phases, namely $(Au,Ni)_3Sn_4$ and $(Au,Ni)Sn_4$. The latter forms a continuous layer, which is responsible for a brittle fracture [40], seen in Figure 6.40 as a long straight crack between the $(Au,Ni)Sn_4$ and $NiSn_4$ layers.

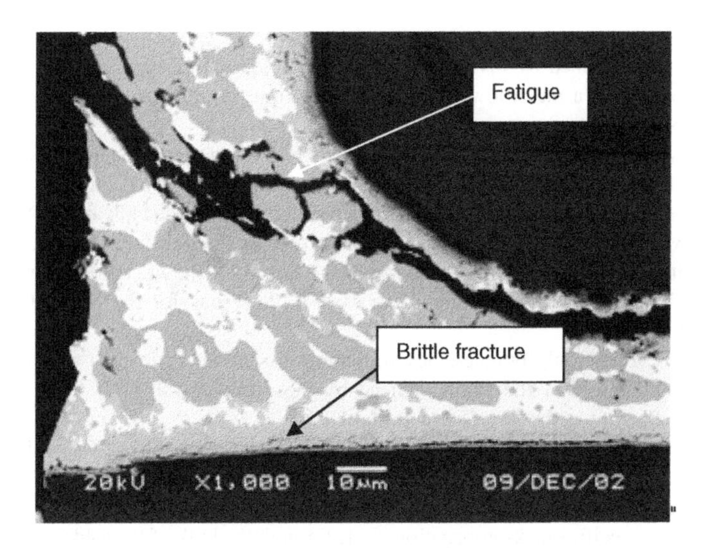

Figure 6.39 Corner Sn-Pb-Ag solder connection, high strain, failed after 1600 cycles (cycled up to 5400 cycles). Two different FMs are present: solder fatigue and brittle interface fracture. Reprinted from Ratchev, P., Vandevelde B. and De Wolf, I., Reliability and Failure Analysis of Sn-Ag-Cu Solder Interconnections for PSGA Packages on Ni/Au Surface Finish, Proceedings of the 10th International Symposium on the Physical and Failure Analysis of Integrated Circuits, 7–11 July 2003, pp. 113–116, Figure 2. © 2003 IEEE [38]

[3] The optimised packages differ only in the over-mould material used, which provides a CTE closer to that of the package polymer body. This results in a five to seven times smaller plastic strain on the critical corner interconnects during thermal cycling [39].

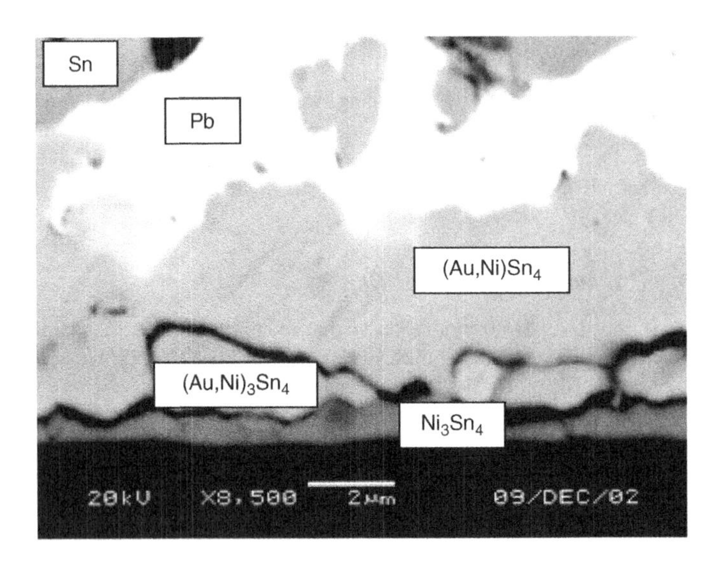

Figure 6.40 The interface between Sn-Pb-Ag (high strain) and under-bump metallisation (UBM) of a joint. Brittle fracture occurs at the interface with the UBM due to the formation of a continuous layer of the brittle (Au,Ni)Sn_4 phase. Reprinted from Ratchev, P., Vandevelde B. and De Wolf, I., Reliability and Failure Analysis of Sn-Ag-Cu Solder Interconnections for PSGA Packages on Ni/Au Surface Finish, Proceedings of the 10th International Symposium on the Physical and Failure Analysis of Integrated Circuits, 7–11 July 2003, pp. 113–116, Figure 3. © 2003 IEEE [38]

- A cross-check was obtained by investigating the next solder connection of the same failed item, a non-corner joint, with a much lower stress concentration. In preview, only the failure by brittle fracture was noticed.
- For Sn-Ag-Cu solder and optimised package, only one FM was noticed, the fatigue-crack formation. Actually, the same intermetallics form at the UBM interface, but not as a continuous layer, because a significant amount of the Au, dissolved in the solder, precipitates in the bulk as (Au,Ni)Sn_4 particles, which naturally reduces the amount of this phase formed at the interface. Therefore a continuous layer of the brittle (Au,Ni)Sn_4 intermetallic cannot be formed at the interface. Consequently, under the same conditions, the brittle-fracture mode is less dangerous for the Sn-Ag-Cu than for the Sn-Pb-Ag solder.
- For optimised packages, only fatigue-crack formation was observed in both solders. This demonstrates the important role of the strain induced by the different values of CTE in initiating the second FM, the fracture at the interface with the UBM.
- However, it is important to mention that, for optimised packages, the Sn-Ag-Cu solder shows a considerably better reliability than the standard Sn-Pb-Ag solder. This can be explained by a new solder fatigue mechanism, which is found to take place in the Sn-Ag-Cu solder: the 'web-like' solder fatigue. This is related to a specific crack-propagation mode through the bulk of the solder, in a weblike way, linking the brittle (Au,Ni)Sn_4 particles. This is beneficial to joint reliability, because it deviates the crack from the shortest path of development and slows down its kinetics of propagation.

6.12.6 Output Data

- The Sn-Ag-Cu solder shows a considerably better reliability than the standard Sn-Pb-Ag solder. Hence Sn-Ag-Cu solder is a valuable solution as a lead-free material for solder joints.

Table 6.1 Synthesis of the 12 case studies, including their relation to previous chapters

Case study no.	Subject	Technological family or operation	FA techniques used during case study/section of Chapter 4	FM identified by the case study/section of Chapter 5
1	$PbZr_xTi_{1-x}O_3$ (PZT) capacitors	Passive electronic parts	Constant voltage stress (CVS) tests/Section 4.3.1 SEM/Section 4.3.5	Time-dependent dielectric breakdown/Section 5.1.1.2
2	Insulated-gate bipolar transistor (IGBT)	Silicon bipolar technology	Decapsulation techniques/Section 4.2.2 X-ray energy-dispersive spectrometry (EDS) inspection/Section 4.3.5 SEM/Section 4.3.5	Interdiffusion of Si into metal/Section 5.1.1.3 Overcurrent/Section 5.3.3
3	8-bit CMOS microcomputer	MOS technology	Decapsulation techniques/Section 4.2.2 Optical microscopy/Section 4.3.2 Techniques for sample preparation/Section 4.2.5 SEM/Section 4.3.5	Electrostatic discharge/Section 5.4.3 Al-Si interdiffusion/Section 5.1.1.3
4	n-channel MOS field-effect transistor (MOSFET)	MOS technology	Optical microscopy/Section 4.3.2 X-ray package inspection/Section 4.3.5 Electrical verification with curve tracer/Section 4.3.1 Decapsulation techniques/Section 4.2.2 Scanning acoustic microscopy (SAM)/Section 4.3.8 Removal of layers/Section 4.2.5	Poor wire-bonding process/Section 5.1.2.3
5	Amorphous silicon (a-Si:H) thin-film transistors (TFT)	Optoelectronic and photonic technologies	Electrical methods/Section 4.3.1	Hot electron degradation/Section 5.1.1.2

(continued overleaf)

Table 6.1 (*continued*)

Case study no.	Subject	Technological family or operation	FA techniques used during case study/section of Chapter 4	FM identified by the case study/section of Chapter 5
6	GaN heterojunction field-effect transistors (HFETs)	Non-silicon technologies	Electrical analysis/Section 4.3.1 Scanning tunnelling electron microscopy (STEM)/Section 4.3.3 Focused ion beam (FIB)/Section 4.2.4	Schottky barrier height (SBH) increase/Section 5.6.2
7	Silicon-based MEMS resonators	Microsystem technology	Electrical analysis/Section 4.3.1	Fatigue in thin-film silicon/Section 5.8.1
8	Micro-cantilever beams achieved by bulk micro-machining	Microsystem technology	Electrical analysis/Section 4.3.1 SEM/Section 4.3.5	Bent fracture of the beam/Section 5.8.1
9	MEMS switches	Microsystem technology	Electron back-scattering diffraction (EBSD) Atomic-force microscopy (AFM)/Section 4.3.3 Nano-indentation/Section 4.3.15	Decrease of gold hardness/ Section 5.8.1
10	Ultra-miniature magnetic switch	Microsystem technology	Electrical analysis/Section 4.3.1 SEM/Section 4.3.5	Contact stiction/Section 5.8.1
11	Lead-free chip-scale package	Packaging	Electrical analysis/Section 4.3.1 SEM with EDS/Section 4.3.5 Micro-hardness tests/Section 4.3.15 Cross-sectioning/Section 4.2.3	Pad-cratering/Section 5.1.2.4 I/O trace fracture/Section 5.1.2.4
12	Solder joints of active electronic components	Mounting	Cross-sectioning/Section 4.2.3 SEM with backscattered electron imaging (BEI mode) and EDS/Section 4.3.5	Solder fatigue/Section 5.1.2.3 Brittle interface fracture/Section 5.1.2.3

6.13 Conclusions

Twelve case studies have been presented, referring to FA of various electronic components described in Chapter 5. In these case studies, some of the FA techniques described in Chapter 4 have been used. For ease of comparison, Table 6.1 contains a synthesis of all the above case studies.

References

1. Bouyssou, E. *et al.* (2007) Extended reliability study of high density PZT capacitors: intrinsic lifetime determination and wafer level screening strategy. 45th Annual International Reliability Physics Symposium, Phoenix, 2007, pp. 433–438.
2. Ey, Wu. and Vollertsen, R.P. (2002) On the weibull shape factor of intrinsic breakdown of dielectric films and its accurate experimental determination – part I: theory, methodology, experimental techniques. *IEEE Trans. Electron Devices*, **49** (12), 2131–2140.
3. Bouyssou, E. *et al.* (2005) Lifetime extrapolation of PZT capacitors. *Integr. Ferroelectr.*, **73** (1), 49–56.
4. Mura, G. and G. Cassanelli (2006) Failure analysis and field failures: a real shortcut to reliability improvements. Therminic 2006, Nice, Côte d'Azur, France, 27–29 September 2006.
5. Cassanelli, G. *et al.* (2005) Reliability prediction in electronic industrial applications. *Microelectron. Reliab.*, **45**, 1321–1326.
6. Pintus, R. *et al.* (2004) Quantitative 3D reconstruction from BS imaging. *Microelectron. Reliab.*, **44**, 1547–1552.
7. Remo, N.C. and Fernandez, J.C.M. (2005) A reliable failure analysis methodology in analyzing the elusive gate-open failures. Proceedings of 12th IPFA 2005, Singapore, pp. 185–189.
8. Palermo, L.B. and Moreno, N. (2004) An effective curve trace testing approach for gate-open failures in DMOS devices. 14th ASEMEP National Technical Symposium, Manila, 2004.
9. Long, K. *et al.* (2006) Stability of amorphous-silicon TFTs deposited on clear plastic substrates at 250 °C to 280 °C. *IEEE Electron Device Lett.*, **27** (2), 111–113.
10. Allee, D.R. *et al.* (2008) Degradation effects in a-Si:H thin film transistors and their impact on circuit performance. Proceedings of the 46th Annual International Reliability Physics Symposium, Phoenix, 2008, pp. 158–167.
11. Roblin, P., Samman, A. and Bibyk, S. (1998) Simulation of hot electron trapping and aging of *n*MOSFETs. *IEEE Trans. Electron Devices*, **35**, 2229–2237.
12. Venugopal, S.M. *et al.* (2006) Threshold-voltage recovery of a-Si:H digital circuits. *J. Soc. Inf. Displays*, **14** (11), 1053–1057.
13. Singhal, S. (2006) Ga-on-Si failure mechanisms and reliability improvements. 44th Annual International Reliability Physics Symposium, San Jose, IRPS 2006, pp. 95–98.
14. Kim, B. *et al.* (2007) Frequency stability of wafer-scale film encapsulated silicon based MEMS resonators. *Sens. Actuators A Phys.*, **136** (1), 125–131.
15. Liu, H.-K., Pan, C.H. and Liu, P.-P. (2008) Dimension effect on mechanical behavior of silicon micro-cantilever beams. *Measurement*, **41**, 885–895.
16. Mandrillon, V., Arrazat, B., Inal, K., Vincent, M. and Poulain, C. (2010) MEMS Switch Gold Ohmic Contact: An Investigation of Possible Failure Mechanisms Linked with Microstructural Evolutions Using EBSD, CEA-LETI Internal Report, June 2010.
17. Rebeiz, G.M. and Muldavin, J.B. (2001) RF MEMS switches and switch circuits. *Microw. Mag.*, **2** (4), 59–71.
18. Coutu, R.A., Kladitis, P.E., Starman, L.A. and Reid, J.R. (2004) A comparison of micro-switch analytic, finite element, and experimental results. *Sens. Actuators A Phys.*, **115** (2–3), 252–258.
19. Coutu, R.A. *et al.* (2006) Microswitches with sputtered Au, AuPd, Au-on-AuPt, and AuPtCu alloy electric contacts. *IEEE Trans. Compon. Packag. Technol.*, **29** (2), 341–349.
20. Chen, L. *et al.* (2007) Contact resistance study of noble metals and alloy films using a scanning probe microscope test station. *J. Appl. Phys.*, **102** (7), 1–7.
21. Yan, X. *et al.* (2003) Finite element analysis of the thermal characteristics of MEMS switches 12th International Conference on Transducers, Solid-State Sensors, Actuators and Microsystems, Vol. 1, June 2003, pp. 412–415.
22. McGruer, N.E. *et al.* (2006) Mechanical, thermal, and material influences on Ohmic-Contact-Type MEMS switch operation. Proceedings of the IEEE Micro Electro Mechanical Systems, pp. 230–233.

23. Gregori G. and Clarke, D.R. (2006) The interrelation between adhesion, contact creep, and roughness on the life of gold contacts in radio-frequency microswitches. *J. Appl. Phys.*, **100** (9), 1.2363745 (10 pages).
24. Humphreys, F.J. (2001) Review grain and subgrain characterization by electron backscatter diffraction. *J. Mater. Sci.*, **36** (16), 212–216.
25. Pharr, G.M., Strader, J.H. and Oliver, W.C. (2009) Critical issues in making small-depth mechanical property measurements by nanoindentation with continuous stiffness measurement. *J. Mater. Res.*, **24** (3), 653–666.
26. Pharr, G.M. and Taljat, B. (2004) Development of pile-up during spherical indentation of elastic– plastic solids. *Int. J. Solids Struct.*, **41**, 3891–3904.
27. Kalidindi, S.R., Bronkhorst, C.A. and Anand, L. (1992) Polycrystalline plasticity and the evolution of crystallographic texture in FCC metals. *Phil. Trans. R. Soc. London*, **341**, 443–477.
28. Vincent, M. *et al.* (2009) An original apparatus for endurance testing of MEMS electrical contact materials. Proceedings of the 55th IEEE Holm Conference on Electrical Contacts, September 2009, pp. 288–292.
29. Coutier, C. *et al.* (2009) A new magnetically actuated switch for precise position detection. Transducers'09, 2009, pp. 861–864.
30. Vincent, M. *et al.* (2008) Electrical contact reliability in a magnetic MEMS switch. 54th IEEE Holm Conference on Electrical Contacts, 2008, pp. 145–150.
31. Poulain, C. *et al.* Improvement of Electrical Contacts Reliability in a Magnetic MEMS Switch. CEA-LETI Internal Report, 2010.
32. Fortini, A. *et al.* (2008) Asperity contacts at the nanoscale: comparison of Ru and Au. *J. Appl. Phys.*, **104** (7), 074320.
33. Hermance, H. and Egan, T. (1958) Organic deposits on precious metal contacts. *Bell Syst. Tech. J.*, **37**, 739–776.
34. Vickers, N. *et al.* (2008) Board level failure analysis of chip scale package drop test assemblies. Proceedings of the 41st International Symposium on Microelectronics, Providence, November 2008.
35. Chong, D.Y.R. *et al.* (2005) Drop impact reliability testing for lead-free & leaded soldered IC packages. Proceedings of 2005 IEEE Electronics Components and Technology Conference, pp. 622–629.
36. Kim, H. *et al.* (2007) Improved drop reliability performance with lead free solders of low Ag content and their failure modes. Proceedings of 2007 IEEE Electronics Components and Technology Conference, pp. 962–967.
37. Pandher, R.S. *et al.* (2007) Drop shock reliability of lead-free alloys – effect of micro-additives. Proceedings of 2007 IEEE Electronics Components and Technology Conference, pp. 669–676.
38. Ratchev, P., Vandevelde, B. and De Wolf, I. (2003) Reliability and failure analysis of Sn-Ag-Cu solder interconnections for PSGA packages on Ni/Au surface finish. Proceedings of the 10th International Symposium on the Physical and Failure Analysis of Integrated Circuits, 7–11 July 2003, pp. 113–116.
39. Vandevelde, B. (2002) Thermo-mechanical modelling for electronic systems. Ph. D. Thesis, Catholic University of Leuven, Belgium, March 2002, pp. 190–201.
40. Lee, J.H. *et al.* (2001) Kinetics of Au-containing ternary intermetallic redeposition at solder/UBM interface. *J. Electron. Mater.*, **30** (9), 1138–1144.
41. Cao, Y. *et al.* (2006) Nanoindentation measurements of the mechanical properties of polycrystalline Au and Ag thin films on silicon substrates: effects of grain size and film thickness. *Mater. Sci. Eng. A*, **427** (1–2), 232–240.

7

Conclusions

The classical three-step approach to a technical presentation, defined by Woelfe [1], describes the necessary structure as follows:

1. Tell them what you're going to tell them.
2. Tell them.
3. Tell them what you told them.

In writing this book, we followed this approach. So, in the preface, the reader was informed about the goals and the plan of the book, and then Chapters 1–6 contained full information about failure analysis (FA), which this chapter now synthesises and presents as our conclusions to each of the previous chapters. The alternative might be to have conclusions at the end of each chapter, to be repeated in this final chapter, but this seems to us to be excessive even for a practical guide. Instead, we preferred to add to the book website sessions of Q&A for each chapter, allowing the reader to self-evaluate whether they have understood the content. The website is intended to be an electronic tool for teaching in FA, maybe one that is more appropriate for the current generation of specialists than the paper version.

In the following, some conclusions are presented, in the order of the previous six chapters.

- The three goals of the book were: (i) to present the basics of FA, as covered by the content of Chapters 4 and 5; (ii) to promote the idea of reliability, as in Chapters 2 and 3; and (iii) to show the beauty of reliability analysis, as shown in the content of Chapter 1 and by describing the details of some case studies in Chapter 6.
- The historical development of reliability as a discipline (presented in Chapter 1) allows the reader to understand the significant role of FA, a role that has increased in recent years, since the physics-of-failure (PoF) has become an essential component of any reliability analysis.
- The state of the art in FA is described in Chapter 1, with three main strands: (i) techniques of FA; (ii) failure mechanisms (FMs); and (iii) models for the PoF. The new trends in FA are linked to device shrinking and complexity growing, from the current domain of microtechnology, towards a new domain, called nanotechnology, with new tools and models for FA, most of them not yet developed.
- Eight possible applications of FA in various fields (industry, research, etc.) have been identified and the details, given in Chapter 2, demonstrate the benefits of using FA: (i) forensic engineering; (ii) reliability modelling; (iii) reverse engineering; (iv) controlling critical input variables;

Failure Analysis: A Practical Guide for Manufacturers of Electronic Components and Systems, First Edition.
Marius I. Bâzu and Titu-Marius I. Băjenescu.
© 2011 John Wiley & Sons, Ltd. Published 2011 by John Wiley & Sons, Ltd.

(v) design for reliability; (vi) process improvement; (vii) saving money by early control; and (viii) a synergetic approach.

- If manufacturers or users of electronic components and systems demand FA for independent laboratories or for specialists within their company, they have to be extremely careful in reading the results. Often nice pictures obtained with expensive tools are used to hide a poor analysis, unable to identify the FMs. By studying the information contained in this book, readers will be able to understand whether the goal of the solicited FA was really accomplished.
- FA has to be used during the whole life cycle of an electronic component or system. The recommended moments for using FA were identified and detailed in Chapter 3: (i) during the development cycle (starting even at the design stage, continuing with the so-called virtual prototyping and ending with reliability testing); (ii) when the fabrication is prepared (e.g. by analysing the reliability of the materials); (iii) during fabrication (by reliability monitoring at wafer and packaged-device level, respectively); (iv) after fabrication (at final testing and during reliability selection); and, finally, (v) during storage and operation (operation failures are essential to the necessary learning process of FA and for developing the preventive maintenance of the electronic systems).
- In all these possible FAs, the manufacturer of the electronic components has to be involved, not to execute the analysis, but to promptly furnish explanations of results and comments on the way the FA was conducted.
- The procedures and tools of FA are detailed in Chapter 4. Some recommended procedures are detailed, which aim to promote a logical use of FA techniques. A large range of possible FA techniques is described, starting with techniques for decapsulating the device and for sample preparation (decapping and decapsulation techniques, cross-sectioning and focused ion beam), and continuing with techniques for FA (electrical techniques, optical microscopy, scanning probe microscopy, microthermographical techniques, electron microscopy, X-ray techniques, spectroscopical, acoustic and laser techniques, holographic interferometry, emission microscopy, atom probe, neutron radiography, electromagnetic field measurements and so on). Each technique is described and the main possible applications are detailed.
- The use of FA techniques for studying the typical FMs for various technologies is described in Chapter 5, starting with the basic distinction between FM (the totality of physical and chemical processes leading to failure) and failure mode FMo (the external symptom of the failed product).
- The FMs must be known for two main reasons: (i) the assessment of the reliability level for a batch of product can be carried out only by modelling the failure rate induced by each FM; and (ii) the process of elaborating efficient corrective actions, able to improve significantly the reliability level, is possible only if the root causes of all FMs are known.
- Each process step may initiate some specific FMs, which are detailed in Section 5.1 and grouped around the two main stages of the technological process: the wafer-level process and the assembly process, respectively.
- At wafer level, the FMs induced by the semiconductors (e.g. crystallographic defects) or by various processes (passivation, metallisation, diffusion, implantation, masking, etching, etc.) have been detailed. For each FM, similar details were mentioned: a short description of the FM, followed by the main techniques used to study the specific FM (among those detailed in Chapter 4) and by a list of corrective actions aiming to diminish the action of the FM.
- The use of test structures was a separate subject, being an important tool for identifying the possible FMs.
- For the assembly process, the FMs induced by various process steps (dicing, die attach, wire connection, sealing, wafer-level packaging, etc.) were detailed, following the same structure as for the wafer-level process.
- FMs arising during operation were mentioned: component mounting in systems and behaviour in a harsh environment (e.g. radiation field).
- The typical FMs of the most important technologies for achieving electronic components were detailed in Sections 5.2–5.8, such as: (i) passive electronic parts (resistors, capacitors, varistors,

connectors, inductive elements, embedded passive components); (ii) silicon bipolar technology (silicon diodes, silicon bipolar transistors, thyristors and other power devices, bipolar integrated circuits); (iii) MOS technology (*n/p* channel technology, MOS transistors, MOS integrated circuits, memories, microprocessors, silicon-on-insulator (SOI) technology); (iv) optoelectronic and photonic technologies (light-emitting diodes, optocouplers, liquid crystal displays, photodiodes, phototransistors, solar cells); (v) nonsilicon technologies (diodes, transistors, integrated circuits); (vi) hybrid technology (thin- and thick-film hybrid circuits); and (vii) microsystem technologies (microsystems and nanosystems).

- We tried to update all the information about each FM detailed in Chapter 5, so more than 450 references (most reported after 2000) were included in this chapter.
- The 12 case studies presented in Chapter 6 are practical applications of the knowledge contained in Chapters 4 and 5. The case studies cover almost all the technologies in Chapter 5 and all the FA techniques described in Chapter 4.
- For any company it is important to use FA to retrieve value from a failure. The information obtained by FA may be extremely valuable to the learning process that must lead to the continuous improvement of the fabrication. In a modern approach, this process must include the customers. For example, Xerox has a network-based system, called Eureka, which performs systematic analyses of small failures in the form of customer breakdowns. This system captures and shares 30 000 repair tips each year, leading to savings estimated at around US$100 million and providing important information for design improvement. On a larger scale, the US Army is known for conducting After Action Reviews that enable participants to analyse, discuss and learn from both the successes and the failures of a variety of military initiatives.
- The recommended sequence of procedures is: (i) visual inspection; (ii) electrical testing; (iii) nondestructive evaluation; and (iv) destructive evaluation (using relevant techniques). This data when properly analysed leads to a viable mechanism for failure.
- Until now, most research has focused on voltage and temperature acceleration effects towards single-failure mechanisms. At the system level, due to the complexity of VLSI circuit and dynamic operating conditions, extrapolation of system voltage and temperature acceleration factors from individual failure mechanisms remains a formidable challenge. Although various models have been proposed to describe the voltage acceleration effect for a single-failure mechanism [2], because of the unique physical process underlying each failure mechanism, for example electromigration (EM), hot carrier injection (HCI), time-dependent dielectric breakdown (TDDB) and negative bias temperature instability (NBTI), no universal voltage acceleration model has been established.
- When analysing a part, board or system, care should be taken with the available data and other factors. If all factors are not carefully considered, much time, effort and money might be wasted.
- FA can be used in a company with important social and organisational benefits. By participating in a discussion about case studies involving FA, the specialists who were not directly involved in the failure will acquire knowledge about the possible failure risks.
- Rigorous analysis of failure requires that people, at least temporarily, put aside their tendency to avoid unpleasant truths and take personal responsibility [3]. Consequently, the human relationships inside the company may be improved, because people will better understand the failure risks and will learn not to attribute too much blame to others or to forces beyond their control.
- FA has an important role in the learning process, resulting from analysing and discussing simple mistakes. Formal processes or forums for discussing, analysing and applying the lessons of failure elsewhere in the organisation are needed to ensure that effective analysis and learning from failure occurs [3].
- Generally, of all of the chronic failures experienced by a company in a given year, 20% represent 80% of the loss. This means that if you investigate the 20% of the failures representing 80% of your losses, you will reap quantum benefits in a short period of time. We call these the 'significant few' failures [4].

Finally, as a sign of the growing importance of the FA field, in the last few years various organisations, such as the Electronic Industries Alliance (EIA) and the Joint Electron Devices Engineering Council (JEDEC)[1] Solid State Technology Association, have proposed a series of standards about FA. Some of these are given below:

- Guidelines for Preparing Customer-Supplied Background Information Relating to a Semiconductor-Device Failure Analysis, EIA/JEP134, September 1998, Electronic Industries Alliance, JEDEC Solid State Technology Division, http://www.jedec.org/download/search/jep134.pdf.
- Component Quality Problem Analysis and Corrective Action Requirements (Including Administrative Quality Problems), JESD671-A (Formerly EIA-671), December 1999, http://www.jedec.org/download/search/jesd671a.pdf.
- Standard for Failure Analysis Report Format, EIA/JESD38, December 1995, http://www.jedec.org/download/search/jesd38.pdf.
- Procedure for Characterising Time-Dependent Dielectric Breakdown of Ultra-Thin Gate Dielectrics, JESD92, August 2003, http://www.jedec.org/download/search/JESD92.pdf.
- Guideline for Characterising Solder Bump Electromigration under Constant Current and Temperature Stress, JEP154, January 2008, http://www.jedec.org/download/search/JEP154.pdf.
- Field-Induced Charged-Device Model Test Method for Electrostatic-Discharge-Withstand Thresholds of Microelectronic Components, JESD22-C101D (Revision of JESD22-C101C, December 2004), October 2008, http://www.jedec.org/download/search/22c101D.pdf.
- Guideline for Residual Gas Analysis (RGA) for Microelectronic Packages, JEP144. July 2002, http://www.jedec.org/download/search/JEP144.pdf.
- Early Life Failure Rate Calculation Procedure for Electronic Components, JESD74, April 2000, Hybrids/MCM JESD93, September 2005 (Reaffirmed: January 2009), http://www.jedec.org/download/search/JESD93.pdf.
- Procedure for the Wafer-Level Testing of Thin Dielectrics, JESD35-A (Revision of JESD35), April 2001, http://www.jedec.org/download/search/jesd35a.pdf.
- Recommended ESD-CDM Target Levels, JEP157, October 2009, http://www.jedec.org/DOWNLOAD/search/JEP157.pdf.
- Recommended ESD Target Levels for HBM/MM Qualification, JEP155, August 2008, http://www.jedec.org/DOWNLOAD/search/JEP155.pdf.

References

1. Woelfe, R.M. (1975) *A Guide for Better Technical Presentations*, IEEE Press.
2. White, M. and Bernstein, J. (2008) *Microelectronics Reliability: Physics-of-Failure Based Modeling and Lifetime Evaluation*, JPL Publication 08-05, Jet Propulsion Laboratory, California Institute of Technology, Pasadena, CA.
3. Edmonson, A. and Cannon, M.D. The Hard Work of Failure Analysis, http://hbswk.hbs.edu/item/4959.html. (Accessed 2010).
4. Latino, R.J. What is Root Cause Failure Analysis? http://www.maintenanceresources.com/referencelibrary/failureanalysis/whatisroot.htm. (Accessed 2010).

[1] *JEDEC* was formed in 1958 as a part of the Electronic Industries Alliance (EIA) – a trade association that represents all areas of the electronics industry in the United States. In 1999, JEDEC was incorporated as an independent association under the name *the JEDEC Solid State Technology Association*. This new organisation then became a recognised sector of the EIA, and maintains that relationship today. *The JEDEC Solid State Technology Association* is the semiconductor engineering standardisation body of the EIA. JEDEC is working with the International Electronics Manufacturing Initiative (iNemi) on a joint-interest group on lead-free issues. All JEDEC standards are available on the Web to download after a free registration.

Acronyms

2D	Bi-Dimensional
2DEG	Two Dimensional Electron Gas
AES	Auger Electron Spectroscopy
AET	Accelerated Environmental Test
AFM	Atomic Force Microscopy
AFP	Atomic Force Probing
ALT	Accelerated Life Test
AOI	Automatic Optical Inspection
AP	Atom Probe
APD	Avalanche Photodiode
AQL	Acceptable Quality Level
ASIC	Application Specific Integrated Circuit
AST	Accelerated Stress Test
ASTM	American Society for Testing and Materials
AVR	Applied Voltage Ratio
BD	Breakdown
BEM	Backside Emission Microscopy
BEOL	Back End of Line
BFA	Bouncing Failure Analysis
BGA	Ball Grid Array
BHT	Bias-Humidity-Temperature
BiCMOS	Bipolar CMOS
BiFET	Bipolar FET
BIR	Building-In Reliability
BISR	Built-In Self-Repair
BIST	Built-In-Self-Test
BJT	Bipolar Junction Transistor
BSE	Backscattered Electron
BT	Bias Temperature
CAD	Computer-Aided Design
CAE	Computer-Aided Engineering
CAFM	Conductive Atomic Force Microscopy
CALCE	Computer-Aided Life-Cycle Engineering
CCD	Charge-Coupled Devices
CCGA	Ceramic Column Grid Array
CCS	Constant Current Stress
CCVC	Capacitive Coupling Voltage Contrast
CDM	Charged Device Model
CE	Concurrent Engineering
CERDIP	Ceramic Dual-In-Line Package
CERQUAD	Ceramic QUAD Flat Package
CFM	Charge Force Microscopy
CFR	Constant Failure Rate
CI	Continuous Improvement (Kaizen, in Japanese); see also CPI
CLSM	Confocal Laser Scanning Microscopy
CMOS	Complementary Metal Oxide Semiconductor
CMP	Chemical-Mechanical Polishing
CMP	Chip Multi-Processor
CNT	Carbon Nano Tube
COTS	Commercial Off-The-Shelf
CPI	Chip-Package Interconnection
CPI	Continuous Process Improvement; see also CI
CPU	Central Processing Unit
C-SAM	C-mode Scanning Acoustic Microscopy

Failure Analysis: A Practical Guide for Manufacturers of Electronic Components and Systems, First Edition.
Marius I. Bâzu and Titu-Marius I. Băjenescu.
© 2011 John Wiley & Sons, Ltd. Published 2011 by John Wiley & Sons, Ltd.

CSM Continuous Stiffness Measurement
CSP Chip-Scale Packaging
CST Constant-Stress Test
CTE Coefficient of Thermal Expansion
CTR Current Transfer Ratio (of an optocoupler)
CVD Chemical Vapour Deposition
CVS Constant Voltage Stress
DB Dielectric Breakdown
DBS Dual Beam Spectroscopy
DFFA Design For Failure Analysis
DFM Design For Manufacturing
DFR Decreasing Failure Rate
DfR Design for Reliability
DFS Design for Safety
DFT Design for Testability
DfY Design for Yield
DG Double Gate
DGSOI Double-Gate Silicon-On-Insulator
DIL Dual In-Line
DIMM Dual In-line Memory Module
DIP Dual In-line plastic Package
DLTS Deep Level Transient Spectroscopy
DMD Digital Micromirror Device
DMOS Diffused Metal-Oxide Semiconductor
DoE Design of Experiment
DPA Destructive Physical Analysis
DPM Defects Per Million
DPPM Defects Parts Per Million
DRAM Dynamic Random Access Memory
DSC Differential Scanning Calorimetry
DTA Differential Thermal Analysis
DUT Device under Test
DUV Deep Ultraviolet
EAPROM Electrically Alterable PROM
EAROM Electrically Alterable Read-Only Memory
EBIC Electron Beam Induced Current
EBL Encapsulated Beam Lead
EBSD Electron Back Scattering Diffraction
EDS Energy Dispersive x-ray Spectroscopy
EDX Energy Dispersive x-ray analysis
EDX Energy-dispersive x-ray
EELS Electron Energy-Loss Spectrometry
EEPROM Electrically Erasable PROM
EEROM Electrically Erasable ROM
EFA Electrical Failure Analysis
EFM Electrical Force Microscopy
EFR Early Failure Rate
EM Electromigration
EMC ElectroMagnetic Compatibility
EMF Electromagnetic Field
EMP ElectroMagnetic Pulse
EOS Electrical OverStress
EPMA Electron-Probe Micro-Analysis
EPROM Erasable ROM
ESC Environmental Stress Cracking
ESCA Electron Spectroscopy for Chemical Analysis
ESD ElectroStatic Discharge
ESEM Environmental SEM
ESS Environmental Stress Screening
ETA Event Tree Analysis
FA Failure Analysis
FAMOS Floating-gate Avalanche-injection Metal-Oxide Semiconductor
FC Flip Chip
FCA Free-Carrier Absorption
FCBGA Flip-Chip Ball Grid Array
FE Finite Element
FEA Finite Element Analysis
FEM Field Electron Microscopy
FEM Finite Element Method
FESEM Field Emission Scanning Electron Microscopy
FET Field-Effect Transistor
FFT Fast Fourier Transform
FI Fault Isolation
FIB Focused Ion Beam
FID Failure Interaction and Dependency
FIM Field Ion Microscopy
FIT Failure In unit Time (1 failure per 10^9 device $\times$ hours)
FM Failure Mechanism
FMA Failure Mode Assessment
FMEA Failure Mode and Effect Analysis

FMECA Failure Mode, Effect and Criticality Analysis
FMEDA Failure Modes, Effects and Diagnostic Analysis
FMMEA Failure Modes, Mechanisms and Effects Analysis
FMo Failure Mode
FP Flat Package
FPBGA Flip-chip Ball Grid Array
FRACAS Failure Reporting, Analysis and Corrective Action System
FTA Fault Tree Analysis
FTIR Fourier Transform Infrared spectroscopy
FTTF First Time To Failure
GTO Gate-Turn-Off thyristor
HALT Highly Accelerated Life Testing
HASS Highly Accelerated Stress Screening
HAST Highly Accelerated Stress Test
HBM Human Body Model
HBT Heterojunction Bipolar Transistor
HC Hot-Carrier
HCD Hot-Carrier Degradation
HCI Hot-Carrier Injection
HEMT High Electron Mobility Transistor
HI Holographic Interferometry
HMOS High-performance, n-channel silicon gate MOS
HRTEM High Resolution TEM
HTOL High Temperature Operating Life
HTRB High Temperature Reverse Bias operating life test current
IC Integrated Circuit
IFR Increasing Failure Rate
IGBT Insulated Gate Bipolar Transistor
IMMA Ion Microprobe Mass Analysis Integrated Circuits
IP Intellectual Property
IPAD Integrated Passive and Active Device
IR Infrared
IRAS Infrared Absorption Spectrometry
IRED Infrared Emitting Diode
IREM Infrared Emission Microscopy
IRLIT IR Lock-in Thermography
IRM Infrared Microscopy
ISS Ion Scattering Spectrometry
ITO Indium–Tin–Oxide
ITP In-Target Probe
ITRS International Technology Roadmap for Semiconductor
JAN Joint Army-Navy (specification)
JEDEC Joint Electron Device Engineering Council
LASER Light Amplification by Stimulated Emission of Radiation
LC Liquid Crystal
LCC Leadless chip carrier
LCD Liquid Crystal Display
LD Laser Diode
LDCC Leaded Chip Carriers
LDD Lightly Doped Drain
LED Light Emitting Diode
LEED Low-Energy Electron Diffraction
LEM Light Emission Microscopy
LGA Land Grid Array
LIGA Lithographie, Galvanoformung, Abformtechnik (In German)
LPCVD Low Pressure Chemical Vapour Deposition
LSI Large Scale Integration
LTPD Lot Tolerance Percent Defective Management
MC Multichip
MCM Multi-Chip Module
MCP Multi-Chip Package
MDM Metal-Dielectric-Metal
MEMS Micro Electro Mechanical Systems
MESFET Metal-Semiconductor FET
MFM Magnetic Force Microscopy
MGT MOS-Gated Thyristor
MI Moiré Interferometry
MIC Mobile Ionic Contamination
MIL Military Electronics
MIL-HDBK Military Handbook
MIL-STD Military Standard
MIM Metal-Insulator-Metal
MIS Metal-Insulator-Semiconductor
MISFET Metal Insulator Semiconductor Field-Effect Transistor
MJ Multijunction

MLCC	Multilayer Ceramic Capacitor
MLE	Maximum Likelihood Estimation
MMIC	Monolithic Microwave
MNOS	Metal-Nitride-Oxide-Semiconductor
MOCVD	Metal-Organic Chemical Vapour Deposition
MOEMS	Micro-Optical Electromechanical Systems
MOS	Metal-Oxide Semiconductor
MOSFET	Metallic Oxide-Semiconductor Field-Effect Transistor
MOST	Metal-Oxide Semiconductor Transistor
MST	Microsystem Technology
MTBF	Mean Time Between Failures
MTTF	Mean Time To Failure
MTTFF	Mean Time To First Failure
MTTR	Mean Time To Repair
NAA	Neutron Activation Energy
Nano-WLP	Nano Wafer Level Packaging
NBTI	Negative Bias Temperature Instability
NDE	Non-Destructive Evaluation
NEMS	Nano-Electro-Mechanical Switches
NF	Noise Figure
NFF	No Fault Found
NMOS	n-channel (type) MOS
NOM	Nomarski Optical Microscopy
NSOM	Near-field Scanning Optical Microscopy, also SNOM
OBIC	Optical Beam-Induced Current
OBIRCH	Optical Beam-Induced Resistance Change
OEICs	Optoelectronic Integrated Circuits
OES	Optical Emission Spectrometry
OFET	Organic FET
OLED	Organic Light-Emitting Diode
OM	Optical Microscopy
OSF	Oxidation-induced Stacking Faults
PAM	Photo-Acoustic Microscopy
PBGA	Plastic-Ball-Grid-Array
PC	Personal Computer
PCB	Printed Circuit Board
PCT	Pressure Cooking Test
PD	Photodetector
PD	Preliminary Design
PDA	Personal Digital Assistant
PDIP	Plastic Dual In-line Package
PDT	Product Development Test
PEALD	Plasma Enhanced Atomic Layer Deposition
PECVD	Plasma Enhanced Chemical Vapour Deposition
PED	Plastic Encapsulated Device
PEM	Photo Emission Microscopy
PEM	Photo Electron Microscopy
PEM	Plastic Encapsulated Microcircuit
PEN	PolyEthylene Naphthalate
PFA	Physical Failure Analysis
PGA	Pin Grid Array
PHEMT	Pseudomorphic High Electron Mobility Transistor
PHM	Prognostics and Health
PIND	Particle Impact Noise Detection/Detector
PL	Photoluminescence
PMOS	p-channel (type) MOS
PoF	Physics-of-Failure
PPoF	Probabilistic Physics of Failure
PQFN	Power Quad Flat-pack No-lead
PQFP	Plastic Quad Flat Pack
PRAM	Phase-change Random Access Memory
PROACT	Preserving failure data; Ordering the analysis; Analysing the data; Communicating findings and recommendations; Tracking for success
PROM	Programmable ROM
PSG	PhosphoSilicate Glass
PSGA	Polymer Stud Grid Array
PV	Photovoltaics
PVD	Physical Vapour Deposition
PVT	Physical Vapour Transport
PWB	Printed Wire/Wiring Board
PWBA	Printed Wiring Board Assembly
Q&RA	Quality and Reliability Assurance
QA	Quality Assurance
QALT	Quantitative Accelerated Life Test
QB	Quasi-Breakdown
QBD	Charge-to-Breakdown
QC	Quality Control

QCT	SPC Quality Control Team
QD VCSELs	Quantum-Dot Vertical-Cavity Surface-Emitting Lasers
QD	Quantum Dot
QFD	Quality Function Deployment
QFN	Quad Flat No-lead
QFP	Quad Flat Package
QLT	Quantitative Life Test
QML	Qualified Manufacturer List
QMP	Quality Measurement Plan
QPL	Qualified Parts List
QRRM	Quick Reaction Reliability Monitor
R&M	Reliability and Maintainability
RA	Reliability Assurance
RAC	Reliability Analysis Centre
RAM	Random Access Memory
RAMS	Reliability, Availability, Maintainability and Safety
RBS	Rutherford Back-Scattering Spectrometry
REPROM	Reprogrammable ROM
RF	Radio Frequency
RGA	Residual Gas Analysis
RH	Relative Humidity
RHEED	Reflected High-Energy Electron Diffraction
RIE	Reactive Ion Etch(ing)
ROHS	Restriction of Hazardous Substances
ROM	Read-Only Memory
RTA	Rapid Thermal Annealing
RVS	Ramped Voltage Stress
SAF	Stuck-At-Fault
SAM	Scanning Acoustic Microscope
SAM	Scanning Auger Microscopy
SAT	Scanning Acoustic Test
SB	Second Breakdown
SB	Soft-Breakdown
SBD	Schottky Barrier Diode
SBE	Single Bit Errors
SBH	Schottky Barrier Height
SCAT	Scanning Acoustic Tomography
SCP	Single Chip Package
SCR	Silicon-Controlled Rectifier
SCS	Single Crystal Silicon
SDRAM	Synchronous DRAM
SEB	Single Event Burnout
SECC	Single Edge Contact Cartridge
SED	Strain Energy Density
SEE	Single Event Effect
SEGD	Single Event Gate Damage
SEGR	Single Event Gate Rupture
SEI	Seebeck Effect Imaging
SEL	Single Event Latch-up
SEM	Scanning Electron Microscopy
SER	Soft-Error Rate
SET	Single Event Transient
SEU	Single Event Upset
SG	Stage Gate
SiC	Silicon Carbide
SILC	Stress-Induced Leakage Current
SILOX	Silicon Oxide
SIMOX	Separation by Implanted Oxygen
SIMS	Secondary-Ion Mass Spectrometry
SiOP	Silicon On Plastic
SIP	Single In-line Package
SIP	System-in-Package
SM	Stress Migration
SMD	Surface Mount Devices
SME	Small-Medium Enterprise
SMT	Surface-Mount Technology
SMU	Source Measure Unit
SNOM	Scanning Near-field Optical Microscopy
SNPEM	Scanning Near-Field Photon Emission Microscopy
SOA	Safe Operating Area
SoB	System on Board
SoC	System on Chip
SOI	Silicon On Insulator
SOIC	Small Outline IC Package
SOJ	Small Outline J-Leaded
SOM	Scanning Optical Microscopy
SOP	Small Outline Package
SOP	System on Package
SOS	Silicon on Sapphire
SPC	Statistical Process Control
SPEM	Spectroscopic Photon Emission Microscopy
SPM	Scanning Probe Microscopy
SQUID	Superconducting QUantum Interference Device
SRAM	Static RAM
SRQAC	Software Reliability and Quality Acceptance Criteria
SSDI	Shadow Spectral Digital Imaging

SSDSCs	Solid-State Dye-sensitized Solar Cells
SSF	Shadow Spectral Filming
SSG	Selective Silicon Growth
SSHE	Scraped Surface Heat Exchanger
SSI	Small Scale Integration
SSL	Solid State Lighting
SSM	Scanning SQUID Microscopy
SSMS	Spark Source Mass Spectrometry
SSR	Step-Stacked-Routing
SST	Step-Stress Tests
SSTL	Stub Series Terminated Logic
SSY	Small-Scale Yielding
ST	Stress Test
STEM	Scanning Transmission Electron Microscopy / Scanning Tunnelling Electron Microscopy
STI	Shallow Trench Isolation
STM	Scanning Tunnelling Microscopy
STM	Scanning Thermal Microscopy
SWAT	SoftWare Anomaly Treatment
SWC	Solderless Wire Wrap Connecting
SWCNT	Single-Walled Carbon Nanotube
SWEAT	Standard Wafer level Electromigration Accelerated Test
SXAPS	Soft x-ray Appearance Potential Spectroscopy
SYRP	Synergetic Reliability Prediction
TAB	Tape Automated Bonding
TAP	Test Access Port
TBC	Thermal Barrier Coating
TBD	Time to Breakdown
TBFD	Trace Based Fault Diagnosis
TBJD	Trench Bipolar Junction Diode
TC	Temperature Coefficient
TC	Temperature Cycling
TCAD	Technology Computer Aided Design
TCCT	Thin Capacitively-Coupled Thyristor
TCR	Temperature Coefficient of Resistance
TCT	Thermal Cycling Test
TD	Thermal Design
TD	Threading Dislocation
TDBI	Test During Burn-In
TDDB	Time-Dependent Dielectric Breakdown
TDI	Test Data In
TDO	Test Data Output
TDR	Time-Domain Reflectometry
TDTR	Time Domain Thermal Reflectance
TED	Transferred Electron Device
TED	Transient Enhanced Diffusion
TEELS	Transmission Electron Energy-Loss Spectrometry
TEGFET	Two-Dimensional Electron gas FET
TEM	Transmission Electron Microscope / Microscopy
TEM-ED	Transmission Electron Microscopy – Electron Diffraction
TFD	Thin Film Delamination
TFH	Thin Film Hybrid
TFO	Thick Field Oxide
TFT	Thin-Film Transistor
TGA	Thermo Gravimetric Analysis
TGO	Thermally Grown Oxide
TH	Lead-Free Through-Hole Flow Soldering
THB	Temperature Humidity Bias
TID	Total Ionising Dose
TIR	Testing-In Reliability
TIVA	Thermally-Induced Voltage Alteration
TLM	Transfer Length Method
TLP	Transmission Line Pulsing
TLPS	Transmission Line Pulse Spectroscopy
TLS	Thermal Laser Stimulation
TLU	Transient induced Latch-Up
TMA	Thermo-Mechanical Analysis
TMAH	TetraMethylAmmonium Hydroxide
TMIC	Transition Metal Ion Chromatography
TMR	Tunnelling Magneto-Restrictive
TMS	Test Mode
TNI	Troubled-Non-Identified
TO	Transistor Outline
TOC	Tactical Operations Centres
TOF	Time-Of-Flight

TOFSIMS	Time-Of-Flight Secondary Ion Mass Spectroscopy
TOSAs	Transmitter Optical Subassemblies
TOV	Temporary Overvoltage
TPR	Technical Plan for Resolution
TQL	Total Quality Leadership
TQM	Total Quality Management
T-RAM	Thyristor-based Random Access Memory
TRM	Thermal Reflectance Microscopy
TRPE	Time-Resolved Photon Emission prober
TRST	Test Reset
TRXRFA	Total Reflection x-ray Fluorescence Analysis
TSC	Two-Step Crystallization
TSI	Top-Surface Imaging
TSOP	Thin Small Online Plastic
TSPD	Thyristor Surge Protection Devices
TST	Thermal Shock Test
TSV	Through-Silicon Vias
TTF	Time To Failure
TTL	Transistor-Transistor Logic
TTL-LS	Transistor-Transistor Logic-Low power Schottky
TTR	Time To Repair
TTS	Transistor-Transistor logic Schottky barrier
TVS	Transient Voltage Suppressor
TXRF	Total Reflection x-ray Fluorescence
TZDB	Time-Zero Dielectric Breakdown
UBM	Under Bump Metallization / Metallurgy
UCL	Upper Confidence Level
UF	Underfilling
UFP	Ultra Fine Pitch
UHV	Ultra High Vacuum
UJT	Unijunction Transistor
ULSI	Ultra Large Scale Integration
UPH	Units Per Hour
UPS	Ultraviolet Photoelectron Spectrometry
UPW	Ultra-Pure Water
UR	Usage Rates
USG	Undoped Silica Glass
USJ	Ultra-Shallow Junctions
UUT	Unit Under Test
UV	Ultraviolet
VC	Voltage Contrast
VCSEL	Vertical-Cavity Surface-Emitting Laser
VDMOSFET	Vertical Diffused MOSFET
VDSM	Very Deep Submicron
VFB	Voltage Flat Band
VHDL	Very High Density Logic
VHSIC	Very High Speed Integrated Circuit
VLSI	Very Large Scale Integration
VMDP	Velocity-Matched Distributed Photodetectors
VMOS	V-groove MOS / Vertical MOS
VPE	Vapour Phase Epitaxy
VSMI	Volumetric System Miniaturization and Interconnection
VTC	Voltage Transfer Characteristics
VTCMOS	Variable Threshold voltage CMOS
WBC	Wafer Backside Coating
WDM	Wavelength Division Multiplexing
WDX	Wavelength Dispersive x-ray
WGPD	Waveguide Photodiode
WIP	Work in Progress
WL	Word Line
WLCSP	Wafer-Level Chip Scale Packaging
WLI	White light interferometry
WLP	Wafer-Level Packaging
WLR	Wafer-Level Reliability
WLTBI	Wafer-Level Testing during Burn-In
WSI	Wafer Scale Integration
WSN	Wireless Sensor Networks
XAES	X-ray-induced Auger Electron Spectrometry
XEDS	X-ray Energy Dispersive Spectroscopy
XIVA	External Induced Voltage Alteration
XPD	X-ray Photoelectron Diffraction
XPS	X-ray Photoelectron Spectrometry / Spectroscopy

XRD X-Ray Diffraction
XRF X-Ray Fluorescence
XRFA X-Ray Fluorescence Spectrometric Analysis
XRM X-Ray Microscopy
XRPM X-Ray Projection Microscopy
XRR X-Ray Reflectivity
XTEM X-ray Transmission Electron Microscopy
ZD Zero Defect
ZIF Zero Insertion Force
ZMR Zone-Melting Recrystallized

Glossary

Terms Related to Electronic Components and Systems

Capacitor	A passive electronic component consisting of a pair of conductors separated by an insulator. When a voltage is applied across the conductors, an electric field appears in the insulator.
Component (or **Part**)	Any smaller, self-contained element of a larger entity [1].
Component, active	A component with gain or directionality and capable of producing energy (e.g. a semiconductor device, vacuum tube).
Component, electronic	A basic electronic element usually packaged in a discrete form with two or more connecting leads or metallic pads. Electronic components are intended to be connected together, forming, together with other elements, electronic systems.
Component, passive	A component without gain or directionality, which consumes, but does not produce energy.
Device	Any component, electronic element, assembly or piece of equipment that can be considered individually. A functional or structural unit, considered an entity for investigations: hardware and/or software, sometimes also including human resources.
Device, discrete	A single device manufactured and encapsulated separately.
Device, optoelectronic	Any device used in optoelectronics, for example a light emitting diode (LED), photodiode or phototransistor, optocoupler or solar cell.
Device, semiconductor	Any device using the properties of semiconductor materials – silicon, germanium, gallium arsenide and so on.
Die (pl. **Dice**, sometimes called **Chip(s)**)	(i) A tiny piece of semiconductor material, broken from a semiconductor slice, on which one or more active electronic component is formed. (ii) A portion of a wafer bearing an individual circuit or device cut or broken from a wafer containing an array of such circuits or devices.

Failure Analysis: A Practical Guide for Manufacturers of Electronic Components and Systems, First Edition.
Marius I. Bâzu and Titu-Marius I. Băjenescu.
© 2011 John Wiley & Sons, Ltd. Published 2011 by John Wiley & Sons, Ltd.

Diode A two-terminal electronic component that conducts an electric current in one direction only.

Hybrid integrated circuit A complete electronic circuit fabricated on an insulating substrate, using a variety of device technologies. The substrate acts as a carrier for the circuit and has interconnecting tracks printed on it using multilayer techniques. Individual devices (comprising chip diodes, transistors, ICs, thick-film resistors and capacitors, which form the circuit function) are attached to the substrate and connected together using the interconnecting tracks.

Integrated circuit (IC) (i) A number of devices (tens of millions of gates or more) manufactured and interconnected on a single semiconductor substrate. (ii) A miniaturised electronic circuit that is manufactured on the surface of a thin substrate of semiconductor material, using bipolar or MOS technology for example.

Microsystems Miniaturised devices that perform non-electronic functions, typically sensing and actuation. For example, microelectromechanical systems (MEMS) combine electrical and mechanical components that can sense, control and actuate on the micro-scale and function either individually or in arrays to generate effects on the macro-scale.

Microtechnology A technology with features of a few micrometres. The term was developed and is mainly used for microsystem technology (MST) but in principle it covers any technology for fabricating electronic components with small features.

Nanotechnology (i) The science of building devices from atoms and molecules or of studying the control of matter at atomic and molecular scale. Generally, nanotechnology deals with structures smaller than 100 nm and involves the development of materials or devices of that size. (ii) Engineering of functional systems at molecular scale.

Nanotube A high-strength cylindrical fullerene graphite structure with attractive physical and chemical properties.

Optoelectronics A technology dealing with the coupling of functional electronic blocks using light beams.

Package The container for an electronic component. Has terminals that provide electrical access to the inside of the container.

Resistor A two-terminal passive electronic component that produces a voltage across its terminals, proportional to the electric current passing through it according to Ohm's law.

System An entity comprising many components and interfaces (e.g. cars, dishwashers, aircraft, etc.), meaning there are many possibilities for failures, particularly across interfaces (e.g. inadequate electrical overstress protection, vibration nodes at weak points, electromagnetic interference, software that contains errors, etc.) [2].

Thyristor A controlled rectifier in which the unidirectional current flow from anode to cathode is initiated by a small signal current from gate to cathode.

Transistor A solid-state semiconductor device used to amplify, control and generate electrical signals, consisting of layers of different semiconductors.

Transistor, bipolar A semiconductor device with three electrodes (emitter, base and collector). A sandwich of two types of doped semiconductor, usually p-type and n-type silicon, meaning it contains two *pn* junctions.

Transistor, field-effect (FET) A transistor with a channel of one type of charge carrier in a semiconductor material, with its conductivity controlled by an electric field. FETs are sometimes called **unipolar transistors** to contrast their single-carrier-type operation with the dual-carrier-type operation of bipolar transistors.

Transistor, metal-oxide semiconductor field effect (MOSFET) A class of voltage-driven devices, with the controlling gate voltage applied to the channel region across an oxide-insulating material. The term can be applied to both transistors in an IC and discrete power devices. The major advantage of a MOSFET is its low power due to its insulation from source and drain.

Wafer A thin semiconductor slice (of silicon, germanium or GaAs) with parallel faces on which matrices of microcircuits or individual semiconductors can be formed. After processing, the wafer is separated into dice or chips containing individual circuits.

Terms Related to Failure Analysis

Defect (i) A fault or malfunction (faulty function) [3]. (ii) A condition that must be removed or corrected. (iii) One or more flaws whose aggregate size, shape, orientation, location and properties do not meet specified acceptance criteria and are rejectable. (iv) A deviation from perfection that can be shown to cause a failure (using a quantitative analysis) that would not have occurred in the absence of the imperfection.

Defect, crystallographic A structural imperfection in a crystal.

Defect, inherent The underlying cause of an intrinsic failure, in the useful life period.

Defect part per million (DPPM) The number of parts returned (owing to defects) by customers for every one million parts delivered. This is a significant measure of the financial success of a product: the lower the DPPM, the higher the profit margin.

Defect, systematic A parametric defect or mask misalignment that is dealt with during the fabrication process.

Degradation (i) A change for the worse in the characteristics of an electric element caused by an external stimulus (e.g. heat, high voltage). (ii) A gradual deterioration in performance as a function of time.

Degradation mode The manner, means or method by which something degrades. Degradation modes have different activation energies and can occur simultaneously.

Error The manifestation of a system fault. For example, a power-company operator enters the wrong account number for cancellation (fault). The system then shuts off power to the wrong node (error) [4]. Both categories, faults and errors, can spread through an electronic system. If a chip shorts out power to ground, it may cause nearby chips to fail as well. Errors can spread because the output of one unit is used as input by other units.

Failure Termination of the ability of an item to perform a required function. A failure occurs when the service delivered by a system or part fails to meet its specification and it is caused by an error. Failures can be classified by two dimensions: the physical location (component level, circuit level, system level or interconnection of systems – which is the highest level) and the stage of occurrence within the lifetime of the system (various stages: design, prototyping, manufacturing, testing, operation) [4].

Failure analysis (FA) (i) A scientific method of determining the root cause of a failure or parameter excursion, so that corrective action can be taken. FA is based on a series of techniques and includes the support of fabrication tool development, processing development, technology development, manufacturing, testing and field-return analysis. (ii) The logical and systematic examination of an item or its diagram(s) to identify and analyse the probabilities, causes and consequences of potential and real failures. (iii) The investigation of products with functional or programming failures, to determine the root cause of failure. (iv) The process of collecting and analysing data to determine the cause of a failure. (v) The objective investigation of material facts associated with a part or system failure.

Failure, catastrophic Sudden and complete failure.

Failure causes Defects in design, process, quality or part application that are the underlying cause of a failure or which initiate a process which leads to failure.

Failure criteria Limiting conditions, relating to the admissibility of a deviation from a characteristic value due to changes after the onset of stress.

Failure, critical Failure that is likely to cause injury to persons or significant damage to material.

Failure, degradation Failure that is both gradual and partial.

Failure distribution The occurrence of failures plotted as a function of time. Usually plotted for a particular group of parts operating in a particular environment.

Failure effect The immediate consequences of a failure on the operation, function/functionality or status of some item.

Failure, extrinsic In essence, any non-intrinsic failure. Typically related to static or dynamic overload events (electrical, thermal, mechanical and radiative) during the component life cycle, or to misapplication (the wrong component for the job).

Failure, gradual A failure that can be anticipated by prior examination or monitoring.

Failure in time (FIT) A measure of reliability equal to 1 failure in 10^9 device operating hours.

Failure, intrinsic A failure that occurs after component delivery, related to component design, materials, processing, assembly, packaging and manufacturing, and provoked under circumstances that are within the design specifications.

Failure mechanism (FM) (i) The physical, chemical or metallurgical process that causes a failure. Identifying the failure mechanism is the final goal of failure analysis and it is important to undertake separate statistical processing of each population affected by the same failure mechanism. For example, one has to process data on components failed by corrosion without looking at components failed by thermal fatigue. (ii) The basic material behaviour that results in a failure, or the chemical, physical or metallurgical process that leads to component failure [5]. (iii) The means by which physical, chemical, mechanical, electrical, human or other actions cause a device to fail. (iv) In failure analysis, a fundamental process or defect responsible for a failure.

Failure mechanisms, extrinsic Mechanisms resulting from the device packaging and interconnection process. The extrinsic conditions affecting the reliability of components vary according to the packaging processes employed and the environment in which the device is operated. As a technology matures and problems in fabrication lines are eliminated, intrinsic failures are reduced, making extrinsic failures all the more important to device reliability.

Failure mechanisms, intrinsic Mechanisms inherent to the semiconductor die itself, including crystal defects, dislocations and processing defects.

Failure mode (FMo) (of an electronic component) (i) The observable consequence of failure. Failure modes usually refer to observable adverse effects (broken structure, cracked surface, etc.) or directly measurable parameter degradation exceeding the prescribed limits. (ii) The manner by which a failure is observed. Generally describes the way in which the failure occurs, for example low or zero output signal, distorted output and so on. (iii) The electrical effect of a failure, which may be: (1) short circuit, (2) open circuit, (3) degraded performance or (4) functional failure [6]. The consequences of a failure mode are synthesised in severity, which considers the worst potential consequence of a failure, determined by the degree of injury, property damage or system damage that could ultimately occur.

Failure precursor An event or series of events that is indicative of an impending failure.

Failure rate The frequency with which a batch of components or systems fails; that is, the number of failures experienced in a given time period. Used as indicator of the reliability level. In systems in particular, the mean time between failures (MTBF) can also be used as a reliability indicator; that is, the numerical inverse of the failure rate for exponentially distributed failures only.

Failure, random A failure associated with the variability of workmanship quality.

Failure (root) cause The circumstances during design, manufacturing or use environment which have lead to failure: lack of design margins, lack of protection against environmental stress, use of defective components, assembly error, use of inappropriate technologies, misuse or abuse of equipment, and so on.

Failure, systematic A deterministic failure, caused by an error or a mistake. The elimination of systematic failures requires a change in the design, production process, operational procedure, documentation or some other area.

Failure, wear-out A failure caused by a mechanism related to the physics of a device and its design and process parameters. Wear-out failures occur in the entire population of parts and can therefore represent a very high percentage of all observable failures.

Failures, early Failures due to randomly distributed weaknesses in material or the manufacturing process (assembly, soldering, etc.). The length of the early-failure period varies from some days to a few thousand hours.

Fault An incorrect system state, in hardware or software, resulting from failures in system components, design errors, environmental interference or operator errors [4].

Fault, intermittent A fault that manifests occasionally, due to unstable hardware or certain system states. For instance, most microprocessors do not perform data-forwarding errors correctly for certain sequences of instructions (injecting a fault in data). As these are discovered, developers add code to compilers to prevent them from generating those specific sequences again [4].

Fault localisation The technique of finding a fault in a failing circuit, functionally and logically.

Fault, transient A fault resulting from a temporary environmental condition. For example, a voltage spike might cause a sensor to report an incorrect value for a few milliseconds before the correct value is reported [4].

Flaw (i) A condition that does not necessarily result in a defective part or failure. (ii) An imperfection or discontinuity that may be detectable by nondestructive testing and is not necessarily rejectable.

Functional testing (i) Testing that ignores the internal mechanism of a system or component and focuses solely on the outputs generated in response to selected inputs and execution conditions. (ii) A test focused on verifying the target-of-test functions as intended, providing the required service(s), method(s) or use case(s). This test is implemented and executed against different targets-of-tests, including units, integrated units, application(s) and systems.

Management, health A new concept, based on understanding *how* and *why* a device fails.

Qualification Verification that a particular component's design, fabrication, workmanship and application are suitable and adequate to assure its operation and survivability under the required environmental and performance conditions. Qualification is applied for components, processes and products.

Qualification, knowledge-based A two-step process that starts by detecting and understanding the failure mechanisms of a specific technology and then applies use-conditions-based qualification plans according to that knowledge.

Quality A measure of the variance of a product from its desired state.

Quality assurance The planned and systematic activities necessary to provide adequate confidence that an item (product or service) will satisfy given requirements for quality.

Reliability The ability of an item to perform a required function under stated conditions for a stated period of time without failures. The numerical value of reliability is expressed as a probability from 0 to 1 and is also sometimes known as the probability of mission success. Reliability is the probability that a system will continue to operate until time t, assuming it was operating at time zero.

Reliability analysis A method of analysing a system design by evaluating reliability-related issues. The goal is to improve the field reliability of the system.

Reliability, intrinsic The reliability a system can achieve based on the types of device and manufacturing process used.

Root cause The first event or condition that triggered, whether directly or indirectly, the occurrence of a failure, for example improper equipment grounding that resulted in electrostatic discharge damage.

References

1. http://en.wikipedia.org/wiki/Component. (Accessed 2010).
2. O'Connor, P.D.T. (2000) Reliability past, present and future. *IEEE Trans. Reliab.*, **49** (4), 335–341.
3. http://en.wiktionary.org/wiki/defect. (Accessed 2010).
4. Siewiorek, D.P. and Swarz, R.S. (1992) *Reliable Computer Systems – Design and Evaluation*, 2nd edn, Digital Press.
5. Jensen, F. (2000) *Electronic Component Reliability*, John Wiley & Sons, Ltd, Chichester.
6. Birolini, A. (1997) *Quality and Reliability of Technical Systems*, Springer, Berlin and New York.

Index

Failure Analysis: A Practical Guide for Manufacturers of Electronic Components and Systems, First Edition.
Marius I. Bâzu and Titu-Marius I. Băjenescu.
© 2011 John Wiley & Sons, Ltd. Published 2011 by John Wiley & Sons, Ltd.

Printed and bound by CPI Group (UK) Ltd, Croydon, CR0 4YY
16/04/2025
14658381-0001